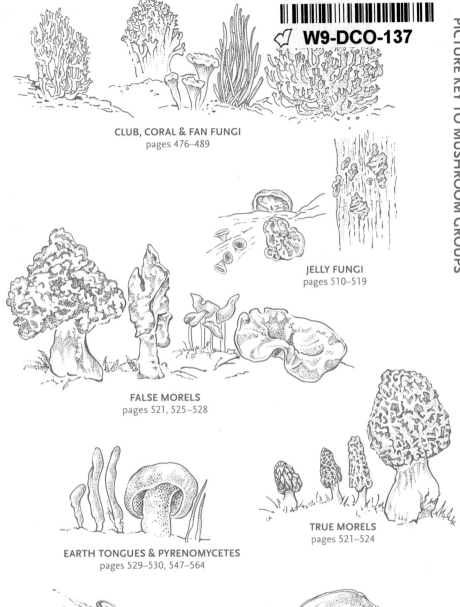

CLUB, CORAL & FAN FUNGI
pages 476–489

JELLY FUNGI
pages 510–519

FALSE MORELS
pages 521, 525–528

TRUE MORELS
pages 521–524

EARTH TONGUES & PYRENOMYCETES
pages 529–530, 547–564

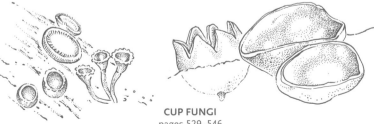

CUP FUNGI
pages 529–546

MUSHROOMS

of the NORTHEASTERN UNITED STATES *and* EASTERN CANADA

Timothy J. Baroni

TIMBER PRESS FIELD GUIDE

This book is dedicated to the following individuals in appreciation of their skills as educators: Howard E. Bigelow and Margaret Elizabeth Barr Bigelow, both now deceased and both formerly professors of mycology and botany at the University of Massachusetts at Amherst, and David Lee Largent, emeritus professor of mycology and biology at Humboldt State University, Arcata, California. During my studies under their guidance, they provided training and inspiration and nurtured in me a lifelong fascination with fungal organisms.

Photo and illustration credits appear on page 575.
Half title page: *Pholiota multifolia.*
Frontispiece: *Tricholomopsis rutilans.*
Contents page: *Laetiporus sulphureus.*

The information in this book is accurate and complete to the best of our knowledge. All recommendations are made without guarantee on the part of the author or Timber Press. The author and publisher disclaim any liability in connection with the use of this information. In particular, eating wild mushrooms is inherently risky. Mushrooms can be easily mistaken, and individuals vary in their physiological reactions to mushrooms that are touched or consumed. Mention of trademark, proprietary product, or vendor does not constitute a guarantee or warranty of the product by the publisher or authors and does not imply its approval to the exclusion of other products or vendors.

Published in 2017 by Timber Press, Inc.

The Haseltine Building
133 S.W. Second Avenue, Suite 450
Portland, Oregon 97204-3527
timberpress.com

Printed in China
Third printing 2020
Text design by Michelle T. Owen based on a series design by Susan Applegate
Cover design by Kristi Pfeffer

Library of Congress Cataloging-in-Publication Data

Names: Baroni, Timothy J., author.
Title: Mushrooms of the northeastern United States and eastern Canada /
    Timothy J. Baroni.
Other titles: Timber Press field guide.
Description: Portland, Oregon: Timber Press, 2017. | Series: Timber Press
    field guide | Includes bibliographical references and index.
Identifiers: LCCN 2016045509 (print) | LCCN 2016047880 (ebook) | ISBN
    9781604696349 (flexibind) | ISBN 9781604698145 (e-book)
Subjects: LCSH: Mushrooms—Northeastern States—Identification. |
    Mushrooms—Canada, Eastern—Identification.
Classification: LCC QK605.5.N85 B37 2017 (print) | LCC QK605.5.N85 (ebook) |
    DDC 579.609713—dc23
LC record available at https://lccn.loc.gov/2016045509

CONTENTS

ACKNOWLEDGMENTS

This book would not have been possible without the donation of a large number of images by Renée Lebeuf. Her generosity in providing images, her enthusiasm for the project, and the many discussions of species concepts for some difficult groups were indispensable. Roy E. Halling also provided images and much needed clarification on the changes in bolete nomenclature.

Roy and I spent many years together exploring for fungi. Ernst Both and I also spent many hours in the field. These two field companions made an enjoyable endeavor even more interesting. I am pleased to be able to provide some of their knowledge to others.

Additionally, many others provided images or discussions on current research including Ron Petersen, Clark Ovrebo, Rick Kerrigan, Andy Methven, D. Jean Lodge, Michael Kuo, Scott Redhead, Michael Goldman, Lance Lacey, Amanda Neville, Joseph Nuzzolese, Jack Ruggirello, and Nina Burghardt. Gerald "Jerry" Sheine is due very special thanks for digitizing all of Howard Bigelow's and my Kodachrome slides.

A significant amount of mushroom biodiversity exploration in the Northeast, that produced collections and images for this work, was made possible by personal funds and also by grants from suny Cortland Faculty Research Program, New York State Albany Museum, and Buffalo Museum of Science. Facilities for field work in the Adirondacks were provided by the suny Cortland Outdoor Education Center at Raquette Lake; by the suny College of Environmental Science and Forestry, Adirondack Ecological Center facility at Huntington Forest in Newcomb, New York; and by Bonnie View Cottages in Wilmington, New York.

I also wish to thank the Department of Biological Sciences, suny Cortland, especially Steven Broyles, chair, for complete support during the writing and assembling of this book, and the college for providing office space, herbarium facilities, the logistical support and microscopic equipment.

Most importantly I want to thank my friend and partner, Robin Marie Wheeler Baroni, for being patient, supportive, and helpful when I needed someone to give a different perspective in solving problems or getting around mental blocks that kept me from progressing forward smoothly. Her support was essential and is constantly priceless.

INTRODUCTION

Let me say this right up front—there are no old, bold mushroom foragers, only old wise, well-trained mushroom foragers. Good doses of caution and common sense are important to remember if you wish to use wild fungi and plants for food. Become an expert first, and then carefully use your knowledge. I have high hopes this field guide and others you may collect over the years, will help you enjoy the beauty and, with great care and caution, the flavors of the wild fungi that can be found in our fields and forests.

But honestly, most fungi covered in this and other field guides are not really edible. The real purpose of a field guide like this one is to help the curious learn more about the species of fleshy fungi in a particular area.

At best guess, several thousand species of mushrooms and other fleshy fungi occur in northeastern North America. Most field guides rarely cover more than 400–500 species. Of course, a good field guide should cover the bulk of the more commonly collected species, and that is the case here. In addition, I present at least a hundred species that are not to be found in any other existing field guide covering North American mushrooms.

Realize, however, that even if you have several good field guides covering a broad range of species, you will still find species not covered. That is one of the exciting things about studying fungi—you might just discover a species from your area that is a new report for the region. You can share that information via several avenues, but a convenient one is MushroomObserver.org. You are allowed to post information on your finds at this site after joining the group.

You will need to learn some photography skills, since this is a site for posting images of your finds as well as information on the features of a species and its ecology. Please take spore prints and provide that information also.

So with all that in mind, bon voyage. Take this field guide and begin the journey. If you began your studies some years ago now, I hope this newest addition will help you continue your journey of learning about the wild mushrooms around us.

Geographical Scope of This Guide

The area covered in this guide is roughly that of the eastern hardwood forests of North America. These territories include New Brunswick, Newfoundland, most of Quebec and Ontario, extending south through Minnesota to Illinois and eastward through the states surrounding the Great Lakes regions to West Virginia, Pennsylvania, Maryland, and then up the coast to Maine. This area includes deciduous and coniferous forests, bogs, and alpine habitats that offer a wide range of species, many of which are only found in this region in North America.

Toxins in Fungi: To Eat or Not To Eat

Not all mushrooms are edible, not all mushrooms are poisonous, but if you wish to use macrofungi for food, you should become an expert at identifying the edible *and* the poisonous species. This approach will serve you well. If you decide to use wild mushrooms in your cuisine, as many people throughout Europe, Asia, Africa, and even North America have done for centuries, I strongly recommend, whether you are a beginner or have some experience, that you thoroughly learn targeted species and use only a small number of these edibles to start: tooth fungi (*Hydnum*

repandum), puffballs (*Lycoperdon*), sulfur shelf polypore (*Laetiporus*), porcini (*Boletus edulis*), and morels (*Morchella*) are easily mastered. I recommend this approach because making a mistake can be very uncomfortable or even deadly.

You might ask, why bother learning about the poisonous species, I am only interested in the edible ones? First of all, it is important to know which ones *not* to consume, but also because you will see these beautiful poisonous species quite frequently on excursions into the forests in late summer and fall. *Amanita bisporigera* and *A. virosa*, the destroying angels, are mycorrhizal with beech and oaks and are rather commonly encountered. They are large, white, and by all accounts taste pretty good. But after ingesting even just a small amount of one of these mushrooms, the cellular toxins absorbed in the blood stream after ingestion begin to disrupt and destroy cells in your liver and that may lead to death after several days of extreme discomfort. There are no known antidotes for such a poisoning.

As a general rule, learn the following genera and avoid using any of the species for food until you become an expert at identification. Some species within otherwise poisonous genera, such as *Entoloma abortivum*, are actually quite good and worth learning. I recommend putting these on your admire-but-do-not-eat list: *Amanita*, red-pored boletes, *Chlorophyllum* (or any mushroom producing a green spore deposit), *Cortinarius*, *Entoloma*, *Galerina*, *Gymnopilus*, *Hebeloma*, *Inocybe*, small *Lepiota* species, and *Omphalotus* (the false chanterelle that glows in the dark, or any mushroom that glows in the dark). This is not a complete account of the poisonous mushrooms, but these genera will be commonly encountered when fruiting conditions are favorable, and you should be able to recognize them.

Not all poisonous mushrooms are deadly poisonous. Most cause gastric distress, diarrhea, or other uncomfortable symptoms, sometimes only if alcohol is also consumed. Recovery from such incidents is usually rapid and complete. Also, to become poisoned you have to actually eat the mushroom. Even the deadliest toxic mushrooms are safe to handle. Contact dermatitis is caused by a few slimy-topped boletes in the genus *Suillus,* and only for those who are sensitive to these mushrooms. Touching a highly toxic mushroom like a destroying angel or a deadly *Galerina* species will not cause any ill effects.

Not very many humans are actually poisoned each year. Most of the suspected poison case reports involve incidents with no symptoms and therefore no real danger.

The real poisoning issues come from individuals who indiscriminately continue to collect and eat mushrooms with little or no knowledge. These are usually recent immigrants who have safely used a species in their former country, that looks very similar to our toxic one. Also a seriously at-risk group are those foolish individuals who are looking to eat psychoactive fungi for recreational purposes. There really is no cure for stupidity.

For a detailed discussion of the different types of toxic mushrooms, and the various sorts of toxic symptoms these mushrooms cause, please visit namyco.org/mushroom_poisoning_syndromes.php, posted by the North American Mycological Association.

Biology of Fungi

Fungi come in many forms, but they are basically microbes, organisms too small to see with the unaided eye unless they form fruit bodies, like mushrooms, to make and release their sexual reproductive spores. We can lump the fungi into two groups: macrofungi, those producing large observable fruit bodies, and microfungi, the ones that require a microscope. Molds are good examples of microfungi.

Until approximately the 1960s, fungi were considered plants and categorized as such. Based on biochemical and DNA studies since then, fungi are now considered a separate kingdom, like animals, plants, protists (for example, algae and protozoa), and bacteria. These studies have found that in fact fungi are more closely related to animals than to plants. The cell walls of fungi are composed of chitin, a substance only found otherwise in the exoskeletons of animals like insects and crustaceans. Plant cell walls are composed of cellulose, a substance not found among fungi, but it is a substance that fungi and bacteria can readily use as a food source. The fungi are clearly a unique group of organisms.

Although some fungi are single-celled—for example, the yeasts that give us bread and alcoholic beverages—by far the majority of fungal species are composed of threadlike filaments called hyphae. One of these cellular units, a hypha, is tubular, long, often branching, and for the most part microscopic. An exception can be found on the moldy strawberries or cantaloupe rind in your garbage can—the large, very long hyphae producing black sporangia and asexual spores at their tips. You can at least see the reflection off the colorless hyphae of these pin molds as they reach up toward the light to release their spores. If your vision is excellent, or you have a magnifying glass, you may actually see the individual long tubular cells.

Hyphae of fungi invade the substrate they will take nutrients from. This mass of hyphae is referred to as a mycelium. The mycelium, or body of the fungus, is exporting enzymes outside its hypha to break down organic and inorganic molecules that can then be absorbed by the hypha as food and used as building blocks by the growing fungus. If you could look into the soil, into all the dead and decomposing organisms around you, and yes, inside the living roots of plants as well, you would see an endless mesh of mycelial networks in those living or decomposing substrates. They are there playing one of three major ecological roles as saprobes, parasites, or symbionts. In biology, nothing is absolute, and occasionally a parasite might take on the role of a saprobe, or vice versa, if the conditions are right, but read on please.

A main function of fungi in our ecosystems is to recycle dead organic matter, along with a host of certain bacteria. These recycling fungi (and bacteria) are saprobes.

Another role fungi play, again like certain bacteria, is that of parasites. These fungi are equipped to attack weakened, injured, or genetically susceptible living plants, animals, and other fungi to absorb nutrients directly from these living organisms.

Undoubtedly key to the health of our planet is the symbiotic relationship land plants entered with fungi to form mycorrhizae.

Mycorrhizae are intimate connections between fungal cells and the root cells of plants. The fungus provides the plant with vital growth nutrients, especially phosphorus. The plant provides the fungus with carbon as it converts sunlight and carbon dioxide into energy-rich sugar compounds. Nearly all plants, more than 90%, have a symbiotic fungal root partner; oftentimes woody plants have many fungal partners.

Certainly some of the very best edible fungi are symbionts with specific trees. If you wish to become a serious mushroom hunter, you must also learn the different kinds of trees. If you wish to find porcini (*Boletus edulis*) for example, your success rate will increase dramatically if you focus on older stands of Norway spruce (*Picea abies*). Sugar and red maple trees do not form ectotrophic mycorrhizae, so you will not find porcini in those types of forests.

Northeastern North American conifer trees that form ectotrophic mycorrhizae are pine (*Pinus*), hemlock (*Tsuga*), spruce (*Picea*), fir (*Abies*), larch or tamarack (*Larix*), and Douglas fir (*Pseudostuga*) which is sometimes planted in this region though it is native to western North America. Conifers that are endotrophic, thus not associated with macrofungi, are northern white cedar (*Thuja*), eastern red cedar or juniper (*Juniperus*), swamp cedar (*Chamaecyparis*), and dawn redwood (*Metasequoia*).

Several hardwood trees also form ectotrophic mycorrhizae. (Hardwoods are often called deciduous trees, but I prefer the former term since larch, a conifer, is deciduous as well.) They are birch (*Betula*); beech (*Fagus*); oak (*Quercus*); linden or lime tree (*Tilia*); rowan or mountain ash (*Sorbus*); poplar or aspen (*Populus*); willow (*Salix*); hornbeam, musclewood, or blue beech (*Carpinus*); American hophornbeam, eastern hophornbeam, or ironwood (*Ostrya*); and American hazelnut (*Corylus*). If you wish to learn more about which plant forms which type of mycorrhiza, visit mycorrhizalonline. com/is-my-plant-mycorrhizal.

What To Call It? Mushroom, Toadstool, Bolete, Stinkhorn, or ?

From a botanical perspective, a mushroom is a fleshy fungus fruit body that has gills, or lamellae, that are covered with the reproductive layer, the hymenium. This hymenium produces basidia, which are the club-shaped cells that produce the sexual spores, the basidiospores. The term *toadstool* has often been used to designate a poisonous mushroom, but that is not a hard-and-fast rule.

The term *mushroom* also has been very loosely applied to the fleshy pore fungi, the boletes or bolete mushrooms. The hymenium covers the inside of each tube and the spores fall out of the opening, or pore, as gravity pulls the released spores down and out of the tubes. Calling boletes mushrooms seems okay, since the true mushrooms and boletes are closely related and you can only tell you have a bolete after you turn it over and see the tubes. However, for accurate and more precise communication, I suggest learning the distinctions between the various groups of fleshy macrofungi and using their common names.

The structures that produce the hymenium help, in an artificial way, to define these groups. For example, coral fungi have the hymenium covering the erect branches, while tooth fungi have the hymenium covering each tooth. These major groups are easily recognized, and when you tell someone you found a stinkhorn, they know exactly what you mean. if you say you found a smelly mushroom, that could just be a bolete that is rotting and smells bad.

All these groups are basidiomycetes except the cup fungi and stromatic fungi or pyrenomycetes, which are ascomycetes. Basidiomycetes produce sexual spores on top of club-shaped cells called basidia in the hymenium. Ascomycetes produce sexual spores in clusters of long tube- or saclike cells called asci (ascus is singular) in the hymenium that, for instance, lines the cup in a discomycete.

Learning these major groups also makes it easier and faster when trying to narrow down the choices for identifying a fungus to genus and species.

Mushroom Structures

Parts of mushrooms and other macrofungi structures are illustrated on the inside front and back covers of this book. These images will help you use the field guide more effectively. I have tried to keep scientific mycological terms to a minimum, but specific terminology is very important when discussing the minute features that help differentiate one fungus from another.

CAP

Although the form of a mushroom fruit body usually changes in shape over time, its features are consistent for each species. For instance, some species have conical-shaped caps while others have

Common Names of Fleshy Macrofungi Groups

Mushrooms Fleshy, fragile, with gills, stipe present or absent

Boletes Fleshy, fragile, tubes instead of gills, tubes separable from cap

Polypores Dry, tough, woody, rarely fleshy, tubes that are not separable from cap, often lacking a stem

Tooth fungi Fleshy or woody, spines or teeth on underside of cap or hanging from a branched system and lacking a cap

Chanterelles Fleshy, fragile, with long decurrent gill-like folds, wrinkles, or ridges, or smooth on underside of often vase-shaped cap

Coral fungi Fleshy, fragile, single or multiple erect branches

Stereoid fungi Mostly tough, leathery, some fragile, stem with fan-shaped cap, underside of cap smooth, bumpy, irregularly poroid, folded

Jelly fungi Moist, gelatinous, variable shapes from globular to coral to toothed

Crust fungi Thin, soft, or tough, flat against the substrate which is usually logs, spore-producing surface smooth, bumpy, poroid, or wrinkled-folded

Puffballs Fleshy at first, dry and papery when mature, white fleshy inside becoming olive-brown dry and powdery, spores ejected through central pore in the sac when the sac is disturbed

Earthstars Fleshy at first, dry and papery or woody with age, in essence a puffball attached in the middle of a thick-fleshed covering that opens and spreads out in a starlike pattern

Earthballs Fleshy at first, then dry and thick-skinned, opening irregularly but not by a discreet central pore, developing black greasy spores that become powdery with age

Bird's nest fungi Felty, woody, or leathery cups filled with lentil-shaped "eggs," on dead wood and wood chips

Stinkhorns Fleshy, fragile, phallic-shaped stalks with slimy, smelly coating over the tip, or cagelike, clawlike, or flowerlike with smelly slime coating on the inner parts, often on woody substrates

Cup fungi, or discomycetes Fleshy, large or small, cuplike fruit bodies, or fruit bodies fingerlike, saddle-shaped, brainlike, spongelike, and all on a stalk or plate- or saucer-shaped, or irregularly rabbit-ear–shaped, and so on

Stromatic fungi, or pyrenomycetes (in part) Fruit body very small (use a hand lens), globular or flask-shaped, with minute pore (a perithecium); fruit bodies clustered on and mostly in a fleshy or dry charcoal-like stroma; stroma is variously colored white, pink, red, green, brown, or black, and variously shaped as erect fingerlike structures, cushions, balls, or crusts covering and parasitizing other fungi, like mushrooms, boletes, or polypores, or on dead woody plant materials convex-shaped caps in the early form, but the caps of each species may both become plane, or flattened, with age.

The mushroom is at first protecting its reproductive surface, the gills, as it pushes its conical or convex cap up through the soil or other substrate. Once in the air currents, the cap fully expands, becomes plane, and the spores are released from the basidia, dropping down between the gills to the wafting air currents below the edges of the gills. The spores are then carried away to find new substrates in which to grow. When you make a spore print, you are catching these spores en masse to determine their color, since this feature is important for identification.

Gills bluntly attached (adnate)

feature at the species level, for example in the case of *Leptonia foliomarginata*. If the gills peel away from the cap easily, that is also a diagnostic feature for genera like *Tapinella* and *Paxillus*.

GILLS

Gills on mushrooms are generally thin and plate- or leaflike. They can be attached to the stem in various ways, or not attached to the stem, in which case they are referred to as free. In this guide, *barely attached* refers to gills that are adnexed, *bluntly attached* is adnate or emarginate, *notched* is sinuate, and *decurrent* is the standard term used for gills running down a stem. Noting attachment and spacing is very helpful when identifying mushrooms.

Gills are generally classified into two types: the lamellae are those that reach the stem, and the lamellulae are the shorter gills that may go halfway or just a short distance from the margin. If the gills are forked repeatedly and equally, dichotomously forked, this feature is characteristic for genera such as *Cantharellula* and *Hygrophoropsis*. When the edge of the gills is differently pigmented than the face of the gills, they are termed marginate. This is a diagnostic

STEM

Mushrooms (but not all fleshy macrofungi) always have a cap and gills, but they may or may not have a stalk, or stem, technically called the stipe. Stalkless mushrooms are referred to as sessile and usually are attached to a substrate like wood, living or dead herbaceous plants, or another fungus (see *Phyllotopsis nidulans*). If the stalk is attached at the margin of the cap, it is called lateral; these are usually found on woody substrates (see *Pleurotus pulmonarius*). If the stalk is off-center, it

Gills decurrent, stem hollow

is called eccentric, and the fruit bodies may or may not be on woody substrates (see *Hygrophoropsis aurantiaca*). These different designs are just different ways to get the gills up into the air currents.

Interestingly, both gills and stalks when present are often gravitropic. That is, they bend in response to gravity to orient the cap and gills to be perpendicular to the earth's surface, so the spores can be precisely dropped from between the gills once they are shot off the basidia. Amanitas often show this gravitropic response. After they have been lying flat in a basket or box for several hours, you will notice the stalk and cap have curved, reorienting to the pull of gravity.

The shape of the stem is important, with the typical shape being equal. But you need to be keenly aware of the other shapes, as they can be diagnostic when trying to make a determination (for example, the rooting stems, or pseudorhiza, of *Phaeocollybia* and *Hymenopellis*). Also, the surface features of stems can be useful diagnostically. For example, the porcini (*Boletus edulis*) always produces a white reticulum, while other look-alike species produce a brown-colored reticulum (for example, *Tylopilus felleus*).

A common feature of many mushroom and bolete stems is to be dotted with fine, variously colored granules. These fine granules are usually at the apex of the stem and white, but on some boletes, such as *Xerocomellus chrysenteron*, they are rhubarb-red and cover the entire stem. If the dots are large and brown or black, they are referred to as scabers and are diagnostic for the bolete genera *Leccinum* and *Leccinellum*. Most mushrooms have unadorned stems; they are glabrous.

VEILS

Protecting the reproductive spores is so important that certain groups of fungi evolved structures, such as the veil, to cover and more effectively protect the hymenium from damage by environmental stresses such as dry conditions and fungus-eating insects.

Some fungi have a partial veil that covers just the gills and then leaves a ring, the annulus, on the stem or a fringed margin on the cap, in which case the cap is described as appendiculate. This partial veil is more effective than not having this covering; however, the ultimate in protection of the hymenium in the early stages of development would be the universal veil that covers the entire fruit body. Evidence of a universal veil can be found as freely removable wartlike patches on the cap and a sheathing volva at the base of the stem after the fruit body has expanded, tearing the universal veil open.

The partial veil usually leaves a ring of tissue, called the ring or the annulus, resembling a skirt around the upper stem, after the cap expands. If it is substantial and membranelike, it is referred to as membranous and can be hanging down, pendant, or sticking out and flaring. If the stem looks as though it is sheathed in this ring tissue from the base of the stem to where the ring is formed, the veil is called peronate or sheathing.

If the ring is spiderweb-like with lots of space between the individual strands and the strands are not confined to a single plane, but crisscrossed from the cap margin to the stem apex, that is a special veil called a cortina and is a diagnostic feature of *Cortinarius*. When a cortina or any fragile veil collapses on

Mushroom emerging from its universal veil

meaning it is soft and easily destroyed and lost. The distinctive collarlike band or circumsessile volva is diagnostic for *A. frostiana*, a species often confused with *A. flavoconia*. The signature of the large orange-yellow-capped *A. muscaria* var. *guessowii* is the concentrically ringed volva at the base of the stem.

Collecting and Identifying Fleshy Fungi

You will need a few items to make good collections, whether for eating or for studying the different species and learning what can be found in your favorite collecting sites. Besides wearing appropriate clothing and footwear such as boots with good traction, the following are essentials for making collections and getting out of the woods safely: basket or bucket, or box with a rope handle; knife (not a penknife but a larger pocket-knife); roll of aluminum foil (heavy duty is best) or waxed paper or paper bags; plastic resealable zipper storage bags with 4-by-5-inch pieces of white paper for spore prints; pen; insect repellent; digital camera (your cell phone camera is sufficient); compass (so you do not become lost; believe me, it can happen).

the stem and is barely noticeable, it is referred to as a ring zone.

The universal veil covers the entire fruit body in the very early stages of development, the button stage. When the universal veil is broken as the mushroom extends upward, parts of the veil are left as the volva surrounding the base of the stem and as warts on the cap in many cases.

The type of volva can be diagnostic for species of *Amanita*. The deadly poisonous *A. phalloides*, the death cap, and *A. bisporigera* and *A. virosa*, the destroying angels, have well-developed cup- or saclike membranous volvas. The very commonly found *A. flavoconia* has yellow conical warts on the cap as remnants of the universal veil, while the volva at the base of the stem is friable,

A sturdy basket is the typical collecting vessel to carry the prizes, but a bucket will work just as well. I recommend not mixing up the collections, but individually wrapping each species, not each specimen, in a single packet using waxed paper or aluminum foil. Other good choices are waxed paper sandwich bags or small brown paper sandwich bags. Do not use plastic wrap or zipper storage bags as these promote increased humidity and temperature, thus bacterial and mold growth that quickly spoils the fruit bodies.

If you find you are collecting the moderate to small species, a good plastic fishing lure or hobby box with multiple compartments is an ideal way to keep each collection separated for later study. Placing a little moss or fern leaf in each compartment keeps the samples fresh and undamaged by preventing them from rolling around in the compartment. Also, if the day is hot and humid and the drive home may be long, take a cooler containing a cold pack to keep the smaller specimens as fresh as possible for the trip back.

A good pocketknife or sheathed knife is also very useful for helping to remove your collections from soil, humus, and wood. In addition, the knife will come in handy if you are absolutely positive of your determination and the mushroom will be used for food. In this case then you can use the knife to clean off the debris or cut off the base of the stem to leave soil and decaying plant litter in the woods, making cleaning your choice edible finds easier back in the kitchen.

Before making a collection, I like to take pictures of the fruit bodies as I find them in their natural settings. Then after collecting, I take another picture in a portrait style, setting them against a gray card, to fully document the features before taking notes and then drying the samples.

To make a good collection, carefully remove the stem from the substrate by digging around the base of the stem, being careful not to break the base off in the soil, humus, or wood. Collect as many fruit bodies as necessary to be able to study all stages of development. If you find a possible species of *Ama-nita*, be certain to dig up the base with the volva intact since that is an important character for identification. Most of the time you can carefully pull away the soil and debris from the base of the stipe so that the entire mushroom can be gently lifted out of its substrate intact. Also, with *Hymenopellis* (formerly *Xerula* or *Oudemansiella*) you need to obtain the long rooting pseudorhiza, so take care to dig down and carefully extract the entire sample.

Try to find the smallest fruit bodies and the largest ones of the same species that appear to come from the same mycelium. Remember the mycelium grows out in a circle, so look for rings of mushrooms. These can be a few feet to many feet in diameter as the fruit bodies are produced upward at the ends of the leading hyphal edges, like apples on the tips of tree branches. Collecting the fruit bodies does not hurt the mycelium, just as it does not hurt the tree to pick the apples.

If you carry 4-by-5-inch pieces of plain white paper in a plastic bag, spore prints can be set up right in the field and placed in the bottom of your basket as you continue to hunt for more collections. Generally, try to make 10–15 good collections on each outing if you intend to study the samples back at home. You certainly can collect more, but if the intent is to document and study each collection in detail, these fleshy fungi only last a day or so, even if you refrigerate them—do not freeze.

If not done in the field, spore prints should be set up immediately after returning to your home. A spore print may take one or up to several hours, but if you leave them overnight in warm humid weather, you will often have nothing but a puddle of putrefying mushroom, sometimes replete with maggots on your white paper in the morning. Check your spore prints every hour or so and remove the cap once you have a reasonable spore drop showing a definite color that you can record.

If your mushroom hunting becomes more than just a passing interest, keep a log book of your finds with dates and, if it is really interesting, a description of each species you have identified. If you get really hooked on the scientific aspect of mushrooming, dry your collections with a food dehydrator on moderate heat setting. High heat makes for samples that are hard to revive and work on using a microscope. High heat also seems to damage the DNA molecules so that stored information becomes less useful for future studies. Once well dried, store your samples in plastic bags in the freezer with your notes and spore prints.

I have provided keys to help you learn the major groups, and then the genera within those groups. The comments section in each species description contains information on similar species and provides diagnostic features used to identify a large number of mushrooms we were unable to include in this field guide because of space limitations. Also, if a species is poisonous or a prized good edible, that information is presented in the comments.

Identifying Your Collections

Once you learn the major groups of fleshy fungi (mushrooms, boletes, corals, and so on) you could just thumb through the pictures in this field guide and try to match up images to your samples; really, that is what most of us do. That may or may not work. Certainly if you think you have a match, please read the description thoroughly and make sure your collection has the key or synoptic features listed at the beginning of the descriptions. If you have a mushroom you have never collected before, make certain you get a spore print to confirm you are in the right group.

BASIDIOMYCETES

These are the large, showy fungi most people see on their lawns and in the woods, representing the mushrooms, polypores, boletes, tooth fungi, puffballs, stinkhorns, coral fungi, chanterelles, and the crust fungi. Basidiomycetes are characterized by spores which are produced by basidia externally on short pegs called sterigmata, where they will be released by force, falling from gills or tubes until air currents below these structures waft the spores away as they fall free from the reproductive tissues. In the case of the gasteromycetes, the spores simply fall off the pegs and make a powdery or slimy mass of spores inside the fruit body that will be exposed to air currents or insects once the structure matures and opens.

There are both deadly poisonous and fine delicious species of basidiomycetes, and they are not so difficult to learn with a little study. The shapes, colors, and forms are highly varied and artistically alluring. Many mycorrhizal fungi very important for forest health are in this group, as are the very destructive tree parasites. The majority of the species, however, play important roles as nutrient recyclers of decaying plant materials and can be found in all habitats where there is decaying plant material.

The fruit bodies only appear when environmental conditions are conducive to fruiting, warm and wet usually. However, like the ascomycetes, species in this group show a seasonality. Some show up early in the spring, for example the dryad's saddle polypore. A wave of fruiting comes in the early summer with species of *Marasmius*, *Gymnopus*, *Russula*, *Lactarius*, *Amanita*, and

so on, then another wave comes in mid- to late summer from the boletes, species of *Entoloma*, *Leptonia*, *Nolanea*, *Mycena*, and so on. A series of different species, like those of *Suillus*, slowly show up as the fall progresses until the fruiting cycle is near its end. The late fall oyster (*Panellus serotinus*) and the honey mushrooms (*Armillaria*) herald the end of the fruiting season. It is bittersweet to see these mushrooms on decaying logs, stumps, or on the ground over decaying roots in the woods.

Some species only fruit sporadically, often with many years between fruit body production, whereas others tend to show up each and every year in the same places at around the same time, like chanterelles, porcini, and oyster mushrooms. If you take the time to learn the patterns and find special places to collect, it is like harvesting a wild, untended vegetable garden.

Presented here first are the true mushrooms, those with gills, arranged by spore print color groupings. The spore print color is the best and most accurate way to identify species. With experience, a person can get good at determining spore colors even before a print is made. As you build knowledge from this exercise, you will find you only have to make spore prints of the collections that are completely new to you. A mushroom that looks like it could produce a white spore print, may give a pale pinkish-buff color—that is an important feature. One with orange gills that looks like it might give an orange spore print, may end up producing a white spore color. Be careful to discriminate spore color from pigments that can leach from the gill edges onto the white paper.

True mushrooms with gills

Gilled Mushrooms with Pale-Colored Spore Prints

Most species of mushrooms are white or pale spored, and the largest number of genera and species will be found in this group. Notice that "pale" includes white, cream, yellow, ochre, grayish-green, grayish-lilac, and pale pinkish-buff. In this group of mushrooms we find the most high profile of the deadly poisonous species, the amanitas. We also find some highly prized gourmet edibles, the oyster mushrooms.

Recently, molecular phylogenetic research has been used to create many new genera; it is this pale-colored group that has received the largest impact from the creation of new generic names for old species of mushrooms, those not found in older field guides. For the most part, the species fall into five main groups which correspond to the families Amanitaceae, Lepiotaceae, Russulaceae, Hygrophoraceae, and Tricholomataceae. Most of these families have been rearranged or changed radically based on molecular studies. This field guide uses the new genera for classification, but for convenience of identification, I use the simpler family organizations.

Gills free

Amanita Volva at stem base and/or warts on cap present
Lepiota, *Chlorophyllum*, *Macrolepiota*, *Leucoagaricus*, *Leucocoprinus* Cap scaly, lacking volva and warts, spore print white, cream, or gray-green
(Note: see also *Cystoderma* and *Cystodermella* with attached gills)

Gills attached, flesh brittle, stem and gills breaking cleanly, spore prints ranging from white, cream, yellow, or up through dark ochre

Lactarius Producing colorless or variously colored latex when injured
Russula Not producing latex when injured, cap often brightly colored

Gills attached, flesh fibrous, gills thick and waxy-looking, fruit bodies white or very brightly colored, small to medium-sized (basidia long, narrow)

Gliophorus Slimy cap and stem, variously colored brown, violet, or green, but not bright yellow
Gloioxanthomyces Slimy cap and stem, bright yellow overall
Cuphophyllus With translucent gills and flesh, not brightly colored
Humidicutis Flesh not translucent, cap and stem brightly colored
Hygrocybe Flesh not translucent, cap not viscid, small, brightly colored
Hygrophorus Thick-fleshed, cap viscid or not, medium or large fruit bodies, some with a ring on the stem

Gills attached in different ways, flesh fibrous, gills not waxy-looking, fruit bodies mostly with dull or muted colors

Tricholoma and allies

Amanita

Amanita bisporigera
G. F. Atkinson

DESTROYING ANGEL
Large entirely white mushroom with enlarged globose base enveloped with white saccate volva; white ring on upper stem also present

CAP 3–10 cm wide, convex, becoming plane, white, glabrous, slightly slippery when fresh, usually lacking patches of universal veil, margin smooth. **GILLS** white, free or barely attached, crowded, broad. **STEM** 6–14 cm long, 7–20 mm broad, equal or tapering upward with globose bulb at base, membranous white saccate volva with free edge around the bulb, surface densely floccose fibrillose-scaly, hollow, white inside and out. **UNIVERSAL VEIL** white, typically absent on cap but forming well-developed sac around stem base. **RING** (annulus) white, near the top of the stem, membranous, hanging down, persistent. **SPORE PRINT** white. **ODOR** and **TASTE** not distinctive.

Habit and habitat single or clustered on soil and duff in hardwood forests with beech, oaks, and birches. Also reported from mixed hardwood and coniferous forests and considered symbiotic with some pine species. July through October. **RANGE** widespread.

Spores subglobose, smooth, inamyloid, 7.5–9.5 × 7–9 µm, basidia 2-sterigmate.

Comments Deadly poisonous. This species is a common large white mushroom in mature forests, but also is found in grassy areas if the host tree species are present. *Amanita virosa* Bertillon is similar, usually more robust, and has 4-sterigmate basidia. Both species turn bright yellow on the cap with aqueous KOH solutions (2–10%).

Amanita phalloides (Vaillant ex Fries) Link

DEATH CAP

SYNONYMS *Agaricus phalloides* Vaillant ex Fries, *Amanita viridis* Persoon, *Fungus phalloides* Vaillant

Large mushroom with green smooth cap, sometimes with large flat white patch; gills faintly yellowish-green in oblique view; stem white or pale yellowish-green with white ring and ample white saclike volva

CAP 6–16 cm wide, convex, becoming plane, streaked or spotted with combinations of green, olive-green, yellow-green, gray, or brown, surface slippery or tacky at first and may have large membranous white patches of universal veil, or patches may be absent, margin not lined. **GILLS** white or cream, but faintly yellowish-green in oblique view, free or barely attached, crowded. **STEM** 5–17(–30) cm long, 9–20 mm broad at the apex, equal with globose bulb at base, surface white or pale yellowish-green below ring, smooth or densely minute cottony scaly, volva white, sac- or cuplike, membranous, ample. **UNIVERSAL VEIL** white, membranous, persistent as a cup. **RING** superior, membranous, white, pendulous. **SPORE PRINT** white. **ODOR** with age repulsively foul or nauseating. **TASTE** mild.

Habit and habitat single or scattered, growing on the ground under oak, but also reported under imported species of pine and even Canadian hemlock and other native conifers. July to November. **RANGE** widespread.

Spores subglobose or broadly elliptical, smooth, amyloid, 8–10 × 6–8 μm.

Comments Deadly poisonous. The much more commonly collected *Amanita citrina* differs by the lemon-yellow cap, flat-topped bulb at the base of the stem that does not have a saccate volva, and the odor of raw potato. *Amanita phalloides* is said to have a pleasant odor and taste when young, but if cooked and eaten, the consequences are serious and too often deadly.

Amanita albocreata
(G. F. Atkinson) E.-J. Gilbert

SYNONYMS *Amanitopsis albocreata* G. F. Atkinson, *Vaginata albocreata* (G. F. Atkinson) Murrill

Tall white mushroom with cream-yellow cap center scattered with white veil patches, also slippery; stem with large globose bulb at base with tightly clasping white cup, lacking a ring

CAP 5–8 cm wide, convex, becoming plane, white with a cream-yellow center, surface slippery, scattered with flat white floccose patches of universal veil, these are easily removed when surface is moist, margin very fine, bumpy, lined at edge. **GILLS** white, free, crowded, broad, edges cottony. **STEM** 10–13 cm long, 6–12 mm broad, equal or tapering upward, large globose bulb at base with membranous cuplike volva tightly adhering and producing a free edge encircling the bulb (ocreate or ensheathing the bulb), surface white and covered with cottony mealy scales, hollow and white inside. **UNIVERSAL VEIL** white, forming soft patches on cap and occasionally as irregular concentric rings on lower stem. **RING** absent. **SPORE PRINT** white. **ODOR** not distinctive. **TASTE** unknown.

Habit and habitat single or clustered on duff in coniferous and hardwood forests with oaks. July and August. **RANGE** widespread, also south to the Gulf Coast.

Spores subglobose or broadly elliptical, smooth, inamyloid, 7–9.5 × 6–8.5 µm.

Comments A striking whitish species that might be poisonous, but that has not been confirmed. However, *Amanita ocreata* Peck is a more robust West Coast species that has an annulus and a cuplike volva, and it is deadly poisonous.

Amanita brunnescens var. brunnescens G. F. Atkinson

CLEFT-FOOTED AMANITA

Large mushroom with brown cap, white flesh, and white stem turning brown when bruised; membranous white ring persistent on stem apex; bulbous stem base vertically cleft

CAP 4–10 cm wide, conical-convex, eventually plane, uniformly brown or grayish-brown overall, becoming paler over margin with expansion, innate brown radiating fibrils obvious, surface glabrous, margin not or faintly lined, with or without white, flat, universal veil patches randomly scattered, flesh discoloring reddish-brown when exposed. GILLS white or creamy-white, free, crowded. STEM 6–15 cm long, 10–20 mm broad at the apex, tapering upward, with large abrupt flat-topped (marginate) bulb splitting vertically in one or more places, glabrous or fibrillose, white but slowly turning brown where handled. UNIVERSAL VEIL white on the pileus, typically not present on the stipe base. RING white, superior, persistent, membranous, pendulous, collapsing on stipe. SPORE PRINT white. ODOR not distinctive, or reminiscent of freshly unearthed potatoes in the stem base and bulb when freshly sectioned. TASTE unknown.

Habit and habitat single or scattered, growing on duff in mixed coniferous and hardwood forests, usually with oaks present. July to October. RANGE widespread.

Spores globose or subglobose, smooth, amyloid, 7–9.5 × 7–8.5 µm.

Comments Poisonous. A variant found mostly in the summer has a white cap, Amanita brunnescens var. pallida L. Krieger. Another similar white-capped species that stains brown on the stipe from handling, A. aestivalis Singer ex Singer, seems to differ by having a slippery cap and a basal bulb without a cleft. We do not know for certain whether these two variants are also poisonous.

Amanita ceciliae (Berkeley & Broome) Bas

STRANGULATED AMANITA

SYNONYMS *Agaricus ceciliae* Berkeley & Broome, *Amanita inaurata* Secretan ex Gillet, *Amanitopsis inaurata* (Secretan ex Gillet) Fayod, *Amanitopsis ceciliae* (Berkeley & Broome) Wasser, *Amanita strangulata* var. *royeri* (L. Maire) Contu

Dark brown, grooved, and lined cap with gray cottony warts; stem lacking a ring, but when fresh, dark gray, flaring, cottony cup edge on base of stem

CAP 5–12 cm wide, conical-convex, becoming plane, brown expanding to paler brown with dark brown center, smooth, glabrous, slippery at first, margin strongly grooved and lined, surface scattered with gray, flat, soft, easily removed patches of universal veil. **FLESH** white, unchanging. **GILLS** white, free, crowded, short gills truncate, ascending straight up to cap. **STEM** 7–18 cm long, 10–20 mm broad at the apex, equal or tapering upward, cottony fibrillose grayish-brown overall, becoming glabrous, white or ashy-white, volva edge gray and flaring, white constricted cottony below, tightly adhering over stem base. **UNIVERSAL VEIL** gray, easily fragmented and destroyed. **RING** absent. **SPORE PRINT** white. **ODOR** not distinctive. **TASTE** unknown.

Habit and habitat single or scattered, growing on duff in mixed coniferous and hardwood forests, usually with oaks or beech present. July to October. **RANGE** widespread.

Spores globose, smooth, inamyloid, various sizes reported: 8–12, 10–12, and 11.5–14 µm.

Comments The range of spore sizes reported appears to show the possibility of cryptic species not yet recognized in this complex. *Amanita inaurata* and *A. strangulata* are European names that have also been applied to these taxa, but the former is a synonym of *A. ceciliae* and the latter has a confused concept and is therefore not a name that can be accurately applied.

Amanita citrina Persoon

CITRON AMANITA

SYNONYMS *Amanita bulbosa* Persoon, *Amanita mappa* (Batsch) Bertillon, *Venenarius mappa* (Batsch) Murrill

Cap greenish-yellow or pale lemon-yellow, sometimes with flat, felted patches; stem white with flat-topped, bulbous base and pale lemon-yellow ring; odor of raw potato

CAP 4–13 cm wide, broadly convex, becoming plane, greenish-yellow or pale lemon-yellow, sometimes with random pale brown discolorations, smooth, glabrous, slippery, or tacky at first, margin not lined, surface scattered with white, buff, or pale yellow, flat, felted, easily removed patches of universal veil. **GILLS** white, free, crowded. **STEM** 6–14 cm long, 10–15 mm broad at the apex, equal or tapering upward from a flat-topped bulbous base (marginate), smooth or cottony fibrillose, becoming glabrous, white, with white, tightly adherent cuplike volva surrounding basal bulb, with uneven projecting margin that may stain pale brown. **UNIVERSAL VEIL** white or pale yellow, soft, thin. **RING** pale lemon-yellow, superior, pendant, edges easily torn, collapsing on stipe. **SPORE PRINT** white. **ODOR** typically of raw potato. **TASTE** unknown.

Habit and habitat single, scattered, or clustered on the ground in mixed pine and hardwood forests, usually with oaks present. July to October. **RANGE** widespread.

Spores subglobose or globose, smooth, amyloid, reported as 6.3–7 × 6–6.3 µm, or elsewhere as 7–10 µm.

Comments It appears our citron amanita consists of two or more species, considering the recorded spore dimensions. DNA evidence supports this indication. A variety with dull lavender cup and lavender on the underside of the annulus goes by *Amanita citrina* var. *lavendula* (Coker) Sartory & Maire and is fairly common in the Northeast.

Amanita farinosa Schweinitz

POWDER-CAP AMANITA

SYNONYMS *Amanitella farinosa* (Schweinitz) Earle, *Amanitopsis farinosa* (Schweinitz) G. F. Atkinson, *Vaginata farinosa* (Schweinitz) Murrill

Small, grayish with strongly lined and gray powdery cap; stem with small bulb at base with well-defined rim, covered with gray powdery sheath

CAP 2–7 cm wide, plano-convex or plane, pale gray or with brown hues, dark on center, strongly grooved and lined from edge to midway, glabrous but covered with dense gray or gray-brown, cottony or powdery universal veil, which is easily removed. **GILLS** white, free, close or subdistant. **STEM** 3–6.5 cm long, 3–10 mm broad at the apex, equal or tapering upward from small basal bulb, gray cottony or powdery overall, becoming glabrous, off-white, with denser gray-brown powdery volva covering basal bulb. **UNIVERSAL VEIL** gray or gray-brown, soft, thin, powdery on cap and rim of stem bulb. **RING** absent. **SPORE PRINT** white. **ODOR** none. **TASTE** unknown.

Habit and habitat single, scattered, or clustered on the ground in mixed conifer and hardwood forests, usually with oaks present. June to October. **RANGE** widespread.

Spores subglobose or globose, smooth, inamyloid, 6.5–9.5 × 4.5–8 µm.

Comments A very distinctive small amanita. No other species in the genus is even similar. It is very fragile and difficult to collect intact unless you are very careful.

Amanita flavoconia
G. F. Atkinson

YELLOW PATCHES AMANITA

Cap yellowish-orange with yellow cottony warts, margin not lined; stem pale yellow or yellowish-orange with oval basal bulb and crumbly-cottony bright yellow volva that is easily lost

CAP 2–8 cm wide, convex, eventually plane, bright yellow or yellowish-orange, glabrous, slippery or tacky, typically with cottony bright yellow warts of universal veil, easily removed, margin not lined. GILLS white, free, close or crowded, broad, short gills rounded or tapered as they ascend to the cap. STEM 5–11 cm long, 5–15 mm broad at the apex, equal or tapering upward from inflated oval basal bulb, pale yellow or yellowish-orange, glabrous or with fine yellowish flattened fibrils over lower half, occasionally with crumbly bright yellow volva material, easily removed, adhering to the stem base. UNIVERSAL VEIL bright yellow or yellowish-orange. RING superior, membranous, pendant, pale yellow or white with yellow edge. SPORE PRINT white. ODOR not distinctive. TASTE unknown.

Habit and habitat single or scattered, growing on the ground in mixed hardwood forests with beech, oak, and birch, also reported with conifers like hemlock. June to November. RANGE widespread.

Spores ellipsoid, smooth, amyloid, 7–9.5 × 4.5–5 μm.

Comments One of the most frequently encountered species of *Amanita* and immediately recognized by the yellowish-orange, unlined cap and the yellow patches of volva on the cap and crumbling at the base of the stem. The stem should be pale yellow or darker yellow-orange. If the stem is white and the cap is pure yellow, lacking orange hues, then you have *A. elongata* Peck. Also compare *A. frostiana*.

Amanita frostiana (Peck)
Saccardo

FROST'S AMANITA
SYNONYMS *Agaricus frostianus* Peck, *Amanitaria frostiana* (Peck) E.-J. Gilbert, *Venenarius frostianus* (Peck) Murrill

Bright orange-yellow cap with lined and grooved margin and yellow cottony warts; stem with globose basal bulb displaying a sharply rimmed cup

CAP 2–8 cm wide, convex, becoming plano-convex, bright orange or orange-yellow, some with reddish-orange on disc, glabrous, with cottony bright yellow or orange-yellow warts of universal veil, margin lined and grooved. **GILLS** white or pale cream, free, close or crowded, short gills truncate, ascending straight up to cap. **STEM** 5–9 cm long, 4–11 mm broad at the apex, slender, equal down to the abruptly globose basal bulb with volva forming a sharp-edged rimmed cup, surface white but sparsely or densely covered with flattened golden-yellow fibrils and scales, may become glabrous, with crumbly volva material adhering around the base of stipe, easily removed. **UNIVERSAL VEIL** bright yellow, crumbly soft, easily lost. **RING** superior, membranous, somewhat bandlike, yellow. **SPORE PRINT** white. **ODOR** not distinctive. **TASTE** unknown.

Habit and habitat single or scattered, growing on the ground in mixed hardwood and coniferous forests, often near oak, beech, or birch, but also under white pine. June to September. **RANGE** widespread.

Spores globose or subglobse, smooth, inamyloid, 8.5–10.5 × 8–10 μm.

Comments *Amanita frostiana* is distinguished in the field from other yellow- or orange-yellow-capped amanitas by its lined and grooved pileus margin, its slender stipe with globose, sharply rimmed volva, and its truncate short gills. The spores are also globose and inamyloid. Compare with the more commonly encountered *A. flavoconia*.

Amanita porphyria Albertini & Schweinitz

PURPLE-BROWN AMANITA
Cap dark grayish or purplish-brown with flat violet-gray universal veil patches; stem with gray persistent ring and a flat-topped basal bulb with membranous cup rim on top of the bulb

CAP 3–12 cm wide, convex, with low broad central bump, dark brownish-gray or purplish-brown, surface glabrous, slippery at first, sometimes with flattened cottony violet-gray patches of universal veil covering significant portions of the cap, margin usually not lined. **GILLS** white, free or barely attached, crowded. **STEM** 5–12 cm long, 6–18 mm broad at apex, equal with abruptly globose flat-topped bulb at base, surface white with grayish flattened fibrils and scales below ring, fine white lines of pubescence on the stipe above ring, volva soft white or grayish, attached to the bulb, projecting a free membranous irregular rim on top of the bulb. **UNIVERSAL VEIL** gray or white becoming gray. **RING** superior, gray or violet-gray, soft, cottony on lower surface, smooth and lined above, eventually collapsing on stipe, persistent. **SPORE PRINT** white. **ODOR** not distinctive. **TASTE** unknown.

Habit and habitat single or scattered, growing on the ground under conifers and in mixed woods. July to October. **RANGE** widespread.

Spores globose or subglobose, smooth, amyloid, 7–10 × 6–10 µm.

Comments Note the similarity in features, except for color, to *Amanita citrina*. It is highly possible that our North American *A. porphyria* is genetically different from the European species and will eventually be described as new.

Amanita muscaria var. *guessowii* Veselý

YELLOW-ORANGE FLY AGARIC

SYNONYM *Amanita muscaria* var. *formosa* Persoon

Large mushroom with dark orange-yellow cap with flattened cream or tan warts and lined, bumpy margin; stem white or cream with characteristic concentric scales and bands running up the stem from a rounded basal bulb; ring white or cream ingrown with stipe, not obvious. **UNIVERSAL VEIL** pale cream or tan, only obvious on cap. **RING** superior, membranous, white or pale yellow, pendulous. **SPORE PRINT** white. **ODOR** not distinctive. **TASTE** mild.

CAP 4–18 cm wide, convex, becoming plane, pale yellow or darker orange-yellow, slippery or tacky at first, covered with flattened creamy-yellow or tan warts of universal veil, margin lined and bumpy with age. **GILLS** white or cream, free, crowded. **STEM** 4–15 cm long, 7–30 mm broad at the apex, equal with tapered subglobose or subrooting bulb at base, white or cream-yellow below ring, with concentric or irregular torn-looking scaly bands covering the upper bulb and lower stem, volva

Habit and habitat single or scattered, growing on the ground under spruce, especially Norway spruce, but also pines, other conifers, and hardwood trees such as poplar. July to November. **RANGE** widespread.

Spores elliptical, smooth, inamyloid, 9–13 × 6–8 µm.

Comments Poisonous to humans and dogs. These big colorful mushrooms fruit prolifically in the wet season and attract attention. The spotted orange-yellow cap and the concentric rings on the base of the stem are diagnostic; however, the yellow or tan warts and the colors of the cap can be washed out in heavy rains. This species also has a pure white form, *Amanita muscaria* var. *alba* Peck.

Amanita rubescens Persoon

BLUSHER

Large mushroom with bronze-brown or brass-yellow cap with small reddish-tan warts; stem white or pale tan with gradually enlarged base, lacking obvious volva; stem and gills slowly turning reddish-brown or wine-red when injured

CAP 4–20 cm wide, convex, becoming plane, bronze-brown or reddish-brown, brass-yellow in some, glabrous, slippery but soon dry, with cottony tan or reddish-tan small warts of universal veil, warts easily lost, margin not lined or faintly so. GILLS white, free or barely attached, close or crowded, bruising reddish-brown. STEM 7–24 cm long, 8–40 mm broad at the apex, equal, base enlarged gradually, lacking any obvious volva, surface white or pale tan with fine fibrils and scales or glabrous, fibrils and flesh bruising wine-red slowly where injured. UNIVERSAL VEIL tan or grayish-red. RING superior, thin, membranous, pendant, white above, white or pale cream-yellow below, shredding, sometimes adhering to cap margin, turning reddish. SPORE PRINT white. ODOR not distinctive. TASTE mild.

Habit and habitat single, scattered, or gregarious, growing on the ground in mixed hardwood and coniferous forests with oak, pine, or fir in the mix. June to October. RANGE widespread.

Spores ellipsoid, smooth, amyloid, 7–10 × 5–6 µm.

Comments Rod Tulloss presents good morphological data to suggest that *Amanita rubescens* of eastern North America is not the same species as *A. rubescens* of Europe and eastern Asia, differing in spore size and several macromorphological features. Our blusher seems to need a new name. One variant has decidedly yellow volva warts in early stages, this may be a different taxon as well. The darker brown-capped form is frequently encountered, but the paler brass-yellow form, imaged here, is not uncommon.

Amanita flavorubens (Berkeley & Montagne) E.-J. Gilbert

YELLOW AMERICAN BLUSHER

SYNONYM *Amanita flavorubescens* G. F. Atkinson

Cap golden-yellow or golden-bronze-brown with cottony yellow warts; ring with golden-yellow tints; stem base enlarged, turning red where injured

CAP 5–10 cm wide, convex, golden-yellow or golden-bronze, some lemon-yellow or with brownish center, glabrous, slippery, with cottony yellow warts of universal veil, these easily lost, margin not lined or faintly so. **GILLS** white or pale cream, free or barely attached, close or crowded. **STEM** 5–15 cm long, 8–17 mm broad at the apex, equal, base ovate enlarged or not enlarged, with or without fragile, easily removed, dark yellow volva patches at the base, surface white with fine yellow fibrils and scales or glabrous, bruising red slowly where injured. **UNIVERSAL VEIL** bright yellow, crumbly soft, easily lost, especially at stem base. **RING** superior, thin, membranous, pendant, white above, yellow below, easily shredding. **SPORE PRINT** white. **ODOR** fragrant or not distinctive. **TASTE** unknown.

Habit and habitat single, scattered, or gregarious, growing on the ground in mixed hardwood and coniferous forests, often with oak, beech, and birch, but also under hemlock. June to October. **RANGE** widespread.

Spores ellipsoid, some elongated, smooth, amyloid, 8–11 × 5.5–7 μm.

Comments Older field guides list this species as *Amanita flavorubescens*. Described from Ohio in 1856, thus the name that must be used, *A. flavorubens* is similar to the red-bruising *A. rubescens* in many aspects, but differs by the yellow colors in the cap, volva, and ring. The red bruising is slow and occurs mostly in the stem base, but it can occur on any part of the fruit body. See also *A. flavoconia* and *A. muscaria* var. *guessowii*; they have yellow colors but lack the bruising reactions.

Amanita wellsii (Murrill) Murrill

SALMON AMANITA

Cap orange-yellow with salmon-pink mixed in, covered with dense yellow cottony patches; stem pale yellow with cream-yellow cottony ring that easily collapses and disappears

CAP 3–10 cm wide, convex, becoming plane, orange-ochre, yellowish-orange, or with salmon-pink hues, glabrous, slightly slippery when moist, lined over margin with age, covered with dense cream or yellow cottony universal veil patches, margin adorned with hanging cottony patches from annulus, all easily removed. **GILLS** pale cream, free or barely attached, crowded. **STEM** 7–16 cm long, 5–25 mm broad at the apex, equal with club-shaped or subglobose basal bulb, pale yellow, cottony above ring but glabrous below, with yellow volva material covering basal bulb. **UNIVERSAL VEIL** cream-yellow, cottony. **RING** superior, delicate cottony, cream-yellow, often collapsing, disappearing. **SPORE PRINT** white. **ODOR** none. **TASTE** mild or unpleasant.

Habit and habitat single, scattered, or clustered, growing on the ground in mixed forests, often with birch present. July to August. **RANGE** widespread.

Spores ellipsoid or somewhat elongated, smooth, inamyloid, 10.5–14 × 6–8 µm.

Comments This pastel-colored species is eye catching and distinctive. Once you remove the veil material from the cap, the colors will remind you of an apricot.

Amanita sinicoflava Tulloss

Cap yellow-brown, margin lined with expansion; stem ringless but with white saclike cup at base, that turns gray with age from the top downward

CAP 2.5–7 cm wide, conical-convex or egg-shaped, becoming plane, often with a broad central bump, yellow-brown, yellowish-olive, or olive-tan, glabrous, slippery-tacky at first, occasionally with pale gray patches of universal veil that are easily removed, margin strongly ribbed or lined with age. **GILLS** white, becoming grayish with age, free or barely attached, close or crowded. **STEM** 6–14 cm long, 4–12 mm broad at the apex, equal or tapering toward apex, surface white with flattened fibrils, volva saclike and attached tightly to stem over lower third to half, loosely sheathing and free from the stem above, white but soon gray from the top downward, edges membranous and lobed. **UNIVERSAL VEIL** white, becoming grayish, membranous, persistent as a sac at the stem base. **RING** absent. **SPORE PRINT** white. **ODOR** not distinctive. **TASTE** mild.

Habit and habitat single or scattered, growing on the ground in mixed hardwood forests, often with oak, beech, or birch, also in mixed coniferous forests including pines or hemlock. June to October. **RANGE** widespread.

Spores subglobose or globose, smooth, inamyloid, 9–12 × 8.5–11.5 µm.

Comments The species is named for the Chinese-mustard color of the cap. Originally it was lumped with the American version of *Amanita fulva* Fries, but that species differs by its orange-brown pileus and orange-staining volva.

Amanita jacksonii Pomerleau

AMERICAN CAESAR'S MUSHROOM

SYNONYM *Amanita umbonata* Pomerleau

Dark red or reddish-orange egg-shaped cap that becomes flat and strongly lined with expansion; stem is yellow with orange fibrils and has a thick, white, saclike cup at base; ring membranous and rich yellow

CAP 5–22 cm wide, conical-convex or egg-shaped, becoming plane, some with a broad central bump, dark reddish or reddish-orange, becoming paler orange-yellow over margin with expansion, glabrous, slippery-tacky at first, usually lacking white universal veil patches, but these are occasionally present, margin strongly grooved and lined to near disc. **GILLS** yellow, free or barely attached, close. **STEM** 9–23 cm long, 5–31 mm broad at the apex, equal or tapering toward apex, surface yellow with orange fibrils and scales scattered sparsely or densely, volva white, deeply saclike, thick and membranous, edges lobed. **UNIVERSAL VEIL** white, not leaving patches on the cap. **RING** superior, membranous, yellow or yellow-orange. **SPORE PRINT** white. **ODOR** not distinctive. **TASTE** mild.

Habit and habitat single or scattered, growing on the ground in mixed hardwood and coniferous forests of oak and pine. July to September. **RANGE** widespread.

Spores elliptical, smooth, inamyloid, 8–10 × 6–7.5 µm.

Comments In older field guides and the technical literature for eastern North America, this species is known as *Amanita caesarea* (Scopoli) Persoon. All current evidence points to our eastern species being distinct from those in Europe and Asia. Our *A. jacksonii*, described in 1984 by René Pomerleau, is considered edible, like *A. caesarea*.

Amanita vaginata (Bulliard) Lamarck

GRISETTE

SYNONYMS *Agaricus vaginatus* Bulliard, *Amanita hyperporea* (P. Karsten) Fayod, *Amanita livida* Persoon, *Amanita strangulata* (Fries) Quélet

Cap gray or gray-brown with grooved margin; stem white with pale gray flattened fibrils, saclike volva white, discoloring gray, ring absent

CAP 4–10 cm wide, conical-convex or egg-shaped, becoming plane, some with a broad central bump, moderate to dark gray or gray-brown, glabrous, slippery-tacky at first, occasionally with white patches of universal veil, these easily removed, margin strongly grooved and lined to near disc. **GILLS** white, or gray with age, free or barely attached, close or subdistant. **STEM** 6–20 cm long, 4–20 mm broad at the apex, equal or tapering toward apex, surface white, glabrous or with gray flattened fibrils or mealy scales, volva saclike and loosely sheathing the stem, white, discoloring gray, edges membranous and lobed. **UNIVERSAL VEIL** white, membranous, persistent. **RING** absent. **SPORE PRINT** white. **ODOR** not distinctive. **TASTE** mild.

Habit and habitat single or scattered, growing on the ground in mixed hardwood forests of oak, beech, or birch, also under mixed coniferous forests including pine or hemlock. In grassy parks with trees in the vicinity. June to October. **RANGE** widespread.

Spores globose, smooth, inamyloid, approximately 8–12 µm.

Comments It appears this "species" consists of many morphological variants in North America, that have not been fully recognized and sorted out. Surely we have more than one species. Use caution if collecting it for eating, or better yet, just avoid it for culinary purposes. See the comments under *Amanita sinicoflava*.

Amanita fulva Fries

TAWNY GRISETTE

SYNONYMS *Amanita vaginata* var. *fulva* (Fries) Gillet, *Amanitopsis fulva* (Fries) W. G. Smith, *Amanitopsis vaginata* var. *fulva* (Fries) Saccardo

Cap dark orange-brown with grooved and lined margin; stem with white saclike volva that often discolors orange-brown; ring absent

CAP 4–10 cm wide, broadly conical-convex or egg-shaped, becoming plane, some with a broad central bump, reddish-brown or orange-brown, becoming paler over margin with expansion, glabrous, slippery-tacky at first, usually lacking white patches of universal veil, but they are occasionally present, margin grooved and lined. **GILLS** white, free or barely attached, close. **STEM** 7–16 cm long, 3–15 mm broad at the apex, equal or tapering toward apex, surface white, glabrous or with fine flattened fibrils or small scales, volva saclike, membranous, edges lobed, white, often discoloring orange-brown. **UNIVERSAL VEIL** white, often with rusty-brown discolorations. **RING** absent. **SPORE PRINT** white. **ODOR** not distinctive. **TASTE** mild.

Habit and habitat single or scattered, growing on the ground in mixed hardwood and coniferous forests. July to September. **RANGE** widespread.

Spores globose or subglobose, smooth, inamyloid, (9–)10–12 × 9–12 µm.

Comments *Amanita fulva* is a European name used widely for our northeastern North American species. The two are morphologically similar, but ours is considered an undescribed species by Rod Tulloss. I will use *A. fulva* for now. Compare with the yellow-brown-capped *A. sinicoflava* that is found in similar habitats.

Amanita multisquamosa
Peck

SYNONYM *Amanita pantherina* var. *multisquamosa* (Peck) D. T. Jenkins
Cap white or pale cream and densely covered with cottony warts; stem white with bulbous base displaying a collarlike band, and with a median flaring, then collapsing ring

CAP 3–11 cm wide, convex, becoming plane with broadly depressed center, white, often with yellow-tan disc, glabrous, slippery or tacky at first, covered with flattened cottony white or pale cream universal veil patches, margin lined. **GILLS** white, free or barely attached, crowded, edges white pubescent fringed. **STEM** 5–14 cm long, 10–20 mm broad at the apex, equal with subglobose or ovoid bulb at base, surface white or sordid cream, becoming watery buff from handling, glabrous or with fine white pubescence above ring, cottony and scaly below ring, volva forming an abrupt white collarlike band (circumsessile) on top of stipe bulb. **UNIVERSAL VEIL** white or pale grayish. **RING** median, white or cream, soft, cottony on lower surface, smooth above, with distinct thickened double-edged rim and sheathing up the stem at first, eventually collapsing on stem, persistent. **SPORE PRINT** white. **ODOR** not distinctive, or of radish. **TASTE** unknown.

Habit and habitat single or scattered, growing on the ground under oaks, pines, or both. July to October. **RANGE** rarely collected.

Spores broadly elliptical, smooth, inamyloid, 9–12 × 6–9 μm.

Comments Poisonous. The pallid colors overall with the densely warted, lined cap, nearly median soft annulus and characteristic bulbous stem base with the collarlike band produced by the volva are features unique to this lovely species. *Amanita cothurnata* G. F. Atkinson may also be a synonym of this species.

Lepiota and Allies

Chlorophyllum molybdites
(G. Meyer) Massee

GREEN-SPORED LEPIOTA
SYNONYMS *Agaricus molybdites* G. Meyer, *Lepiota molybdites* (G. Meyer) Saccardo, *Macrolepiota molybdites* (G. Meyer) G. Moreno, Bañares & Heykoop, *Chlorophyllum morganii* (Peck) Massee, *Agaricus morganii* Peck

Large mushroom growing in grassy areas, with brown scales on the white cap; gills free, white, but soon greenish; stem white, bruising brown, with double shaggy ring; flesh white, sometimes turning dingy reddish

CAP 10–30 cm wide, convex, becoming plane, smooth and uniformly pale pinkish-buff or pinkish-brown when unopened, breaking up into scattered pinkish-brown flattened scales with expansion, exposing white fibrillose ground color, the center remains evenly pinkish-brown, dry and soft, margin with small white fibrillose scales. **FLESH** white, unchanging or reddish when injured, thick. **GILLS** white, turning grayish-green, free, crowded, broad. **STEM** 5–25 cm long, 15–30 mm broad, equal or club-shaped, white, discoloring brownish from handling or aging, glabrous. **RING** white above, brown or greenish below, shaggy fibrillose, large, double, persistent and often movable. **SPORE PRINT** grayish-green. **ODOR** and **TASTE** not distinctive.

Habit and habitat usually scattered or clustered, and frequently forming large fairy rings, growing on lawns, in disturbed areas, in the open. August to October. **RANGE** widespread.

Spores ellipsoid or almond-shaped, end truncate, smooth, colorless, thick-walled with apical pore, dextrinoid, 9–13 × 6–9 μm.

Comments Poisonous. This species causes severe gastrointestinal distress. Take a spore print if you are contemplating eating large mushrooms off lawns. If it is green, do not try to cook and eat this fungus. See also *Chlorophyllum rachodes*.

Chlorophyllum rachodes
(Vittadini) Vellinga

SHAGGY PARASOL

SYNONYMS *Agaricus rachodes* Vittadini, *Lepiota rachodes* (Vittadini) Quélet, *Macrolepiota rachodes* (Vittadini) Singer

Large mushroom growing on the ground, with large brown scales on the white shaggy fibrillose cap; gills free, white, bruising brown; stem white, bruising brown, with double shaggy ring; flesh white, turning reddish, then brown

CAP 6–20 cm wide, globose then convex with expansion, becoming plane, smooth and evenly brown or grayish-brown at first, with expansion exposing white fibrillose ground color, breaking up into large shaggy brown flattened or upturned scales, center evenly brown or grayish-brown, dry and soft, margin with shaggy pale gray fibrillose scales. **FLESH** white or pale gray-brown, turning peach-red when exposed, then brown, thick. **GILLS** white, turning brown with bruising or age, free, close or crowded, broad. **STEM** 10–20 cm long, 10–30 mm broad, club-shaped with a basal bulb, off-white, discoloring from handling or aging brownish, glabrous. **RING** white above, brown below, shaggy fibrillose, large, double, usually persistent and often movable. **SPORE PRINT** white. **ODOR** not distinctive. **TASTE** mild.

Habit and habitat single, scattered, or clustered, sometimes forming fairy rings, growing on lawns, in disturbed areas, in the open, under trees, often near spruce. June to October. **RANGE** widespread.

Spores ellipsoid or somewhat almond-shaped, end truncate, smooth, colorless, thick-walled with apical pore, dextrinoid, 8–13 × 5.5–8 μm.

Comments *Chlorophyllum rachodes* does not turn reddish when dried. Also the gills and stipe do not turn bright ochre-yellow when bruised, as do those of *Leucoagaricus americanus*. *Chlorophyllum rachodes* is known to cause gastrointestinal upset in some mycophagists and thus should be avoided. See also *C. molybdites* and *Macrolepiota procera*.

Macrolepiota procera
(Scopoli) Singer

PARASOL
SYNONYMS *Agaricus procerus* Scopoli, *Lepiota procera* (Scopoli) Gray, *Amanita procera* (Scopoli) Fries, *Leucocoprinus procerus* (Scopoli) Patouillard

Tall, slender, mostly white mushroom with brown scales on the cap and stem; gills free, white; stem with double-edged, movable ring; flesh white and unchanging

CAP 7–20 cm wide, oval then convex or bell-shaped, eventually plane with a rounded bump on the center, smooth and uniformly brown or reddish-brown, with expansion the surface breaks up into scattered brown scales, exposing the white fibrillose ground color, dry and soft, margin shaggy fibrillose. FLESH white, unchanging or with a dingy reddish tint, soft. GILLS white, darkening to pinkish or tan with age, free, close, broad. STEM 15–40 cm long, 8–13 mm broad, equal with bulbous base, slender and tall, white but covered with small brown scales. RING white, shaggy fibrillose, double-edged, persistent and freely movable. SPORE PRINT white. ODOR and TASTE not distinctive.

Habit and habitat single, scattered, or clustered on the ground under conifers or hardwoods, along footpaths or in open woods. August to October. RANGE widespread.

Spores ellipsoid, smooth, colorless, thick-walled with apical pore, dextrinoid, 12–18 × 8–12 µm.

Comments Considered a good edible, but it has some similarities to species that cause problems. Please use caution if you decide to make this one part of your harvest. Compare with *Chlorophyllum rachodes*, the shaggy parasol, which is a stockier mushroom with reddening flesh when exposed, and the stem lacks the small brown scales characteristic of *Macrolepiota procera*. If you find specimens with smaller caps and white scales on the stem, and they produce a pinkish spore print, you have found the not-so-commonly-collected *M. prominens* (Saccardo) M. M. Moser.

Leucoagaricus americanus
(Peck) Vellinga

REDDENING LEPIOTA
SYNONYMS *Agaricus americanus* Peck, *Lepiota americana* (Peck) Saccardo, *Leucocoprinus americanus* (Peck) Redhead

Large mushroom growing on stumps, wood chips, sawdust piles, with numerous flat reddish-brown scales on the white cap; gills free, white, bruising ochre-yellow; stem white, bruising bright yellow, with double ring

CAP 3–15 cm wide, convex, then bell-shaped, becoming plane with low broad central bump, in youngest stages evenly pinkish-buff or reddish-brown, exposing white ground color with expansion, covered with brown flattened or upturned scales, center evenly pinkish or grayish-brown, dry, margin finely lined. **FLESH** white, bruising yellow-orange, turning vinaceous. **GILLS** white, turning ochre-yellow when bruised, free, close, broad. **STEM** 7–15 cm long, 8–22 mm broad, equal or swollen over lower part, then tapered at the base (ventricose), white, discoloring or aging reddish-brown, but turning quickly bright yellow when bruised, then becoming reddish, glabrous. **RING** white, large, double, usually persistent. **SPORE PRINT** white. **ODOR** not distinctive. **TASTE** mild.

Habit and habitat scattered or clustered on stumps, sawdust piles, wood chips. June to October. **RANGE** widespread.

Spores ellipsoid, smooth, colorless, thick-walled with apical pore, dextrinoid, 9–14 × 4–5 μm.

Comments This mushroom turns wine-red when dried. It is listed as edible and good. Compare with *Chlorophyllum rachodes*, which looks similar and does make some people ill.

Leucoagaricus leucothites
(Vittadini) Wasser

SMOOTH LEPIOTA
SYNONYMS *Lepiota leucothites* (Vittadini) P. D. Orton, *Lepiota naucina* (Fries) P. Kummer, *Leucoagaricus naucinus* (Fries) Singer, *Lepiota holosericea* Gillet

All-white, growing on grassy lawns in the fall; gills may become pinkish-gray; ring is obvious and collarlike

CAP 5–10 cm wide, oval then broadly convex, becoming plane, often with broad central bump, white or ashy-white or pale gray, typically glabrous and smooth, but occasionally fibrous-scaly, dry, surface outline can be lumpy. **FLESH** white, unchanging. **GILLS** white, occasionally developing pinkish-gray tints, free, close, broad. **STEM** 5–15 cm long, 5–15 mm broad, equal but often with bulbous base, white but may discolor pale yellow then slowly pale brown, glabrous, dry. **RING** white, double-edged, collarlike, movable. **SPORE PRINT** white. **ODOR** not distinctive, or unpleasant. **TASTE** not distinctive.

Habit and habitat scattered or clustered on soil among grasses on lawns typically, occasionally also found in forest or thin woodlands with conifers present. September to October. **RANGE** widespread.

Spores ellipsoid, smooth, colorless, thick-walled with an apical pore, dextrinoid, 7–9 × 5–6 µm.

Comments Two vaguely similar white species with a ring should be compared to *Leucoagaricus leucothites*. The edible *Agaricus campestris* has gills that turn pink, then chocolate, and the spores are dark chocolate-brown. The deadly poisonous *Amanita bisporigera* grows near beech trees and has a distinct cuplike volva at the base of the stem. The grayish form of *L. leucothites* has been implicated in gastrointestinal upset poisonings.

Leucocoprinus cepistipes
(Sowerby) Patouillard

ONION-STALKED LEPIOTA

SYNONYMS *Agaricus cepistipes* Sowerby, *Lepiota cepistipes* (Sowerby) P. Kummer

White, soft, and flabby mushroom with white granular cap and upper stem; cap margin lined; gills free, crowded; stem with white bandlike, flaring ring; often staining yellow when handled

CAP 2–8 cm wide, oval with flattened top, then bell-shaped, eventually plane with a rounded central bump, white at first with a brown center, becoming yellowish or even purplish drab overall with age, dry and very soft, covered with mealy white granules or scales that are easily rubbed off, margin finely lined. **FLESH** white, sometimes yellowing, soft. **GILLS** white, free, crowded, narrow. **STEM** 4–14 cm long, 3–6 mm broad, equal with a gradually swollen base, white but often yellowing where handled, covered with white granules above ring, glabrous below. **RING** white, sheathing below, flaring above, membranous, persistent. **SPORE PRINT** white. **ODOR** and **TASTE** not distinctive.

Habit and habitat usually clustered on the ground in rich soil, compost, wood chips, sawdust under conifers or hardwoods. June to September. **RANGE** widespread.

Spores ellipsoid, smooth, colorless, thick-walled with small apical pore, dextrinoid, $6–12 \times 6–8$ µm.

Comments This species is named for the shape of the stem base, resembling a scallion or green onion (*Allium cepa* is the scientific name of the commonly consumed onion bulbs, thus *cepistipes*). This flabby little mushroom is reported to cause gastrointestinal upset for some.

Lepiota cristata (Bolton)
P. Kummer

MALODOROUS LEPIOTA
SYNONYMS *Agaricus cristatus* Bolton, *Lepiota colubrina* var. *cristata* (Bolton) Gray

Small white mushrooms with reddish-brown crusty scales on the cap; gills white, free; stem white with a fragile, easily torn ring; odor pungent but hard to describe, not pleasant

CAP 1–7 cm wide, convex, then plane with a rounded bump on the center, uniformly brown or reddish-brown in early stages, soon breaking into rings of small brown scales on white ground color with a brown center, dry. **FLESH** white, thin, fragile. **GILLS** white, barely free, close, narrow. **STEM** 2–8 cm long, 2–6 mm broad, mostly equal, white but may be darker toward base, glabrous or silky-covered below ring. **RING** white, very fragile and easily lost.

SPORE PRINT white. **ODOR** pungent, disagreeable, but with a fragrant spiciness. **TASTE** not distinctive.

Habit and habitat scattered or clustered on humus or leaf litter under various types of trees, along unimproved roads, in grassy areas, or in disturbed sites. June to October. **RANGE** widespread.

Spores wedge- or bullet-shaped with a spur on the flattened end, smooth, colorless, thin-walled and lacking an apical pore, dextrinoid, 5–8 × 3–5 μm.

Comments One of the many small to medium-sized *Lepiota* species. Most can only be accurately identified with microscopic features.

Russula and *Lactarius*

Russula aeruginea Lindblad ex Fries

TACKY GREEN RUSSULA
Moderately large cap, smooth, green, and slightly slippery; gills creamy-white and fragile; stem white; taste mild

CAP 5–9 cm wide, convex, becoming plane with sunken center, dull green or tinted with olive, yellow, or even gray hues, color persistent, glabrous, slightly slippery when wet, quickly drying, smooth or minutely velvety, may have radially raised bumpy lines over margin (tuberculate-striate) upon expansion. **FLESH** white, brittle. **GILLS** white or pale yellow, may become brown-spotted, attached, close, often forked near stem, brittle, snapping when bent. **STEM** 4–8 cm long, 10–20 mm broad, equal or enlarged up or down, white or pale cream-yellow, discoloring brown over the base, glabrous, chalky, snapping cleanly. **SPORE PRINT** cream or pale yellow. **ODOR** not distinctive. **TASTE** mild.

Habit and habitat single, scattered, or clustered on humus, under hardwoods, especially oak, may also be found under conifers. July to September. **RANGE** widespread.

Spores broadly ellipsoid or subglobose, small isolated conical warts with few connections, colorless but ornamentations amyloid, 6–8.5 × 5–7 μm, excluding ornamentation, warts to 0.8 μm high.

Comments This species is considered a good edible. If the specimens are smaller, with caps only 4–6 cm wide, with a green, dry cap surface that has bluish-metallic tints and is composed of cracked, felty, or velvety patches, they are most likely *Russula parvovirescens* Buyck, D. Mitchell & Parrent. If the specimens have cracked cap surfaces that are yellowish-brown with olive hues, are slippery over the center, and have larger caps measuring 5–12 cm, then *Russula crustosa* Peck is the most likely candidate. Compare also with *Russula variata* that can have green colors on the cap.

Russula variata Banning

VARIABLE RUSSULA

Large cap mottled green, purple, and yellow, can become cracked with age; gills and stem white; taste mild but also younger caps and gills can be slowly acrid hot

CAP 5–15 cm wide, convex, becoming plane with center sunken or vase-shaped on some, extremely variable in color, usually mottled with green, purple, olive, yellow, and pink in various mixtures, occasionally one color dominating, glabrous, smooth, wrinkled, or cracked, slightly slippery when wet, quickly drying, may have low radial lines over margin (striate) upon expansion. **FLESH** white, brittle. **GILLS** white, may become brown-spotted, bluntly attached or subdecurrent on some, close or crowded, forked near stem or cap margin, sometimes repeatedly so, when young soft and flexible, with age brittle and snapping

when bent. **STEM** 3–10 cm long, 10–30 mm broad, equal, white, discoloring brown irregularly, glabrous or felted above, chalky, snapping cleanly. **SPORE PRINT** white. **ODOR** not distinctive. **TASTE** of cap and gills mild or younger ones acrid slowly.

Habit and habitat single, scattered, or clustered on humus under hardwoods, especially oak and aspen. June to October. **RANGE** widespread.

Spores broadly ellipsoid, small isolated conical warts with few connections, colorless but ornamentations amyloid, 7–11.5 × 6–8 μm, excluding ornamentation, warts to 1 μm high.

Comments Compare with *Russula cyanoxantha* (Schaeffer) Fries that has similar cap-color variations among green, purple, olive, and pink, but the gills are not forked, at least not as frequently, and the taste is always mild.

Russula brevipes Peck

SHORT-STALKED WHITE RUSSULA
Large white, often vase-shaped, dry cap; gills white, subdecurrent, close; stem white, short, very solid; all surfaces can discolor brownish as well

CAP 10–20 cm wide, convex, broadly and sometimes deeply sunken centrally, with inrolled smooth margin at first, white or creamy-white, some with ochre-tan colors, irregularly developing brownish discolorations, smooth or somewhat felted, dry. **FLESH** white, thick, brittle. **GILLS** white, becoming cream-yellow, sometimes slowly discoloring cinnamon-brown after bruising, bluntly attached or subdecurrent, close or crowded, brittle, making a subtle snapping feeling when bent. **STEM** 2–8 cm long, 25–40 mm broad, equal or tapering downward, very hard and sturdy, white, sometimes discoloring brown when handled, chalky, snapping cleanly.

SPORE PRINT off-white to buff. **ODOR** not distinctive, or somewhat unpleasant. **TASTE** not distinctive.

Habit and habitat single, scattered, or clustered in the needle duff under conifers, frequently hemlocks, or in mixed woods. July to October. **RANGE** widespread.

Spores broadly ellipsoid, warted and partially reticulate, colorless but ornamentations amyloid, 8–11 × 6.5–8.5 μm.

Comments *Russula brevipes* var. *acrior* Shaffer, shown here, differs by its very short tapered stem with a blue-green-colored ring where the gills attach to the stem. The taste can be quite acrid hot as well. *Russula brevipes* often occurs with *Lactarius deceptivus* under hemlocks. The copious white latex, the fluffy, cottony roll of tissue on the cap margin, and the slowly but strongly acrid hot taste will differentiate *L. deceptivus*. Also compare with *R. densifolia* and *R. compacta*.

Russula compacta Frost

FIRM RUSSULA

Medium to large mushroom with very firm flesh, white but soon discolored rusty-brown, turning orange-brown where bruised; flesh white but slowly turning reddish-brown when exposed; odor strong, unpleasant

CAP 3–18 cm wide, convex, becoming plane or shallowly vase-shaped, white or pale buff at first, with discolored areas of rusty-brown or tawny, turning ochre-brown where bruised, smooth or somewhat felted, slippery or sticky when wet. **FLESH** white, discoloring slowly yellowish to reddish-brown when exposed, thick, firm. **GILLS** white, becoming cream-yellow, turning reddish-brown after bruising, bluntly attached, close or subdistant with age, brittle, making a subtle snapping feeling when bent. **STEM** 3–12 cm long, 10–45 mm broad, equal, firm, white, staining tawny slowly when handled, chalky, snapping cleanly. **SPORE PRINT** white. **ODOR** strong, fishy, disagreeable, and intensifying with age or drying, smell remains on your fingers after handling. **TASTE** not distinctive, or slightly bitter acrid.

Habit and habitat single, scattered, or clustered under conifers. July to October. **RANGE** widespread.

Spores broadly ellipsoid or subglobose, warted and partially or completely reticulate, colorless but ornamentations amyloid, 7–10 × 6–8.5 μm excluding ornamentation, warts to 1.2 μm high.

Comments Several white *Russula* species turn various colors when bruised. See comments under *R. densifolia*. I avoid picking *R. compacta* because of the very fishy and disagreeable odor it leaves on my hands.

Russula densifolia Secretan ex Gillett

Medium to large, white, with shallow vase-shaped, slippery cap that becomes gray to black; flesh white turning red, then black when exposed; gills white, close, turning red when bruised; stem white, turning red, then black when bruised

CAP 4–15 cm wide, convex, becoming plane, shallowly vase-shaped with inrolled smooth margin at first, white, discoloring to ash-gray, brown, or black, smooth or somewhat felted, slippery or sticky when wet. FLESH white then slowly reddish, eventually black, thick, firm. GILLS white, becoming cream-yellow, turning reddish, then black after bruising, bluntly attached or subdecurrent, close or crowded, brittle, making a subtle snapping feeling when bent. STEM 2–9 cm long, 10–35 mm broad, equal, firm, white, turning red then slowly black when handled, chalky, snapping cleanly. SPORE PRINT white. ODOR not distinctive. TASTE not distinctive, or slightly acrid.

Habit and habitat single, scattered, or clustered under hardwoods or conifers. July to October. RANGE widespread.

Spores broadly ellipsoid or subglobose, warted and partially or completely reticulate, colorless but ornamentations amyloid, 7–11 × 6–8 μm excluding ornamentation, warts 0.1–0.7 μm high.

Comments Several white *Russula* species turn color when bruised. For *R. dissimulans* Shaffer the flesh turns red, then black, and the gills are subdistant (spores 8–11 × 6.5–9 μm). If the gills are distantly spaced and the spores are smaller (6.5–8 × 5–7 μm) then it is *R. nigricans* Fries. If the exposed flesh goes directly black and it has a slightly acrid taste, then you have *R. albonigra* (Krombholtz) Fries; its spore ornamentations are 0.1 μm high. If the taste is mild and the collection was from under conifers, it is *R. adusta* (Persoon) Fries; its spore ornamentations measure 0.4 μm high.

Russula granulata Peck

Moderately sized, yellow-brown, slimy cap with pale crustlike patches; gills and stem white, becoming pale orange-yellow, stem reddish-brown over base; odor fragrant but not completely pleasant, taste acrid hot

CAP 4–7 cm wide, subglobose, soon convex, becoming plane with center sunken, pale brown, yellowish-brown, dark grayish-yellow, with paler yellow-brown crustlike patches mostly over the center, slimy when wet, quickly drying, with raised bumpy ridges over margin. **FLESH** pale yellow, brown around larval tunnels, firm. **GILLS** white, becoming yellow or orange-yellow, attached, close or subdistant, brittle, snapping when bent. **STEM** 3–6 cm long, 10–26 mm broad, equal or enlarged downward, pale yellow or pale orange-yellow, becoming reddish-brown basally, longitudinally bumpy, chalky, snapping cleanly. **SPORE PRINT** pale orange-yellow. **ODOR** fetid but with a fragrant component. **TASTE** cap flesh oily, unpleasant and slightly acrid hot, the lamellae not acrid or only slightly so.

Habit and habitat single, scattered, or clustered on humus, mosses, or well-rotted wood under hardwoods, conifers, or mixed forests. July to October. **RANGE** widespread.

Spores broadly ellipsoid, warts mostly not connected, colorless but ornamentations amyloid, 6–8 × 4.5–6.5 µm excluding ornamentation, warts to 1 µm high; cylindrical cells in crustlike patches are smooth.

Comments Several yellowish-brown, slippery-capped *Russula* species with fetid or bitter almond smells are found in the Northeast. All are members of the genus *Russula* section *Ingratae* subsection *Foetentinae*. Microscopic analysis is typically needed to make accurate identifications. However, if the fruit bodies are large and lack crustlike patches, see *R. grata*. If you have a microscope and the cylindrical cells in the crustlike patches are warted, you have *R. pulverulenta* Peck. If moderately sized like *R. granulata*, but lacking crustlike patches, the odor is only slightly fetid or spermatic, and the lamellae taste acrid hot, you have *R. pectinatoides* Peck.

Russula grata Britzelmayr

ALMOND-SCENTED RUSSULA

SYNONYMS *Russula laurocerasi* Melzer, *Russula foetens* var. *grata* (Britzelmeyer) Singer, *Russula foetens* var. *laurocerasi* (Melzer) Singer

Moderate to large, pale yellow, slimy cap with obvious bumpy lines on the margin; gills and stem white becoming pale orange-yellow, stem discoloring brown after handling; odor fragrant of bitter almonds, taste nauseating acrid hot

CAP 3–13 cm wide, globose with incurved margin pressing the stem, soon convex, becoming plane and then with center sunken, light yellow, orange-yellow, or yellowish-brown, with expansion paler and more yellowish, glabrous, slimy when wet, quickly drying, with radially raised bumpy lines over margin upon expansion. **FLESH** pale yellow, brown around larval tunnels, firm. **GILLS** white, becoming yellow or orange-yellow, attached, close or subdistant, brittle, snapping when bent. **STEM** 2.5–11 cm long, 9–30 mm broad, equal or enlarged up or down, pale cream-white or pale orange-yellow, discoloring dark yellow or reddish-brown where handled, fuzzy over apex, glabrous elsewhere, chalky, snapping cleanly, hollow with maturity. **SPORE PRINT** pale orange-yellow. **ODOR** fragrant like benzaldehyde or bitter almonds. **TASTE** nauseating acrid hot.

Habit and habitat single, scattered, or clustered on humus under hardwoods or mixed with conifers, frequently near beech, maple, and hemlock. July to September. **RANGE** widespread.

Spores broadly ellipsoid, broadly ovoid, or subglobose, large conical warts often connected to form a partial reticulum, colorless but ornamentations amyloid, 7.5–11 × 7.5–9 µm excluding ornamentation, warts to 2.6 µm high.

Comments If you find really large specimens with the cap reaching 20 cm, and the odor is more fetid than bitter almond, especially with the older more mature individuals, then you most likely have *Russula fragrantissima* Romagnesi. If the spores have warts only 1 µm high, that will confirm the identification.

Russula ventricosipes Peck

Medium to large, caps orange-brown, slippery, retaining sand and plant debris as slime dries; gills creamy, staining brown; stem pale with wine-red patches over the base; smell of bitter almonds, taste slowly acrid hot

CAP 4–13 cm wide, convex, becoming plane with center sunken, with a faint peachy-orange or brownish felted surface in button stage that is quickly lost, surface yellowish-brown or mostly orange-brown, slippery or slimy, quickly drying, glabrous but sand and plant debris sticking to cap, mostly with radially raised bumpy lines over margin (tuberculate-striate) with expansion. FLESH pale cream, becoming darker with exposure, firm. GILLS pale cream, staining or spotting brown, attached, close or crowded, some forked near stem, brittle. STEM 2–10 cm long, 15–50 mm broad, equal or tapered at both ends, off-white under a layer of reddish to reddish-brown patches and scales, especially over the base, very hard but breaking cleanly. SPORE PRINT cream. ODOR of bitter almonds or benzaldehyde. TASTE slowly acrid hot.

Habit and habitat scattered or clustered in sandy soil under pines, especially pitch pine, in sand dunes around lakes or oceans, or inland on sand eskers. August to September. RANGE widespread.

Spores ellipsoid-oblong, obscurely warted with very few connections, colorless but ornamentations amyloid, 7–10 × 4.5–6 μm excluding ornamentation, warts to 0.1 μm high.

Comments A species you will only find if you visit sandy soil and pine habitats, such as those in pine barrens and lake or ocean dunes.

Russula subdepallens Peck

Medium to large, red cap with yellow blotches; flesh white, very fragile, crumbly; gills white, easily breaking; taste mild, nutty; growing in hardwood or mixed forests

CAP 5–14 cm wide, convex, becoming plane, center broadly and shallowly sunken, margin becoming elevated, bright rosy-red, with yellowish blotches eventually developing centrally, bumpy lined over the margin, minutely wrinkled elsewhere, slippery or sticky when wet, quickly drying. **FLESH** white, or slight grayish, very fragile, crumbly. **GILLS** white, attached, broad near cap margin, narrowed near stem, subdistant, very fragile, easily broken. **STEM** 4–10 cm long, 10–30 mm broad, equal, white, chalky, snapping cleanly. **SPORE PRINT** white. **ODOR** not distinctive. **TASTE** mild, nutty.

Habit and habitat single, scattered, or clustered on humus under hardwoods of beech, yellow birch, red oak, and poplar or in mixed forests with hemlock. June to August. **RANGE** widespread, but not commonly reported.

Spores subglobose or ovoid, warted-spiny and with a few broken or partial connectives, colorless but ornamentations amyloid, 7–9.5 × 6.5–8 μm excluding ornamentation, warts 0.8(–1) μm high, apiculus colorless, large 1.6–3 × 1–1.6 μm.

Comments This species fruits early in the season, thus it is missed by those who wait for July and August to begin their explorations. The very crumbly fragile gills are an excellent field character to identify this mild-tasting red russula. If the stem has any red or pink, you probably have *Russula peckii* Singer, which is also somewhat fragile and mild tasting.

Russula silvicola Shaffer

Medium to large, red or orange-red cap; white gills and stem; taste acrid hot; found growing in hardwood or mixed forests

CAP 2–8 cm wide, convex, becoming plane, often with center shallowly sunken, somewhat variable in color but mostly evenly dark red or with mixtures of dark pink or reddish-orange, also with areas of orange-yellow, yellow, or white, smooth or minutely bumpy, slippery or sticky when wet, but quickly drying, with raised bumpy lines over margin. **FLESH** white, firm but brittle. **GILLS** white or pale yellowish, attached, close or subdistant, brittle, making a subtle snapping feeling when bent. **STEM** 2–8 cm long, 4–20 mm broad, equal or enlarged downward, white, longitudinally bumpy, chalky, snapping cleanly. **SPORE PRINT** white or pale cream. **ODOR** not distinctive. **TASTE** slightly to sharply acrid hot.

Habit and habitat single, scattered, or clustered on humus or well-rotted wood under hardwoods or conifers or mixed forests in well-drained, nonboggy areas. July to September. **RANGE** widespread.

Spores broadly ellipsoid, warted and partially or completely reticulate, colorless but ornamentations amyloid, 6–10 × 5–9 µm excluding ornamentation, warts to 1.2 µm high; cystidia of pileus and stipe lacking cross walls or having up to three per cystidium.

Comments The other common red-capped russula with hot acrid taste is *Russula emetica* (Schaeffer) Persoon that is found on *Sphagnum*, typically in bogs in our area. The white stipe of *R. emetica* is usually longer and the cap colors more uniformly deep apple-red. Of the many red-capped species of *Russula*, only a few of the more common ones are covered here. See also, *R. subdepallens*, a mild-tasting species.

Russula decolorans (Fries) Fries

SYNONYM *Agaricus decolorans* Fries

Cap orange-red or copper-red, sometimes with yellows, slippery at first; gills pale cream becoming ochraceous, staining gray; stem white, turning gray when injured or handled; taste mild

CAP 5–12 cm wide, subglobose then convex, with center sunken, mostly orange-red or burnished copper, sometimes ochre on center but reddish around the margin, sometimes additionally mixed with yellows and reds, slippery but soon dry, glabrous, smooth, sometimes with subtle short raised lines over margin (striate) with expansion. **FLESH** white, quite firm at first, turning slowly gray or black with exposure. **GILLS** pale cream, turning toward ochre, staining or spotting slowly gray, bluntly attached or subdecurrent, close, often forked near stem, brittle. **STEM** 4–12 cm long, 10–30 mm broad, equal, white, staining dark gray with age or handling, smooth, firm when young, becoming fragile. **SPORE PRINT** pale ochre. **ODOR** not distinctive. **TASTE** mild.

Habit and habitat single, scattered, or clustered on the ground under conifers, also in conifer-dominated bogs or fens. July to September. **RANGE** widespread.

Spores ellipsoid, conical warts isolated, colorless but ornamentations amyloid, $9–12 \times 7–9$ µm excluding ornamentation, warts to 1.5 µm high.

Comments The cap colors can be somewhat variable, but always with reddish-orange dominating. *Russula paludosa* Britzelmayr can be much larger and does not turn gray on the gills, stem, or flesh.

Russula fragilis Fries

FRAGILE RUSSULA

Small or medium, very fragile, colors of the slippery cap variable, ranging from purplish to red or pink with yellow or olive tints, margin bumpy-striate; gills and stem white, very fragile; taste acrid hot

CAP 2–6 cm wide, convex, becoming plane and then with center shallowly sunken; highly variable in color but with purplish, reddish, grayish, or pinkish hues intermixed, also after expansion and fading, developing touches of yellow, or pale pink with olive hues especially over the center; smooth, slippery or sticky when wet, but quickly drying, with raised bumpy lines over margin. **FLESH** white, very thin and easily crumbled. **GILLS** white or pale yellowish, attached, close or subdistant, brittle, snapping when bent. **STEM** 2–7 cm long, 5–15 mm broad, equal or enlarged downward, white, longitudinally bumpy, chalky, snapping cleanly. **SPORE PRINT** white or pale cream. **ODOR** not distinctive. **TASTE** moderately to sharply acrid hot.

Habit and habitat single, scattered, or clustered on humus, mosses, or well-rotted wood under hardwoods or conifers or both. July to October. **RANGE** widespread.

Spores broadly ellipsoid or broadly ovoid, warted and partially or completely reticulate, colorless but ornamentations amyloid, $6–9 \times 5–8$ µm excluding ornamentation, warts to $1(–1.6)$ µm high.

Comments This small, multicolored, very fragile species of *Russula* seems to be very common in the Northeast, but you may have difficulty accurately identifying it until you have seen all its extremes in color variation.

Russula claroflava Grove

GRAYING YELLOW RUSSULA

SYNONYMS *Russula flava* Romell, *Russula ochroleuca* var. *claroflava* (Grove) Cooke

Cap clear golden-yellow, dry; gills pale cream or creamy-ochre, staining sooty-gray; stem white or pale yellow, turning sooty-gray when injured or handled; taste mild

CAP 4–12 cm wide, subglobose then convex, with center sunken, clear golden-yellow, slippery at first, soon dry, glabrous, smooth, but with short radially raised lines over margin with expansion. **FLESH** white, turning slowly ash-gray or black with exposure. **GILLS** pale cream, becoming darker creamy-ochre, staining or spotting gray or sooty-gray, attached, close, often forked near stem, brittle. **STEM** 3–10 cm long, 10–20 mm broad, equal, white or very pale yellowish, staining sooty-gray, smooth. **SPORE PRINT** ochre. **ODOR** not distinctive. **TASTE** mild.

Habit and habitat scattered or clustered on the ground under conifers or in woods mixed with hardwoods. July to September. **RANGE** widespread.

Spores ellipsoid, conical warts connected by ridges forming a reticulum, colorless but ornamentations amyloid, 8.5–10 × 7.5–8 µm excluding ornamentation, warts to 1 µm high.

Comments *Russula claroflava* is considered edible. See also *R. decolorans*, which often grows in areas frequented by *R. claroflava*.

Russula simillima Peck

Medium-sized, ochre-yellow, and slightly slippery cap; gills creamy or yellow-ochre, close; stem pale ochre-yellow; taste slowly acrid hot

CAP 3–6(–8) cm wide, convex, becoming plane, center sunken, ochre-yellow with darker ochre center, color persistent, glabrous, but under a lens densely pubescent on center and minutely dotted orange-brown, especially over the margin, slippery when wet, quickly drying, smooth, with faint radially raised bumpy lines over margin (tuberculate-striate) with expansion. **FLESH** cream. **GILLS** pale cream to pale yellowish-ochre, attached, close or crowded, some forked near stem, brittle. **STEM** 3–7 cm long, 5–18 mm broad, equal or narrowly club-shaped, pale ochre-yellow, pruinose overall at first, but soon glabrous, chalky. **SPORE PRINT** white. **ODOR** not distinctive, or vaguely pungent. **TASTE** slowly acrid hot.

Habit and habitat scattered or clustered on soil and humus under hardwoods with beech present. August to September. **RANGE** widespread.

Spores broadly ellipsoid, conical warts with some connections forming a partial reticulum, colorless but ornamentations amyloid, 6.5–8 × 5–6.5 μm excluding ornamentation, warts to 0.8 μm high.

Comments Some suggest this ochre-yellow russula described by Peck from near the Adirondack region is the same as *Russula fellea* (Fries) Fries, a species of Europe with a fruity smell of stewed apples or geraniums. It is also associated with beech trees. *Russula simillima* differs by its lack of a distinctive odor, based on Peck's description and my collections, and by its slowly acrid taste, not the bitter taste of *R. fellea*. The cap surface on *R. simillima* is also finely orange-brown punctate under a hand lens, especially so in the dried condition, another feature not noted for *R. fellea*. *Russula ballouii* Peck is darker rusty-brown, and the cap and stem are covered with scalelike pigmented patches.

Russula ochroleucoides
Kauffman

Medium to large, yellow or orange-yellow, and when young with reddish hues on the velvety cap; gills white, discoloring yellow over the edges; stem white, discoloring erratically yellow; flesh firm; found growing under hardwoods

CAP 6–12 cm wide, convex, becoming plane and broadly sunken, from pale straw-yellow at first to ochre-yellow with darker ochre center, sometimes with reddish-ochre center in early stages, soft, finely velvety, dry, smooth or center eventually cracked, margin smooth, not lined. FLESH white, thick. GILLS white but with yellow over edges near the cap margin from bruising, attached or barely so, close or subdistant, some forked near stem, with short gills infrequent, brittle. STEM 4–6 cm long, 15–20 mm broad, equal or tapering downward, mostly white, but some with irregular splashes of yellow, especially around the lower half, glabrous or finely

pruinose, chalky. SPORE PRINT white. ODOR faintly aromatic or none. TASTE slowly somewhat bitter acrid.

Habit and habitat scattered or clustered on soil and humus, under hardwoods, especially with oak, yellow birch, beech, poplar, and suckering American chestnut. July to September. RANGE widespread but not common.

Spores subglobose or broadly ellipsoid, conical warts with some connections forming a partial reticulum, colorless but ornamentations amyloid, 6–8 × 5–7 µm excluding ornamentation, warts to 0.5 µm high.

Comments An attractive but rarely collected species. If you have a yellow-capped collection with a cream or darker yellow spore deposit, then you have one of several yellow species not covered in this field guide. *Russula simillima* has some similarities to this species, but it is smaller and does not discolor yellow on the gills and stem.

Lactarius deceptivus Peck

DECEPTIVE MILKY

A large white mushroom with sunken cap and a cottony roll over the margin; latex abundant, white; taste slowly but strongly acrid hot

CAP 5–24 cm wide, convex, sunken with strongly inrolled cottony margin that covers the gills at first, becoming deeply sunken over the center with expansion, white but soon with brown stains, becoming tan with age, dry, glabrous at first, but becoming torn into large thick scales and patches of dense soft cottony tissue, especially over the margin. **FLESH** white, firm, thick. **GILLS** white becoming cream color, staining brown where injured, subdecurrent, close or subdistant, moderately broad, brittle. **STEM** 4–9 cm long, 10–35 mm broad, equal or tapering downward, white, staining brown, glabrous or becoming matted fibrillose, very firm, solid. **LATEX** white, abundant, turning tissues brownish. **SPORE PRINT** white or pale ochraceous-buff. **ODOR** not distinctive when young, pungent with age. **TASTE** slowly but strongly acrid hot, occasionally only peppery.

Habit and habitat scattered or clustered on soil or humus under hemlock, sometimes oak may also be present. July to October. **RANGE** widespread.

Spores broadly ellipsoid, well-developed isolated warts and spines, colorless, but ornamentations amyloid, 9–13 × 7.5–9 µm excluding ornamentation, which measures to 1.5 µm high.

Comments This species can look like *Russula brevipes*, which is found in similar habitats at the same time, but *R. brevipes* does not produce latex and the taste, although not agreeable, is not acrid hot. Other all-white species, such as *Lactarius piperatus*, have crowded gills, while *L. subvellereus* lacks a cottony roll of tissue on the cap margin.

Lactarius subvellereus Peck

SYNONYM *Lactifluus subvellereus* (Peck) Nuytinck

Cap white, velvety soft; gills pale cream, spotting brown where injured; stem surface finely velvety and soft, but flesh firm; taste very acrid hot; found growing under oaks

CAP 4–15 cm wide, convex with inrolled margin, becoming plane, shallowly sunken over the center, some nearly vase-shaped, white or with yellow or yellow-brown spots or staining, dry, velvety and downy soft. **FLESH** white, turning yellow when exposed. **GILLS** pale cream, turning brown where injured, bluntly attached or subdecurrent, close or somewhat crowded, broad. **STEM** 2–5 cm long, 12–35 mm broad, equal or tapered downward, colored as cap, sometimes bruising brown, finely velvety, dry, solid. **LATEX** pale creamy-yellow, abundant. **SPORE PRINT** white. **ODOR** not distinctive. **TASTE** very acrid hot.

Habit and habitat scattered or clustered on soil or humus under hardwoods, especially with oak. July to October. **RANGE** widespread.

Spores ellipsoid, mostly with scattered fine small warts, colorless, but ornamentations amyloid, 7.5–9 × 6–7 µm excluding ornamentation, which measures to 0.2 µm high.

Comments Compare with *Lactarius deceptivus* which is also white, but its cap margin has a roll of cottony white tissue. It also develops large flat scales on the pileus and is found under hemlock with which it is a mycorrhizal partner. The velvety stem and more-separated gills distinguish *L. subvellereus* from *L. piperatus*.

Lactarius piperatus (Linnaeus) Persoon

PEPPERY MILKY

SYNONYM *Agaricus piperatus* Linnaeus

Completely white, very firm, cap smooth, dry; gills tightly crowded, narrow; latex white, unchanging; taste immediately very hot

CAP 5–15 cm wide, convex, sunken over the center, becoming vase-shaped, white, but often with pale brown or pinkish-brown stains, dry, glabrous, smooth or uneven. FLESH white, unchanging. GILLS white, then pale cream, bluntly attached or sub-decurrent, tightly crowded, narrow, often forked. STEM 2–8 cm long, 10–25 mm broad, equal, white, dry, very firm, solid. LATEX white, abundant, sometimes staining gills yellow. SPORE PRINT white. ODOR not distinctive. TASTE immediately very acrid hot.

Habit and habitat scattered or clustered on soil in mixed forests. July to October. RANGE widespread.

Spores ellipsoid or subglobose, fine lines and scattered small warts, colorless, but ornamentations amyloid, $5–7 \times 5–5.5\ \mu m$ excluding ornamentation, which measures to $0.2\ \mu m$ high.

Comments Compare with *Lactarius deceptivus* that is also white and slowly but intensely acrid hot when tasted, but its cap margin is cottony, its cap surface is not firm and not glabrous, and it grows under conifers. *Lactarius piperatus* var. *glaucescens* (Crossland) Hesler & A. H. Smith, now *Lactifluus glaucescens* (Crossland) Verbeken, differs by the latex that dries a pale green color.

Lactarius deliciosus var. deterrimus (Gröger) Hesler & A. H. Smith

ORANGE-LATEX MILKY

Medium- to large-sized orange mushroom with orange latex, cap and gills staining green slowly after injury; growing in cold wet coniferous forests

CAP 3–13 cm wide, convex and sunken over the center, becoming plane and more deeply sunken; apricot-orange or pale orange-buff, sometimes alternating with zones of paler rings, some with a sheen, turning green slowly in areas where injured; slippery or even thinly slimy when wet, glabrous. **FLESH** orange-buff. **GILLS** orange-salmon, staining reddish-vinaceous then dull green when injured, bluntly attached or subdecurrent, close, broad. **STEM** 2.5–7 cm long, 15–25 mm broad, equal or slightly tapered downward, same color as cap, glabrous and uneven, but not pitted, moist or scarcely slippery but not slimy, brittle, becoming hollow, orange inside but slowly dull wine-red. **LATEX** bright orange, slowly staining tissue purplish-vinaceous then tissue turning dull green. **SPORE PRINT** pale buff. **ODOR** not distinctive. **TASTE** mild or tardily peppery in young caps.

Habit and habitat scattered or clustered on moss beds or humus in cold wet coniferous forests with uniform stands or mixtures of spruce, fir, hemlock, and pine. July to October. **RANGE** widespread.

Spores broadly ellipsoid, isolated warts and partially formed ridges producing a broken reticulum, colorless, but ornamentations amyloid, 7.5–9 × 6–7 μm excluding ornamentation, which measures to 0.8 μm high.

Comments This green-staining orange-latex milky is commonly collected. DNA evidence suggests this mushroom and others found in North America and described as varieties of *Lactarius deliciosus* are genetically separate from similar-looking mushrooms found in Europe. We will use this species name for now since it is used in Hesler and Smith's monograph on *Lactarius* for North America.

Lactarius thyinos A. H. Smith

Medium- to large-sized orange mushroom with orange latex, base of stem slowly turning wine-red when cut open; never with green staining; found growing in cold water swamps and bogs with arborvitae

CAP 3–9 cm wide, convex, becoming deeply sunken but with margin spreading or arched, carrot-orange or salmon-orange alternating with zones of paler orange or yellow, some bands with silvery sheen, slippery or even thinly slimy when wet, glabrous. **FLESH** orange-buff. **GILLS** bright orange-salmon, becoming paler with age, vinaceous-brown or red when injured, bluntly attached or subdecurrent, close, broad. **STEM** 4–8 cm long, 8–25 mm broad, equal or slightly tapered downward, same color as gills or with white sheen at first, glabrous, slippery or thinly slimy, orange inside filled with white cottonlike stuffing at first, becoming hollow and slowly dull wine-red. **LATEX** orange, slowly staining tissue wine-red or brown. **SPORE PRINT** pale yellow. **ODOR** faintly fragrant. **TASTE** mild.

Habit and habitat scattered or clustered on moss beds or humus under arborvitae (*Thuja*) in cold swamps and bogs. July to October. **RANGE** widespread.

Spores subglobose or broadly ellipsoid, isolated warts and partially formed ridges producing a broken reticulum, colorless, but ornamentations amyloid, 9–12 × 7.5–9 µm excluding ornamentation, which measures to 1 µm high.

Comments *Lactarius deliciosus* is similar in color and can be found in the same habitats, but it turns green slowly after injury. Both are edible, but only of fair quality. See further comments under *L. deliciosus* var. *deterrimus*.

Lactarius subpurpureus Peck

VARIEGATED MILKY

Cap subzonate and mottled with wine-red and silvery-pink spots; gills wine-red; stem colored as cap; latex dark wine-red, scanty; associated with hemlock

CAP 3–10 cm wide, convex and sunken over the center, becoming nearly vase-shaped, spotted or mottled wine-red and silvery-pink, subzonate, fading to pinkish-buff, staining greenish or ochraceous with age, dry or slightly tacky, glabrous. FLESH white or pinkish, turning red quickly from the latex when exposed. GILLS wine-red, developing green stains or spots with injury or age, bluntly attached or subdecurrent, subdistant, broad. STEM 3–8 cm long, 8–15 mm broad, equal, colored as cap but with darker reddish mottling, also often with silvery hoary sheen, slightly slippery when moist. LATEX dark wine-red, scanty. SPORE PRINT cream. ODOR not distinctive. TASTE mild or slightly peppery.

Habit and habitat scattered or clustered on soil or humus under hemlock in mixed woods. July to October. RANGE widespread.

Spores ellipsoid or subglobose, with coarse and fine lines forming a partial reticulum, colorless, but ornamentations amyloid, $8–11 \times 6.5–8$ µm excluding ornamentation, which measures to 0.5 µm high.

Comments *Lactarius paradoxus* Beardslee & Burlingham, which is found mainly in the southeastern and gulf coastal areas of the United States, has similar wine-red-colored latex, but it differs by its grayish-blue colors in the cap and stem.

Lactarius chelidonium Peck

Cap blue and yellowish-brown mottled, staining green, center sunken; gills yellow becoming orange-brown; stem colored as cap; latex pale yellow-brown, like grasshopper juice, scanty; associated with pines

CAP 3–8 cm wide, broadly convex and sunken over the center, shallowly vase-shaped; spotted or mottled grayish-blue or deeper sky-blue, but also with dingy orange or yellowish-brown, greenish from staining, subzonate, fading to orange-brown or dull reddish-brown, staining green or olive, sometimes the entire cap becomes greenish with age; slippery at first, soon dry, glabrous. **FLESH** grayish-blue or sky-blue. **GILLS** pale yellow becoming dull orange or orange-brown, bluntly attached or subdecurrent, close, broad. **STEM** 2–6 cm long, 10–25 mm broad, equal, colored as cap or paler, dry. **LATEX** dull yellow or pale yellow-brown, very

scanty. **SPORE PRINT** pale buff. **ODOR** not distinctive, or slightly fragrant. **TASTE** mild or slightly peppery.

Habit and habitat scattered or clustered on soil or humus under pines, especially eastern white pine and red pine. August to October. **RANGE** widespread but not common.

Spores ellipsoid, with warts and lines forming bands and a partial reticulum, colorless, but ornamentations amyloid, 7–9 × 5–7 µm excluding ornamentation, which measures 0.5–1 µm high.

Comments The blue colors of the cap remind one of a pale *Lactarius indigo*, while the green staining suggests *L. deliciosus* var. *deterrimus*; please see each of those species and their descriptions. *Lactarius chelidonium* is an exceptionally attractive species in its early developmental stages.

Lactarius indigo (Schweinitz) Fries

INDIGO MILKY

SYNONYM *Agaricus indigo* Schweinitz
Cap dark blue, slippery at first, also zoned; gills dark blue; stem dark blue and slippery at first; latex blue, staining all parts greenish

CAP 5–15 cm wide, convex and sunken over the center, becoming vase-shaped, dark blue at first, becoming silvery or pale grayish mixed with dark blue, developing green stains where injured, zoned around margin, slippery or sticky, glabrous. **FLESH** white but quickly blue from latex, then green. **GILLS** dark blue, changing colors as cap, staining green where injured, bluntly attached or subdecurrent, close, broad. **STEM** 2–8 cm long, 10–25 mm broad, central but may be eccentric, equal or narrowed downward, dark blue but with a hoary sheen, slippery at first, soon dry, sometimes with spots, rather hard, becoming hollow. **LATEX** dark blue, slowly turning green. **SPORE PRINT** cream. **ODOR** not distinctive. **TASTE** mild or slightly bitter.

Habit and habitat scattered or clustered on soil under conifers, especially pines, and mixed hardwoods, especially oaks. July to October. **RANGE** widespread.

Spores ellipsoid or subglobose, banded with ridges forming a distinct or broken reticulum, colorless, but ornamentations amyloid, 7–9 × 5.5–7.5 μm excluding ornamentation, which measures to 0.5 μm high.

Comments This is an easily identified species, and it is edible. *Lactarius paradoxus* Beardslee & Burlingham, a southern species, has pale silvery-blue colors on the cap, but the lamellae are cinnamon-colored and the latex is vinaceous-brown. *Lactarius chelidonium* also has blue colors at first on the cap and in the cap flesh, but the cap develops orange-brown colors, the gills are yellow-brown, and the latex is yellowish-brown.

Lactarius uvidus (Fries) Fries

COMMON VIOLET-LATEX MILKY
SYNONYM *Agaricus uvidus* Fries

Cap drab purple, slippery; gills cream, turning lilac from injury; stem white with ochre-yellow, turning lilac when injured; latex abundant, white, then cream, turning the flesh lilac

CAP 3–10 cm wide, convex, becoming plane, slightly sunken and occasionally with low broad bump over the center, pale grayish-lilac or purplish-drab, center often dark brown, slimy or only slippery in drier conditions, glabrous. **FLESH** white, turning lilac when exposed to air. **GILLS** pale cream, turning lilac where injured, bluntly attached or subdecurrent, close, broad. **STEM** 3–7 cm long, 10–16 mm broad, equal, white or with yellow-ochre colors irregularly scattered, staining lilac where injured, glabrous, slippery. **LATEX** white, abundant, becoming pale creamy-yellow, staining injured surfaces lilac. **SPORE PRINT** white. **ODOR** not distinctive. **TASTE** slowly bitter, then slightly acrid hot.

Habit and habitat scattered or clustered on soil or humus under hardwoods, especially with aspen or poplar, birch, and pine. July to October. **RANGE** widespread.

Spores ellipsoid, mostly with ridges but not truly reticulate, colorless, but ornamentations amyloid, 7.5–11 × 6.5–8.5 μm excluding ornamentation, which measures to 1 μm high.

Comments There are other lilac- or purple-staining *Lactarius* species, but this is the most frequently encountered one in the Northeast. A drop of 3% KOH solution will turn the cap surface olive-green. *Lactarius maculatus* Peck and *L. subpalustris* Hesler & A. H. Smith are two other, less common northeastern species that also have white- to cream-colored latex that stains the tissues lilac. *Lactarius maculatus* has white gills and is acrid hot to the taste, while *L. subpalustris* has tan or cinnamon-buff-colored gills and is mild when tasted. Neither reacts green with KOH solution.

Lactarius sordidus Peck

DIRTY MILKY

Large, heavy mushrooms with yellow-brown, slippery caps with olive tints; gills staining olive-brown from milky-white latex; stem with spots; taste acrid hot

CAP 5–10(–15) cm wide, convex, with shallowly sunken center or somewhat vase-shaped, dull honey-brown with olive tints over the center, slippery or tacky at first, soon dry, smooth, glabrous, or with margin slightly cottony at first. FLESH white or yellow, slowly olive-brown, thick. GILLS white or pale creamy-yellow, bluntly attached or subdecurrent, crowded, staining olive-brown from injury. STEM 4–12 cm long, 10–25 mm broad, equal or enlarged downward, pale over apex or completely colored as cap, usually covered with darker, greenish-brown, sunken spots (scrobiculate), dry, hollow. SPORE PRINT dull white.

LATEX white, milky, staining tissues olive-brown. ODOR not distinctive. TASTE slowly but intensely acrid.

Habit and habitat single, scattered, or clustered on soil and humus associated with conifers, especially spruce and fir. July to October. RANGE widespread.

Spores ellipsoid, broken or complete reticulum of heavy banding, colorless, but ornamentations amyloid, 5.5–7.5 × 5–6 μm excluding ornamentation, which measures 0.5 μm high.

Comments If you like to use the term *toadstool*, here is the perfect mushroom for it. This unattractive species might well cause some problems if eaten, because of the acrid hot taste. It is not currently known to be edible. *Lactarius atroviridis* Peck is similar looking and also produces milk-white latex, but the fruit bodies are dark green.

Lactarius psammicola
A. H. Smith

Cap zoned ochraceous-orange and buff, margin densely hairy; gills white, decurrent, close or crowded; stem white or pale grayish, often with some orange wet spots; latex white, unchanging; taste very hot

CAP 4–14 cm wide, convex and sunken over the center, becoming vase-shaped, margin inrolled and hairy, ochraceous-orange or ochraceous-buff and zoned with alternating lighter buff bands, slippery or sticky, densely hairy at first, then matted fibrillose. **FLESH** dingy buff. **GILLS** light buff or off-white, sordid ochre with age, decurrent, close or crowded, narrow. **STEM** 1–3 cm long, 10–20 mm broad, tapering to base, white or pale grayish, dry, sometimes with dark orange-ochre slippery spots, rather hard, becoming hollow. **LATEX** white, abundant, staining white paper slowly yellow. **SPORE PRINT** yellow. **ODOR** not distinctive. **TASTE** very acrid hot.

Habit and habitat scattered or clustered on soil under mixed hardwoods and conifers. July to October. **RANGE** widespread.

Spores broadly ellipsoid or subglobose, with short ridges and isolated warts forming a broken reticulum, colorless, but ornamentations amyloid, 7–9 × 6–7.5 µm excluding ornamentation, which measures to 0.5 µm high.

Comments This rather diverse group of *Lactarius* species with zonate pileus colors is not well understood in North America. This particular species shown here will be found as *L. zonarius* (Bulliard) Fries in older field guides, but the common northeastern species appears to be *L. psammicola*. Some field guides show a paler colored species, but the original description by Smith (1941) states "ochraceous-buff to ochraceous-orange" alternating with light buff bands as shown here.

Lactarius vinaceorufescens
A. H. Smith

YELLOW-LATEX MILKY
Cap pallid, becoming wine-red, zoned; gills pale buff, eventually spotted with vinaceous stains; stem staining vinaceous, eventually vinaceous-brown; latex abundant, white, turning rapidly bright yellow

CAP 4–12 cm wide, convex, becoming plane, often slightly sunken over the center, pale buff, with sordid cream or buff concentric zones, becoming pinkish-buff or pinkish-cinnamon or wine-red, but always watery concentrically zoned, slippery at first, becoming dry, glabrous but minutely pubescent on margin at first. **FLESH** white, turning bright yellow from latex. **GILLS** off-white or pale orange-buff, spotting or discoloring pinkish-brown, turning reddish-brown with age, bluntly attached or subdecurrent, close, narrow. **STEM** 4–7 cm long, 10–25 mm broad, equal or enlarged downward, off-white or pale pinkish-buff, staining dark pinkish-cinnamon or vinaceous and eventually completely vinaceous-brown, glabrous. **LATEX** white at first, abundant, turning rapidly bright yellow. **SPORE PRINT** white or pale yellow. **ODOR** not distinctive. **TASTE** slowly acrid hot.

Habit and habitat scattered or clustered on soil or humus under pine, especially white pine. July to October. **RANGE** widespread.

Spores subglobose or ellipsoid, bumpy ridges that are sometimes reticulate, colorless, but ornamentations amyloid, 6.5–9 × 6–7 μm excluding ornamentation, which measures to 0.8 μm high.

Comments *Lactarius chrysorrheus* Fries has a much paler cap (off-white or pale buff, zonate or not, with pale buff, not spotted gills) and occurs under hardwoods, especially oak. Both species are slowly but exceedingly hot to the taste.

Lactarius chrysorrheus

Lactarius helvus (Fries) Fries
BURNT-SUGAR MILKY
SYNONYMS *Lactarius aquifluus* Peck, *Lactarius helvus* var. *aquifluus* (Peck) Peck
Large rosy-cinnamon-brown caps with darker watery spotting; latex colorless, watery; odor strongly of burnt sugar, maple syrup, or curry; found in wet areas, usually in *Sphagnum*

CAP 3–15 cm wide, convex, becoming plane with shallow sunken center, some with low central bump, dull rosy-brown or fawn- or cinnamon-colored, often mottled with darker spots looking like water marks, dry, smooth, glabrous or becoming finely velvety or fibrillose-scaly. **FLESH** colored as cap, thick. **GILLS** white, becoming pale pinkish-cinnamon, bluntly attached, close. **STEM** 3–10 cm long, 10–20 mm broad, equal, variable in color but often mixtures of orange, pink, and pale browns, usually covered with thin white bloom that is easily rubbed off and thus darker colors show through, dry, white mycelium over the base. **SPORE PRINT** buff or creamy-white. **LATEX** colorless, watery.

ODOR strong, of burnt sugar, maple syrup, or curry. **TASTE** mild or slightly acidulous, slowly a little acrid.

Habit and habitat single, scattered, but more often clustered in wet mossy areas, usually with *Sphagnum*, associated with conifers. July to October. **RANGE** widespread.

Spores ellipsoid, broken or irregular reticulum and spines, colorless, but ornamentations amyloid, 6.5–8 × 5–6.5 μm excluding ornamentation, which measures 1 μm high.

Comments Previously known as *Lactarius aquifluus*, it seems to be indistinguishable morphologically from the European *L. helvus*, at least by current interpretations. We shall see what the DNA comparisons eventually uncover. We may go back to Peck's original name. Although notice that Peck changed his mind 10 years after describing it and placed it as a variety of *L. helvus* in 1885. In Europe, its counterpart has been involved in cases of mild poisoning.

Lactarius hibbardae Peck

Cap dark grayish or vinaceous-brown with faintly zoned margin; gills cream or buff; stem buff, becoming similar in color to cap; odor of anise, taste slowly acrid

CAP 2–8(–10) cm wide, broadly convex or plane and then shallowly sunken over the center, sometimes with small conical or rounded bump on the center; dark gray, vinaceous-gray, or vinaceous-brown, may be faintly zoned around margin; dry, radiately hairy or with fine scales. **FLESH** white. **GILLS** pale cream, becoming ochre-buff, bluntly attached or subdecurrent, close or crowded. **STEM** 2–5(–7) cm long, 4–10(–17) mm broad, equal, colored as gills at first, then as cap or paler, white over base, dry, with fine white dusting overall, becoming hollow. **SPORE PRINT** white or creamy-white. **LATEX** white, milky, becoming waterier with age. **ODOR** fragrant, faintly of anise. **TASTE** slowly acrid.

Habit and habitat scattered or clustered on humus or soil under conifers and mixed hardwoods, such as balsam fir, birches, and alder. July to October. **RANGE** widespread.

Spores broadly ellipsoid, banded with ridges forming a distinct reticulum, colorless, but ornamentations amyloid, 6–9(–10) × 5–6.5 µm excluding ornamentation, which measures to 0.4 µm high.

Comments Usually found in northern areas and at higher altitudes.

Lactarius glyciosmus (Fries)
Fries

SYNONYMS *Agaricus glyciosmus* Fries, *Galorrheus glyciosmus* (Fries) P. Kummer

Small to medium-sized with pale pinkish or vinaceous-gray colors on the cap and stem, the cap is silky fibrillose, the tissues are somewhat fragile with an odor of coconut

CAP 2–6 cm wide, convex, becoming plane then shallowly sunken in center or somewhat vase-shaped, sometimes with small papillate bump on center, pinkish-gray or vinaceous-gray and remaining that color, dry, silky radially hairy, or with fine scales, sometimes concentrically zoned from color or ridges. **FLESH** pale buff, thin. **GILLS** off-white or pinkish-buff or pale pinkish-cinnamon, with cream tints, subdecurrent or decurrent, close or crowded. **STEM** 2–5(–10) cm long, 10–15 mm broad, equal, colored as cap or paler, dry, with tiny dots and fibrils over upper half or becoming entirely glabrous, hollow, fragile. **SPORE PRINT** pale cream or pinkish buff. **LATEX** white, milky, unchanging. **ODOR** fragrant of coconut. **TASTE** disagreeable, slightly acrid.

Habit and habitat scattered or clustered on humus or on well-decayed, moss-covered conifer logs, near fir, birches, and alder. July to September. **RANGE** widespread.

Spores broadly ellipsoid, warty and with some short ridges forming a broken reticulum, colorless, but ornamentations amyloid, 6–9(–10) × 5–7 µm excluding ornamentation, which measures to 0.8 µm high.

Comments This very pale, delicate pinkish species is always interesting to find, just to smell the fragrant odor. It just might put you in the mood for a piña colada. The mushroom is listed as edible, but it is unusual to find it in any quantity.

Lactarius cinereus var. *fagetorum* Hesler & A. H. Smith

Caps ash-gray with slight olive tints, fragile, gills cream, subdecurrent; stem same ash-gray as cap, but with characteristic white band at apex; taste of flesh slightly acrid

CAP 2.5–6 cm wide, convex, with shallowly sunken center or somewhat vase-shaped, pale ash-gray or with pale olive tints, pale brown over the center, slippery or tacky at first, but soon dry, smooth, glabrous. **FLESH** white, fragile. **GILLS** white, becoming pale cream, subdecurrent, crowded. **STEM** 4–7 cm long, 8–15 mm broad, equal or enlarged downward, colored as cap except for a white band at very apex, thinly viscid at first, soon dry, glabrous, hollow. **SPORE PRINT** pale yellow. **LATEX** white, milky, scant. **ODOR** not distinctive. **TASTE** mild then slowly somewhat acrid, latex slowly burning acrid.

Habit and habitat single, scattered, or clustered on soil and humus associated with beech. August to October. **RANGE** widespread.

Spores ellipsoid, warty or with some short ridges or banding, colorless, but ornamentations amyloid, 6–7.5 × 5–6 µm excluding ornamentation, which measures to 0.7 µm high.

Comments A nondescript but distinctive species once the field characters are recognized. Look for this mushroom in mature beech forests. Compare with the much smaller, but somewhat similar-looking *Lactarius griseus*.

Lactarius griseus Peck

Small and fragile, with gray fibrillose-scaly caps; gills cream-colored, subdecurrent; stem with creamy-white band at apex; found on moss-covered logs and humus in wet locations

CAP 1.5–5 cm wide, convex, with well-sunken center or somewhat vase-shaped with conical bump on center, gray or violet-brown flattened fibrils over paler ashy-gray surface, center darker than margin, eventually with ochre undertones, dry, fibrillose or scaly, especially over the center. **FLESH** white at first, soon cream, very fragile. **GILLS** white, becoming pale cream or buff, bluntly attached or subdecurrent, close or subdistant. **STEM** 2–6.5 cm long, 3–6 mm broad, equal, colored as cap except for a pale band at very apex, dry, glabrous, hollow, white hairy at base, very fragile. **SPORE PRINT** pale yellow.

LATEX white, milky, scant, drying yellowish in droplets or on white paper. **ODOR** not distinctive. **TASTE** slowly somewhat acrid on young specimens, often mild on mature ones.

Habit and habitat single, scattered, or more often clustered on moss-covered decaying conifer logs or thick humus, also in bogs or very wet seepage areas. June to October. **RANGE** widespread.

Spores subglobose or broadly ellipsoid, warty or with some short ridges, colorless, but ornamentations amyloid, 7–8 × 6–7 μm excluding ornamentation, which measures to 1 μm high.

Comments This small fragile gray mushroom growing on wet well-decayed mossy logs is very commonly found. The slowly acrid taste seems to vary with age and gives the sharpest bite when first picked.

Lactarius camphoratus
(Bulliard) Fries

AROMATIC MILKY

SYNONYM *Agaricus camphoratus* Bulliard

Small to medium-sized with reddish brown or brick-red colors, cap with conical bump; latex white; odor fragrant, often like maple syrup

CAP 1.5–5 cm wide, broadly conical, becoming plane then sunken with well-developed conical bump on center, dark reddish-brown, becoming orange-brown or brick-red over margin as it fades, center darker than margin, moist becoming dry, glabrous, often bumpy. **FLESH** cap color, brittle. **GILLS** off-white or pale buff, becoming pinkish-cinnamon or wine-red and developing a frosted sheen, bluntly attached or subdecurrent, close or crowded. **STEM** 1.5–6 cm long, 3–11 mm broad, equal or enlarged downward, colored as cap but with silvery hoary coating, dry, white long hairs at base, hollow, fragile. **SPORE PRINT** pale yellow in heavy deposit. **LATEX** white, milky, thin and watery for older specimens. **ODOR** fragrant when fresh, faintly of maple syrup, stronger upon drying. **TASTE** disagreeable, bitter.

Habit and habitat single, scattered, or clustered on humus in coniferous and mixed forests. June to November. **RANGE** widespread.

Spores subglobose or broadly ellipsoid, warty or with some short ridges, colorless, but ornamentations amyloid, 7–8.5 × 6–7.5 μm excluding ornamentation, which measures to 1 μm high.

Comments The fragrant odor is often not immediately noticeable when first picked. However, if fruit bodies are placed in a wax or plain paper bag and left for a half hour or so, the odor of curry, maple syrup, or burnt sugar is strong. If your sample has this odor but the gills are yellow and the latex is watery, even in the younger ones, you probably have *Lactarius fragilis* (Burlingham) Hesler & A. H. Smith. See also *Lactarius volemus*.

Lactarius volemus (Fries) Fries

VOLUMINOUS-LATEX MILKY

SYNONYMS *Agaricus camphoratus* Fries, *Agaricus lactifluus* Linnaeus, *Lactarius lactifluus* (Linnaeus) Quélet

Cap orange-brown, minutely velvety; gills cream-white, staining dark brown; latex milky-white, abundant; unpleasant odor of bad fish

CAP 3–13 cm wide, convex with inrolled margin, becoming plane then sunken and somewhat vase-shaped, light or dark orange-brown, minutely velvety, dry, smooth or faintly cracking, often a bit wrinkled over the margin. **FLESH** white, turning brown. **GILLS** cream-white, staining dark brown, bluntly attached or subdecurrent, close. **STEM** 4–7(–10) cm long, 5–25 mm broad, equal or enlarged downward, orange-yellow or pale orange-brown, mostly paler than the cap, dry, hollow. **SPORE PRINT** white. **LATEX** white, milky, abundant, staining mushroom tissues and paper (and your fingers) brown.

ODOR faintly to strongly fishy, unpleasant. **TASTE** mild.

Habit and habitat single, scattered, or clustered on humus in hardwood forests. June to September. **RANGE** widespread.

Spores subglobose or broadly ellipsoid, warty and with ridges forming a reticulum, colorless, but ornamentations amyloid, 7.5–9(–10) × 7.5–8.5(–9) μm excluding ornamentation, which measures mostly 0.5–0.8(–1) μm high.

Comments If the gills are widely spaced, the fruit bodies lack the fishy odor, but the colors are in this range and the copious white milky latex does not stain everything brown, you have *Lactarius hygrophoroides* Berkeley & M. A. Curtis. Both species are considered good to eat. *Lactarius corrugis* Peck also produces copious white latex that turns gills and tissues brown, but it differs by its dark brown or reddish-brown wrinkled cap, dark brown stem, and lack of a strong fishy odor.

Lactarius rufus (Scopoli) Fries

RED-HOT MILKY

SYNONYMS *Agaricus rufus* Scopoli, *Agaricus variabilis* Persoon, *Lactarius boughtonii* Peck, *Lactarius rubescens* (Saccardo) Britzelmayr

Medium-sized brick-red mushroom with copious white latex; taste instantly and strongly acrid hot—be careful about the quantity tried, the smaller the better.

CAP 4–10(–12) cm wide, convex, becoming plane, then sunken with central conical umbo, often strongly incurved margin, dark bay-red or brick-red after white bloom disappears, paler with aging, fine ribbing on margin, dry, glabrous. **FLESH** pale vinaceous-buff or purplish. **GILLS** white becoming vinaceous-buff or vinaceous-tan, subdecurrent, crowded, narrow. **STEM** 5–11 cm long, 9–17 mm broad, equal or enlarged downward, surface white hoary at first, colored as cap except for white base once the bloom disappears, glabrous, dry, cottony white filled. **LATEX** milk white, abundant, turning white paper yellow.

SPORE PRINT cream-buff or pale pinkish-buff. **ODOR** not distinctive, or pungent with age. **TASTE** instantly and strongly acrid hot, only try a tiny piece.

Habit and habitat scattered or clustered on soil or humus under conifers, especially pine, spruce, and larch, also known to grow under birch, often on *Sphagnum* and other mosses, also in bogs. July to October. **RANGE** widespread.

Spores broadly ellipsoid, ridged and forming a partial reticulum, colorless, but ornamentations amyloid, 8.5–12 × 6–8 µm excluding ornamentation, which measures to 0.5 µm high.

Comments There appear to be several variations of this species in North America. The one outlined here with the small conical bump on the cap center is the one most commonly collected in the Northeast, especially in bogs or poorly drained soils. Although *Lactarius rufus* is considered edible in some European countries, it is listed as poisonous in North America.

Lactarius lignyotus Fries

CHOCOLATE MILKY

Cap and stem dark chocolate-brown, velvety, cap with sharp conical central bump; gills white, staining pink from dried latex; latex milky-white, abundant

CAP 2–10 cm wide, broadly convex with sharp conical bump on the center, becoming shallowly sunken, very dark chocolate-brown or almost black, becoming paler with age, dry, soft velvety, radially wrinkled. **FLESH** white, often becoming pinkish with exposure. **GILLS** white, becoming ochre buff, staining pink from drying latex, bluntly attached, close or subdistant. **STEM** 4–12 cm long, 4–20 mm broad, equal, colored as cap or slightly paler, white over base, dry, minutely velvety overall. **SPORE PRINT** bright orange-yellow. **LATEX** white, milky, abundant, usually drying pinkish. **ODOR** and **TASTE** not distinctive.

Habit and habitat scattered or clustered on humus and mosses under conifers, often on well-decayed logs. July to October. **RANGE** widespread.

Spores broadly ellipsoid or subglobose, banded with ridges forming a reticulum, colorless, but ornamentations amyloid, 9–10.5 × 9–10 μm excluding ornamentation, which measures to 1.5 μm high.

Comments There are several varieties of this species in the Northeast. If the gill edges are brown, or marginate, it is *L. lignyotus* var. *canadensis* A. H. Smith & Hesler. If the flesh turns dark violet when injured and the gills are not marginate it is *L. nigroviolascens* G. F. Atkinson. If the flesh turns dark violet and the gills are brown marginate it is *L. nigroviolascens* var. *marginatus* A. H. Smith & Hesler [syn. *L. lignyotus* var. *marginatus* (A. H. Smith & Hesler) A. H. Smith & Hesler]. If the cap is yellow-brown, the lamellae are not marginate, the latex stains damaged tissues red, then it is *L. fumosus* Peck. If the gills are distantly spaced and the cap is dark yellowish-brown, then *L. gerardii* Peck. This group includes many other described species, but these are the most common morphological forms encountered.

Wax Caps

Cuphophyllus lacmus
(Schumacher) Bon

SYNONYMS *Hygrophorus lacmus* (Schumacher) Fries, *Hygrocybe lacmus* (Schumacher) P. D. Orton & Watling, *Hygrophorus subviolaceus* Peck, *Hygrocybe subviolacea* (Peck) P. D. Orton & Watling, *Cuphophyllus subviolaceus* (Peck) Bon

Cap thinly slippery, violet-gray; gills paler violet-gray, decurrent, broad and thick; stem white with tints of violet-gray, dry, tapered and curved; odor unpleasant

CAP 2–6 cm wide, convex, becoming plane, uniformly violet-gray, ashy-gray or bluish-gray or brownish-violaceous, hygrophanous and becoming paler with brown tints from center outward, thin slippery layer, glabrous, translucent-lined when wet. **FLESH** same color as cap, thin at margin. **GILLS** white then pale smoky violet-gray or darker gray on some collections, decurrent, subdistant or distant, moderately broad, thick and waxy-looking. **STEM** 3–7 cm long, 4–11 mm broad, equal or tapering downward and curved at base often, white or colored as cap but paler, sometimes pale cream in base, not slippery, glabrous, solid becoming hollow. **SPORE PRINT** white. **ODOR** unpleasant. **TASTE** at first mild, then bitter or nauseous, finally acrid burning.

Habit and habitat scattered or clustered, on wet soil or humus under mixed hardwoods. July to September. **RANGE** widespread.

Spores broadly elliptical or oblong, colorless, inamyloid, smooth, 6–8 × 4–6 µm.

Comments The taste of this mushroom is unpleasant, but you have to chew a little while before expectoration to get the effect. Some collections are darker gray in the cap and gills than others, but otherwise have similar features.

Gliophorus irrigatus (Persoon) A. M. Ainsworth & P. M. Kirk

SYNONYMS *Hygrophorus irrigatus* (Persoon) Fries, *Hygrocybe irrigata* (Persoon) Bon, *Hygrophorus unguinosus* (Fries) Fries, *Hygrocybe unguinosa* (Fries) P. Karsten, *Gliophorus unguinosus* (Fries) Kovalenko

Very slippery or slimy small black or dark grayish-brown mushroom with bright white gills

CAP 1–5 cm wide, convex or conical-convex, becoming plane, with a low broad bump over the center, dark grayish-brown or black, becoming pale smoke gray over the margin, slimy when wet, translucent-lined, glabrous. **FLESH** white or grayish, soft. **GILLS** white or pale grayish, contrasting sharply with cap and stem, bluntly attached or slight decurrent, subdistant, broad. **STEM** 3–9 cm long, 3–5 mm broad, equal or tapered at base, same color as cap or more often pale grayish-brown, translucent, slippery or slimy layer sheathing stem, surface glabrous, shiny varnished looking when dry. **SPORE PRINT** white. **ODOR** none. **TASTE** mild.

Habit and habitat scattered or clustered, on humus and soil under conifers or in mixed woods, also in swampy areas on mosses. August to October. **RANGE** widespread.

Spores ellipsoid, colorless, inamyloid, smooth, 7–10 × 4–6 μm.

Comments This is a dull and darkly colored slippery little mushroom.

Gliophorus laetus (Persoon) Herink

SYNONYMS *Hygrophorus laetus* (Persoon) Fries, *Hygrocybe laeta* (Persoon) P. Kummer

Small very slimy cap, stem and edge of gills, variable in colors ranging from reddish-brown, violet-gray or pink; growing in wet mucky areas or on *Sphagnum*, often in bogs

CAP 1–3.5 cm wide, convex, becoming plane or with slightly sunken center, colors highly variable but often with reddish-brown or orange colors, also pale violaceous-gray, olivaceous-orange, sometimes olive-brown or buff or pinkish-buff, slimy, glabrous, faintly translucent-lined at margin. **FLESH** same color as cap, thin but tough. **GILLS** variable like cap, but usually paler and often pinkish or vinaceous-gray, bluntly attached and often with decurrent tooth, subdistant, moderately broad, waxy-looking, but edges viscid. **STEM** 3–12 cm long, 2–4 mm broad, equal, may be same color as cap or can be paler, often with pale violaceous-gray, very slimy, glabrous. **SPORE PRINT** white. **ODOR** none or faintly fishy (disagreeable). **TASTE** not distinctive.

Habit and habitat usually clustered, on wet earth or humus or more frequently in *Sphagnum* in boggy areas, under hardwood or mixed hardwoods and conifers. July to October. **RANGE** widespread.

Spores elliptical, colorless, inamyloid, smooth, 5–8 × 3–5 µm.

Comments This species is almost impossible to pick by grabbing the stem. It is so slimy and slippery, you must also bring up part of the substrate in which it is growing.

Gliophorus psittacinus
(Schaeffer) Herink

PARROT MUSHROOM

SYNONYMS *Hygrophorus psittacinus* (Schaeffer) Fries, *Hygrocybe psittacina* (Schaeffer) P. Kummer

Small very slimy cap and stem, variable in colors but cap typically dark green entirely at first, then margin becoming pinkish or yellow; gills and stem fading to yellow

CAP 1–3 cm wide, conical-convex, becoming bell-shaped or eventually plane, dark green at first, fading from margin inward to center to pale reddish-brown, pinkish-flesh, yellow-ochre, yellow, or olivaceous-orange, center often remaining greenish, slimy, glabrous, translucent-lined. **FLESH** same color as cap, thin. **GILLS** pale greenish becoming yellow, bluntly attached, subdistant, moderately broad, waxy-looking. **STEM** 3–7 cm long, 2–5 mm broad, equal, may be same color as cap or yellow and translucent, very slimy, glabrous, hollow. **SPORE PRINT** white. **ODOR** and **TASTE** not distinctive.

Habit and habitat scattered or clustered, on soil or humus under hardwood or mixed hardwoods and conifers. July to October. **RANGE** widespread.

Spores elliptical, colorless, inamyloid, smooth, 6.5–8 × 4–5 µm.

Comments Another species that is almost impossible to pick by grabbing the stem, it is so slimy and slippery. The green color is sometimes described as parrot green, hence the common name.

Gloioxanthomyces nitidus
(Berkeley & M. A. Curtis) Lodge, Vizzini, Ercole & Boertmann

SYNONYMS *Hygrophorus nitidus* Berkeley & M. A. Curtis, *Hygrocybe nitida* (Berkeley & M. A. Curtis) Murrill, *Gliophorus nitidus* (Berkeley & M. A. Curtis) Kovalenko

Small very slimy cap and stem, dark yellow overall, cap deeply sunken with lined margin; cap and stem becoming shiny when dried

CAP 1–4 cm wide, convex, becoming plane and with a deeply sunken center, rich dark golden-yellow, may fade to pale yellow or cream color, slimy, glabrous, lined or grooved over margin. FLESH same color as cap, thin. GILLS same color as cap or paler, bluntly attached or subdecurrent, subdistant, moderately broad, waxy-looking, fragile. STEM 3–8 cm long, 2–5 mm broad, equal, same color as cap, fading to near white with age, very slimy, glabrous, hollow. SPORE PRINT white. ODOR and TASTE not distinctive.

Habit and habitat scattered or clustered, on wet soil or humus under hardwood or mixed hardwoods and conifers, also on *Sphagnum* in bogs. July to October. RANGE widespread.

Spores elliptical, colorless, inamyloid, smooth, 6.5–8 × 4–5 µm.

Comments Another species that is almost impossible to pick by grabbing the slimy and slippery stem. These small mushrooms are like glowing little yellow lights in the dark woods. *Nitidus* means bright, shining.

Humidicutis calyptriformis
(Berkeley) Vizzini & Ercole

PINK WAX CAP
SYNONYMS *Hygrocybe calyptriformis* (Berkeley) Fayod, *Hygrophorus calyptriformis* (Berkeley) Berkeley, *Godfinia calyptriformis* (Berkeley) Herink, *Porpolomopsis calytriformis* (Berkeley) Bresinsky

Cap bright pink and sharply conical, splitting after expansion, slippery when wet; gills pinkish, subdistant; stem white with a pink flush, very fragile, splitting

CAP 2.5–7 cm wide, narrowly and sharply conical, expanding to near plane and splitting to near center, retaining central conical bump, dull pinkish-red, bright shell-pink, or some pale lilac, conical bump may become white, slippery when wet, becoming dry, with radiately flattened fibrils. **FLESH** pinkish. **GILLS** pale pink at first, attached or barely so, close or subdistant, moderately broad. **STEM** 5.5–16 cm long, 4–8 mm broad, equal but often with tapered base, white but tinted with pale pinkish or pale rose over upper half, dry, glabrous, splitting easily, hollow. **SPORE PRINT** white. **ODOR** and **TASTE** not distinctive.

Habit and habitat scattered or clustered, on soil and humus in conifer or mixed woods. August to October. **RANGE** in the southern and western part of the Northeast (Maryland and Ohio). May be more widely spread.

Spores elliptical, colorless, inamyloid, smooth, 6.5–8 × 4.5–6 μm.

Comments This beautiful mushroom prefers warmer climates and is found more frequently in southeastern areas in North America.

Humidicutis marginata var. *marginata* (Peck) Singer

ORANGE-GILLED WAX CAP

SYNONYMS *Hygrophorus marginatus* Peck, *Hygrocybe marginata* (Peck) Murrill, *Tricholoma marginatum* (Peck) Singer

Cap orange-yellow, fading to pale yellow, dry; gills brilliant orange, not fading like cap, or at least edges remaining dark orange; stem golden-yellow, dry

CAP 1–5 cm wide, rounded conical, then convex or becoming bell-shaped, some with low broad bump over center, rich orange-yellow or golden-orange, hygrophanous and becoming pale yellow or nearly white with age, dry or moist, glabrous, some faintly translucent-lined at margin when fresh. **FLESH** same color as cap, thin. **GILLS** brilliant orange or dark orange, color persistent at least on edges, bluntly attached, subdistant, moderately broad, waxy-looking. **STEM** 4–10 cm long, 3–6 mm broad, equal or tapered above and below a swollen middle, pale orange-yellow or bright lemon-yellow or rich golden-yellow, dry, glabrous, hollow, easily splitting. **SPORE PRINT** white. **ODOR** none. **TASTE** mild.

Habit and habitat single, scattered, or clustered on humus and leaf litter under hardwood or mixed hardwoods and conifers. July to October. **RANGE** widespread.

Spores elliptical or oblong, colorless, inamyloid, smooth, 7–10 × 4–6 μm.

Comments This species has two main varieties, one in which the lamellae are the same golden-yellow color as the cap surface, *Humidicutis marginata* var. *concolor*, and one that has a strongly olive-colored cap and shades of olive-green on the stem, *H. marginata* var. *olivacea*.

Humidicutis pura (Peck)
E. Horak

SYNONYMS *Hygrophorus purus* Peck, *Hygrocybe pura* (Peck) Murrill

Entirely white, cap and stem slimy, cap conical, at least on the center

CAP 4–7 cm wide, conic, becoming broadly bell-shaped with conical center and eventually uplifted margin, pure white, may become pinkish where wounded, slippery or slimy, glabrous, margin faintly translucent-lined when fresh, becoming opaque. **FLESH** white, thin, waxy. **GILLS** white, attached, subdistant, broad. **STEM** 4–8 cm long, 3–8 mm broad, equal or tapered either upward or downward, white but base may turn pinkish, slimy, glabrous, hollow, easily splitting. **SPORE PRINT** white. **ODOR** none. **TASTE** mild.

Habit and habitat single, scattered, or clustered on rich humus and leaf litter under hardwood or mixed hardwoods and conifers. August to October. **RANGE** widespread but not common.

Spores ellipsoid, colorless, inamyloid, smooth, 7–9 × 4–5 μm.

Comments The more commonly collected white wax caps are smaller, with caps 1–5 cm broad: *Cuphophyllus virgineus* (Wulfen) Kovalenko has a dry cap while *C. borealis* (Peck) Bon ex Courtecuisse has a slippery cap. They also have in common a dry stem and short decurrent gills. *Cuphophyllus* and *Humidicutis* were taken out of *Hygrophorus* based on some morphological features, like hyphal arrangement of the gill tissue, and now this move is supported by DNA phylogenetics. *Cuphophyllus* tends to be colorless with translucent flesh and intervened gills, while *Humidicutis* has brightly colored species, or at least not with translucent flesh, and the caps are often sharply conical.

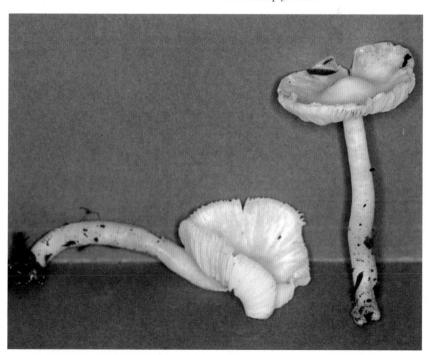

Hygrocybe cantharellus
(Schweinitz) Murrill

CHANTERELLE WAX CAP

SYNONYMS *Agaricus cantharellus* Schweinitz, *Camarophyllus cantharellus* (Schweinitz) Murrill, *Hygrophorus cantharellus* (Schweinitz) Fries

Small bright reddish-orange mushroom with tiny scales on the cap; gills decurrent, yellow, distant, waxy-looking; stem colored as cap, dry, smooth

CAP 1–4 cm wide, convex then plane, shallowly sunken in center, scarlet or dark orange-red when young, fading to orange-ochre or paler, finely silky or scurfy-scaly (use a lens), opaque. **FLESH** yellow or orange, thin. **GILLS** orange-yellow, paler than cap, decurrent, subdistant or distant, broad, waxy-looking. **STEM** 2.5–10 cm long, 1.5–5 mm broad, equal, color of cap or becoming paler and more orange, white over base, glabrous, hollow. **SPORE PRINT** white. **ODOR** not distinctive. **TASTE** mild.

Habit and habitat scattered or clustered, on the ground, on mosses, on rich humus, on decaying logs, in bogs under hardwoods or conifers. July to October. **RANGE** widespread.

Spores elliptical, colorless, smooth, 7–12 × 4–8 µm.

Comments The several other reddish-orange species of *Hygrocybe* do not have decurrent gills. Compare with *H. miniata*, which has reddish-orange gills that are bluntly attached. It is mostly smaller in stature but does have a scaly cap like *H. cantharellus*.

Hygrocybe coccineocrenata
(P. D. Orton) M. M. Moser

SYNONYMS *Hygrophorus coccineocrenatus* P. D. Orton, *Pseudohygrocybe coccineocrenata* (P. D. Orton) Kovalenko, *Hygrophorus miniatus* var. *sphagnophilus* Peck, *Hygrophorus turundus* var. *sphagnophilus* (Peck) Hesler & A. H. Smith

Cap reddish orange with brown scales and fibrils, also with scalloped margin; gills off-white, decurrent, broad and thick; stem colored as cap; on *Sphagnum*

CAP 0.5–2 cm wide, convex, becoming broadly convex then plane, with a shallowly sunken center and scalloped (crenate) margin, scarlet or orange or yellowish and covered with dark brown or blackish-brown fibrils and scales making the cap very dark when young, dry, minutely scaly with expansion. **FLESH** same color as cap, thin. **GILLS** white or cream-buff, decurrent, subdistant or distant, broad, thick and waxy-looking. **STEM** 1.5–7 cm long, 1.5–4 mm broad, equal, colored as cap or more reddish-orange, not viscid, glabrous. **SPORE PRINT** white. **ODOR** and **TASTE** not distinctive.

Habit and habitat scattered or clustered in humus and *Sphagnum* beds in high elevation bogs. July to September. **RANGE** widespread but restricted to bogs or at least wet areas with *Sphagnum*.

Spores elliptical or somewhat bean-shaped, colorless, inamyloid, smooth, 9–14 × 5–8 μm.

Comments This small delicate mushroom is very distinctive. If you frequent areas with *Sphagnum*, such as bogs in the Adirondack Mountains, you will run into this small colorful species.

Hygrocybe miniata (Fries)
P. Kummer

FADING SCARLET WAX CAP
SYNONYM *Hygrophorus miniatus* (Fries) Fries

Small bright scarlet capped mushroom with scurfy-scaly cap; gills bluntly attached, scarlet at first, then orange, waxy-looking; stem colored as cap or yellow, dry, smooth

CAP 2–4 cm wide, convex then plane, shallow and broadly sunken over center, scarlet, fading to orange or paler, smooth or minutely fibrillose or scurfy-scaly on the center, faintly translucent-lined or opaque. **FLESH** same as cap, thin. **GILLS** scarlet at first, then orange-yellow, paler than cap, bluntly attached, close or subdistant, broad, waxy-looking. **STEM** 3–5 cm long, 3–4 mm broad, equal, color of cap or becoming yellow, white over base, glabrous, hollow. **SPORE PRINT** white. **ODOR** not distinctive. **TASTE** mild.

Habit and habitat scattered or clustered, on the ground, on mosses, on rich humus, on moss covered decaying logs, under hardwoods or mixed hardwoods and conifers. July to November. **RANGE** widespread.

Spores elliptical or pear or bean-shaped, with at least some spores constricted in the middle, colorless, smooth, 6–10 × 4–6 μm.

Comments *Hygrocybe squamulosa* (Ellis & Everhart) Arnolds is similar in form and color but has larger fruit bodies. The caps reach 5 cm broad and are noticeably scaly, and the cap flesh is thicker. It is also usually associated with older mature forests.

Hygrocybe flavescens
(Kauffman) Singer

GOLDEN WAX CAP
SYNONYMS *Hygrophorus puniceus* var. *flavescens* Kauffman, *Hygrophorus flavescens* (Kauffman) A. H. Smith & Hesler

Cap bright yellow or slight orange-yellow, slippery; gills pale yellow; stem same color as cap, dry

CAP 2.5–7 cm wide, broadly convex, becoming plane or with slightly sunken center, bright yellow or faintly orange-yellow, slippery at first but becoming dry and shiny, glabrous, faintly translucent-lined at margin. **FLESH** same color as cap, thin. **GILLS** pale yellow, paler than cap, barely attached, close or subdistant, broad, waxy-looking. **STEM** 4–7 cm long, 5–15 mm broad, equal or tapered at base, often flattened, color of cap except white over base, dry, glabrous, hollow, easily splitting. **SPORE PRINT** white. **ODOR** and **TASTE** not distinctive.

Habit and habitat single or scattered or clustered, on the ground, in thick humus or amongst mosses in wet areas of hardwood or mixed hardwoods and conifers. July to October. **RANGE** widespread.

Spores elliptical, colorless, inamyloid, smooth, 7–9 × 4–5 µm.

Comments *Hygrocybe chlorophana* (Fries) Wünsche is very similar in form and color, but the stem of the fruit bodies is slippery. It may be that these two are the same species, in which case the name *H. chlorophana* must be used. Be aware that handling the stipe of this dry species long enough will make it slightly slippery from injuring the cells of the stem surface.

Hygrocybe parvula (Peck)
Murrill

SYNONYMS *Hygrophorus parvulus* Peck, *Pseudohygrocybe parvula* (Peck) Kovalenko

Cap uniformly yellow, dry; gills yellow, broad; stem orange-yellow or with salmon tints, dry, not slippery, under hardwoods

CAP 1–3 cm wide, convex, becoming plane, uniformly yellow or wax-yellow, sometimes with a little orange mixed in, may fade to pale yellow or cream color, moist, glabrous, translucent-lined over margin. **FLESH** same color as cap, thin. **GILLS** same color as cap or paler, bluntly attached or subdecurrent, subdistant, moderately broad, waxy-looking, fragile. **STEM** 3–6 cm long, 2–3 mm broad, equal, uniformly dark orange-yellow or salmon-ochre, glabrous, moist or dry, not viscid, hollow. **SPORE PRINT** white. **ODOR** and **TASTE** not distinctive.

Habit and habitat scattered or clustered, on wet soil or humus under hardwood or mixed hardwoods and conifers. June to October. **RANGE** widespread.

Spores elliptical, colorless, inamyloid, smooth, 5–7 × 3.5–5 μm.

Comments This small species has bright contrasting colors, the stem often more brightly colored than the cap.

Hygrocybe ovina (Bulliard) Kühner

SYNONYMS *Hygrophorus ovinus* (Bulliard) Fries, *Neohygrocybe ovina* (Bulliard) Herink, *Camarophyllus ovinus* (Bulliard) P. Kummer, *Hygrocybe nitiosa* (A. Blytt) M. M. Moser, *Hygrophorus nitiosus* A. Blytt

Cap and stem dry and uniformly grayish-brown; gills paler grayish-brown, broad and thick; all parts including the flesh turning red when injured

CAP 2–5 cm wide, convex, becoming plane, uniformly gray-brown or darker sooty-brown, sometimes clay-brown on the center and becoming pinkish-buff over the margin, moist or dry, not slippery, glabrous or becoming scaly over center with age. **FLESH** pallid, turning red when damaged, thin. **GILLS** gray-brown but paler than cap, turning pinkish-red when bruised, bluntly attached, close or subdistant, moderately broad, thick and waxy-looking. **STEM** 4–7 cm long, 5–10 mm broad, equal, colored as cap or paler, turning pinkish or vinaceous where handled, not slippery, glabrous, hollow. **SPORE PRINT** white. **ODOR** faintly fruity when cut or crushed. **TASTE** not distinctive, or alkaline.

Habit and habitat scattered or clustered, on wet soil or humus under mixed hardwoods. July to September. **RANGE** widespread.

Spores elliptical or oblong, colorless, inamyloid, smooth, 7–9 × 4.5–6 µm.

Comments The dark colors make this species difficult to see in a dense forest. The odor, as described for the European collections, is sometimes slightly nitrous when cut and unpleasant in older specimens. Our collections are distinctly fruity in odor.

Hygrophorus amygdalinus
Peck

Cap gray with scalloped margin, slippery; gills white, subdistant; stem grayish and minutely dotted scaly, slippery; odor of bitter almonds

CAP 2.5–4 cm wide, convex with low conical bump over the center, becoming plane with conical bump still present, margin rolled in and often scalloped (crenate), gray or gray-brown, somewhat slippery when wet, glabrous. **FLESH** white, thick. **GILLS** white, bluntly attached or subdecurrent, subdistant, broad. **STEM** 5–15 cm long, 4–6 mm broad, equal but often with tapered base, similar color as cap, white at base becoming yellowish, slippery, minutely dotted scaly over upper half. **SPORE PRINT** white. **ODOR** of bitter almonds. **TASTE** mild.

Habit and habitat scattered or clustered on soil and humus in conifer woods, especially pines. August to October. **RANGE** widespread but not commonly collected.

Spores elliptical, colorless, inamyloid, smooth, 8–12 × 4.5–6 µm.

Comments This slender gray mushroom is distinctive because of its odor. Compare with the more commonly collected *Cantharellula umbonata* that lacks such an odor, has repeatedly forking gills, and grows on hair cap moss beds. Repeatedly forked gills are not present in *H. amaydalinus*.

Hygrophorus ponderatus
Britzelmayr

Large white mushroom with slimy cap and stem; ring on stem very wispy, delicate, easily lost

CAP 5–14 cm wide, convex, then plane, becoming broadly sunken, white, slippery or slimy when wet, glabrous, margin with cottony flecks from the ring. **FLESH** white, thick. **GILLS** white, bluntly attached or decurrent, subdistant, broad. **STEM** 2.5–9 cm long, 15–35 mm broad, equal or tapered up or downward, white, slippery or slimy, silky-fibrillose under slime, with thin wispy, delicate white ring near apex, but that soon collapsing, solid and white inside. **RING** white, lacey fibrillose. **SPORE PRINT** white. **ODOR** and **TASTE** not distinctive.

Habit and habitat scattered or clustered on soil often associated with mounds of moss, in conifer or mixed woods. August to October. **RANGE** widespread but not commonly encountered in the Northeast.

Spores ellipsoid, colorless, inamyloid, smooth, 6.5–9 × 4–5.5 μm.

Comments This large white species occurs in the forests of western Massachusetts, near Amherst. Compare with *Hygrophorus sordidus* Peck, another large white species. It is not slippery or slimy on the cap or stem, and it does not produce a ring.

Hygrophorus speciosus Peck

Very slippery or slimy large mushroom with bright orange-red cap and creamy-white gills; under conifers, often on *Sphagnum*

CAP 2–5 cm wide, convex or broadly conical-convex, some with low rounded bump over the center, orange-red or orange at first, fading to golden-yellow, slimy when wet, glabrous. **FLESH** white or yellowish, soft. **GILLS** white or cream-yellowish, bluntly attached or short decurrent, subdistant, broad. **STEM** 4–10 cm long, 4–8 mm broad, equal or tapered up or downward, white or with pale orange tints, slippery or slimy layer sheathing stem to near apex, surface irregularly bumpy, white-dotted over apex, soft and filled with cottony white tissue. **RING** colorless, thick or thin slime. **SPORE PRINT** white. **ODOR** and **TASTE** not distinctive.

Habit and habitat scattered or clustered, on humus under various types of conifers, pine, cedar, spruce, but most often under tamarack and in or near *Sphagnum* bogs. August to October. **RANGE** widespread.

Spores ellipsoid, colorless, inamyloid, smooth, 8–10 × 4.5–6 μm.

Comments This species has two versions. If a white inner membranous ring is present under the slimy veil that covers the stipe, then the name *Hygrophorus speciosus* var. *kauffmanii* Hesler & A. H. Smith may be applied and is the image shown here.

Tricholoma and Allies

This group is large with many genera. I have separated them mostly based on ecology and gill attachment. Some genera can be found on the ground but are attached to buried wood, others grow solely from obviously decaying woody substrates. The gill attachment feature is also helpful because the difference between decurrent gills and bluntly or barely attached gills is usually obvious. You may have to examine several descriptions before making a decision for your collection. Making a spore print is very helpful for some members of this large group of "white"-spored species.

Almost all these genera can be clearly characterized with macroscopic features; however, in several cases, identification can be accurately accomplished only with microscopic features. I have noted those features in parentheses in the key.

Growing on other mushrooms

Asterophora On large fruit bodies of *Russula* or *Lactarius* species, either cap brown and powdery with gills absent, or cap white and silky with gills well-developed
Collybia Very small, cap and stem white, stem long, thin, on blackened, decayed, mummified mushroom fruit bodies that are barely recognizable

Growing on conifer cones or magnolia fruits

Baeospora myosura In the fall on cones of Norway spruce and white pine, cap brown (spores amyloid, cystidia thin-walled, not encrusted)
Strobilurus esculentus In the early spring on cones of Norway spruce, cap brown (spores not amyloid, cystidia thick-walled and encrusted)
Strobilurus conigenoides On decaying magnolia fruits, cap white or pale tan

Growing on *Sphagnum*, hair cap, or other mosses

Arrhenia gerardiana On *Sphagnum*, cap small and vase shaped with dark brown fibrillose dots, gills long decurrent
Sphagnurus paluster On dense *Sphagnum* beds, cap gray and translucent-lined
Cantharellula umbonata On hair cap moss (*Polytrichum commune*), cap pale gray with conical umbo, gills white and repeatedly forked, staining reddish
Rickenella Small, delicate, convex with sunken center, stem minutely fibrillose overall (long projecting cystidia on cap, stem, and gills)
Mycena sanguinolenta On mosses with conifer needles, delicate, fragile, producing red juice when injured, gills reddish-brown on edges

Growing on soil, humus, small twigs, or decaying leaves

GILLS NOTCHED OR BLUNTLY ATTACHED, OR SHORT DECURRENT WITH AGE
Tricholoma Gills notched, mostly large fleshy (spores smooth, not amyloid)
Pseudotricholoma Gills notched, gray-brown, scaly (spores smooth, amyloid)
Lepista Spore print pinkish-buff (spores with cyanophilic bumps)
Leucopaxillus Large fleshy, binding substrate with thick white mycelium (spores amyloid warted)
Melanoleuca With strict-straight stem and rigid stature, gills crowded (spores amyloid warted, cystidia often present and sharp-pointed, some with crystals adorning the pointed apex)

Lyophyllum Watery brown or black cap colors, often staining black, sometimes in bouquet-like clusters (basidia with siderophilous and cyanophilous bodies)

Rugosomyces On lawns, cap and stem pink or pinkish-brown (basidia with siderophilous and cyanophilous bodies)

Armillaria On buried roots or on stumps, cap with fine black hairs or scales, stem with ring, but one species lacks a ring and has short decurrent gills

Laccaria Gills flesh-pink, violet, or purple, thick and waxy-looking, cap and stem mostly orange-brown, but some violet or purple, fruit bodies small, medium, or large

Cystoderma Cap orange-brown or yellow-brown, densely granular, stem with distinct ring or ring zone (spores amyloid)

Cystodermella Cap reddish-orange, granular, large, ring granular and soon collapsing (spores not amyloid, cystidia pointed with crystal-encrusted tips)

Hymenopellis With long rooting stem, or pseudorhiza, cap with a rubbery texture when fresh (spores large, ovoid, lemon- or almond-shaped, not amyloid)

Gymnopus Small to medium-sized, caps convex or plane, margin often inrolled

Connopus In bouquetlike clusters, cap dark purplish-brown, gills white and crowded

Rhodocollybia Spore print pinkish-buff, small or large, fleshy, variously colored (many spores red-brown, dextrinoid in Melzer's reagent)

Marasmius Stem mostly dark, shiny, horsehairlike, caps small, variously colored, convex or plane, sometimes bell-shaped, often corrugate-lined (spores not amyloid, tissues not dextrinoid)

Mycetinis scorodonius Marasmius-like, cap red-brown, then paler, odor of garlic

Mycena Small, delicate, caps variously colored, bell-shaped, stem fragile (spores often amyloid, tissues often dextrinoid)

Crinipellis Marasmius-like but caps with brown hairs or fibrillose scales (hairs and scales turning reddish-brown in Melzer's reagent)

Resinomycena Small, white, with sticky gill edges and sticky stem

Tetrapyrgos Stem black but white-frosted with tiny hairs, cap small, white, wrinkled

GILLS DECURRENT, BUT MAY BE BLUNTLY ATTACHED AT FIRST

Ampulloclitocybe Medium-sized, gray-brown cap and stem, stem swollen club-shaped, gills cream-colored, growing under conifers

Clitocybe Small to large, variously colored (spores small, hyaline, not amyloid)

Gerhardtia highlandensis Moderately large, white with hoary-canescent cap, odor fragrant, growing under white pine (spores with random cyanophilic bumps; basidia with siderophilous or cyanophilous bodies)

Hygrophoropsis Orange overall, gills repeatedly forked, soft-fleshed, growing under conifers

Singerocybe Cap deeply funnel-shaped, slippery and gray-brown, turning off-white from moisture loss, medium-sized, odor and taste unpleasant

Rickenella Small, delicate, convex with sunken center, stem minutely fibrillose overall (long projecting cystidia on cap, stem, and gills)

Lichenomphalia Small, delicate, umbrella-like caps with scalloped margins, green algae at base of stems

Growing on wood, such as logs, stumps, or large branches on the ground

STEM ABSENT OR STRONGLY OFF-CENTER, OR STEM AT CAP MARGIN

Pleurotus Cap white, tan, or brown, gills decurrent on short stem, spore print lilac-gray or white, medium to large (spores cylindrical, not amyloid)

Hypsizygus On elms and various hardwoods, cap large, white, smooth, cracking with age, gills bluntly attached or notched, stem long, frequently off-center (spores globose or subglobose, not amyloid, but cyanophilic)

Ossicaulis On maples, chalk-white, gills bluntly or barely attached, crowded, stem well-developed, off-center (spores ellipsoid, not amyloid, not cyanophilic)

Phyllotopsis Fan-shaped, orange, woolly, gills orange, odor of propane

Plicaturopsis Caps small overlapping, fan-shaped with lobed margins, fuzzy and zoned, gills wavy-crinkled, growing on smaller decaying branches, fruit bodies pliant tough

Panellus Caps small or large, kidney- or fan-shaped, tan or olive-green, gills crowded and lateral, stem often broad at gill attachment, tough (spores sausage-shaped and amyloid)

Panus neostrigosus Reddish-purple fuzzy caps, gills off-white or tan, crowded, decurrent, tough

Pleurocybella Milk-white, spatula-shaped caps, gills white, growing on conifers, medium to large, fleshy-fragile (spores globose, not amyloid)

Cheimonophyllum Small, cap white, shell-shaped, gills white with granular fringed edges, fleshy-fragile

Schizophyllum Gray overall, caps shell-shaped, hairy, margin scalloped, gills split lengthwise

Resupinatus Tiny, dark gray-brown inverted cups with corrugate-grooved margins, gills distant

STEM CENTRAL OR ONLY SLIGHTLY OFF-CENTER AND GILLS DECURRENT, BUT MAY BE BLUNTLY ATTACHED AT FIRST

Pseudoclitocybe Cap hygrophanous, dark brown fading to pale gray-brown, deeply sunken and smooth, gills watery brown, also growing on the ground (spores smooth and amyloid)

Arrhenia epichysium Small, dark brown or black hygrophanous cap with low raised lines over inrolled margin, becoming pale gray-brown (spores smooth and not amyloid)

Xeromphalina Small, orange-brown cap and stem, gills yellow, stem base covered with fuzzy orange tufted hairs (spores amyloid)

Gerronema Cap vase-shaped, yellow with brown flattened fibrils, gills pale yellow

Clitocybula In bouquetlike clusters, cap pale ash-white or pale brown

Cyptotrama Bright golden-yellow granular or scaly cap and stem, gills white

Panus neostrigosus Stem can be central, slightly off-center, and even lateral at cap margin; see also stem absent or strongly off-center

Neolentinus White with flat brown scales on cap, gill edges saw-toothed, large

Omphalotus Bright orange overall, growing in bouquetlike clusters, large

STEM CENTRAL AND GILLS BLUNTLY ATTACHED, NOTCHED, OR BARELY ATTACHED

Megacollybia Cap gray-brown with white, lens-shaped fissures, stem white with white cords radiating out from base

Flammulina Cap orange-brown, slimy, stem densely dark brown velvety, often in clusters on standing dead trees, especially elm

Tricholomopsis Cap with yellow ground color, many also with black, brown, olive, or brick-red fibrils on cap (cheilocystidia inflated, abundant)

Leucopholiota decorosa Cap and stem brown scaly, gills white

Mycena Small, delicate, variously colored, cap often bell-shaped

Baeospora myriadophylla Cap lavender then ochre-brown centrally, gills lavender, crowded, small

Hypsizygus Large white with long stem, growing on wounds of standing dead trees; see also stem absent or strongly off-center

Tricholoma aurantium
(Schaeffer) Ricken

VEILED TRICH

SYNONYMS *Agaricus aurantius* Schaeffer, *Armillaria aurantia* (Schaeffer) Quélet, *Melanoleuca aurantia* (Schaeffer) Murrill

Moderately large orange or reddish-orange scaly cap; gills white, staining rust-brown, notched attached, close; stem densely mottled with flattened orange scales except for stark white band at apex

CAP 3–9 cm wide, convex, becoming plane, orange or dull reddish-orange or tawny-ochre, slimy at first, densely covered with flattened scales and fibrils, opaque. **FLESH** white, thick, unchanging. **GILLS** white, developing rust-brown stains, attached and mostly notched (sinuate), close, narrow. **STEM** 4–8 cm long, 10–20 mm broad at apex, equal, densely mottled with flattened scales that have similar colors as the cap, stark white above the zone at the apex where the scales abruptly stop, white, solid inside. **SPORE PRINT** white. **ODOR** farinaceous. **TASTE** farinaceous or rancid farinaceous.

Habit and habitat scattered or clustered on the ground under. August to October. **RANGE** widespread.

Spores elliptical, smooth, colorless in KOH, 4–6 × 3–4 μm; hymenial cystidia and clamp connections absent.

Comments The sharp demarcation between the pigmented part of the stem and the white apex looks like the mushroom should have a ring, but it does not. This look is prominent enough, however, to have earned the common name for the veiled trich, referring to its nonexistent partial veil.

Tricholoma caligatum
(Viviani) Ricken

SYNONYMS *Agaricus caligatus* Viviani, *Armillaria caligata* (Viviani) Gillet

Medium to large, solid mushroom with dry brown fibrillose and scaly cap; stem white above flaring membranous ring, brown fibrillose and scaly below ring

CAP 4–2 cm wide, convex, becoming plane or shallowly sunken, white around the margin but with reddish-brown, cinnamon-brown or grayish-brown flattened scales, patches and fibrils over most of the surface, margin with white ring tissue, moist or dry, smooth, opaque. **FLESH** white, thick, unchanging. **GILLS** white, developing brown spots with age, attached and mostly notched (sinuate), close, broad. **STEM** 4–10 cm long, 15–30 mm broad, equal, white above ring, colored as cap below ring from fibrils and scales, white and solid inside. **RING** white above, cinnamon-buff below, flaring, thin, cottony to membranous. **SPORE PRINT** white. **ODOR** not distinctive, or spicy-fruity or strongly disagreeable. **TASTE** bitter or nutty or mild.

Habit and habitat scattered or clustered on leaf litter under hardwoods, especially oak. July to October. **RANGE** widespread.

Spores broad ellipsoid, smooth, colorless in KOH, not amyloid, 6–8 × 4.5–5.5 μm; hymenial cystidia and clamp connections absent.

Comments In older guides this would belong to the genus *Armillaria*. There are probably several cryptic species hidden under this current concept as depicted by the description. *Armillaria viscidipes* Peck, which has yet to be transferred validly to *Tricholoma*, differs from *T. caligatum* by its slippery or slimy stem, pungent odor of potato, and symbiosis with hemlocks.

Tricholoma equestre
(Linnaeus) P. Kummer

CANARY TRICH

SYNONYMS *Agaricus equestris* Linnaeus, *Melanoleuca equestris* (Linnaeus) Murrill, *Tricholoma flavovirens* (Persoon) Lundell

Medium to large with bright yellow, slippery cap with center becoming reddish-brown; gills yellow, notched at attachment; stem with yellow color; mostly under pine

CAP 3–12 cm wide, convex, becoming plane, some with low broad central bump, bright yellow overall or only around the margin when developing olive-brown or reddish-brown colors over the center, slimy at first, smooth, or with some flattened scales and fibrils over the center, opaque. **FLESH** white or pale lemon-yellow near cap, thick, unchanging. **GILLS** bright yellow, sometimes pale yellow, attached and mostly notched (sinuate), close, broad. **STEM** 2–10 cm long, 10–25 mm broad at apex, equal but often with bulb at base, white or nearly so over the apex, yellow downward, smooth or with fine fibrils, white and solid inside. **SPORE PRINT** white. **ODOR** and **TASTE** faintly farinaceous or not distinctive.

Habit and habitat scattered or clustered on leaf litter under or near conifers, typically pines. August to November. **RANGE** widespread.

Spores ellipsoid, smooth, colorless in KOH, not amyloid, 6–8 × 3–5 µm; cystidia and clamp connections absent.

Comments This mushroom is considered a good edible, but it is among several species you would not want to confuse, since they may cause gastrointestinal problems or just taste bad. Compare with *Tricholoma subluteum* and *T. sejunctum*. If the cap colors are paler yellow with more brown on the center, and the gills are white, you most likely have *T. intermedium* Peck. If the pileus is dry and the taste is bitter, you most likely have *T. aestuans* (Fries) Gillet.

Tricholoma grave Peck

Large heavy mushroom with pale orange-brown, dry cap with inrolled margin; gills pale orange-tawny; stem pale grayish-buff, with distinctive scales or branlike patches overall, the base penetrating the soil deeply

CAP 12–20 cm wide, convex with inrolled margin, grayish orange-brown, becoming paler, dry, covered with minute grayish-white silky fibrils and also intermixed with tawny spots and streaks, opaque. **FLESH** pale gray, unchanging, thick. **GILLS** white becoming pale ochre-tawny, attached-notched (sinuate), subdistant, broad. **STEM** 7.5–10 cm long, 25–38 mm broad, equal with tapered base penetrating soil about a third of length, grayish-white or pale gray, covered with scales or small branlike patches, stout, white mycelioid over base, solid. **SPORE PRINT** apparently white. **ODOR** and **TASTE** not known.

Habit and habitat single or scattered on soil under pine and oak. September. **RANGE** Massachusetts and New Hampshire.

Spores ellipsoid, smooth, colorless in KOH, not amyloid, 7.5 × 5 μm.

Comments This species is very rarely collected. A recent published account (Bessette et al., 2013) indicates the taste is unpleasant, slightly bitter, and the odor is disagreeable. Also the stem flesh is off-white, turning reddish-brown when exposed, and extremely thick (30–60 mm). The stem can reach extreme lengths of 20 cm. It would be hard to pass up such a massive mushroom without at least having a look, a smell, and a picture; thus, it must be a very rare fungus since it has only recently been included in field guides and the technical literature. For this book I have adapted the description originally given by Peck in 1890.

Tricholoma myomyces
(Persoon) J. E. Lange

SYNONYM *Agaricus myomyces* Persoon
Cap gray to blackish, dry, densely fibrillose-scaly; gills pale gray or white; stem white or pale grayish, silky fibrillose coated; under conifers

CAP 1–6 cm wide, rounded conical or convex with low broad central bump, the margin incurved and in youngest stages and attached to the stem with a cobwebby veil that is lost with expansion, gray, dark gray or almost black, becoming paler with age, dry, densely fibrillose centrally, fibrillose-scaly elsewhere, opaque. **FLESH** pale gray, thick. **GILLS** pale gray, becoming white, attached-notched (sinuate), close. **STEM** 1.5–8 cm long, 5–10 mm broad, equal or enlarged downward, white or pale gray, covered with white silky fibrils, solid or hollow. **RING** white to pale gray cobwebby fibrils (a cortina), leaving remnants on the upper stem. **SPORE PRINT** white. **ODOR** and **TASTE** not distinctive.

Habit and habitat scattered or clustered on leaf litter under conifers, usually spruce or pine. August to October. **RANGE** widespread.

Spores ellipsoid, smooth, colorless in KOH, not amyloid, 6.5–7.5 × 4–5 µm; cheilocystidia and clamp connections absent; with swollen cells below the cylindrical cells of the cap surface (a pseudoparenchymatous subpellis).

Comments Some consider this species to be the same as *Tricholoma terreum* (Schaeffer) P. Kummer, but that species lacks the cobweblike veil, and perhaps the colors tend to be more brown on the cap. If the cap is paler grayish with widely spaced scales, and the odor and taste are farinaceous, you most likely have *T. sculpturatum* (Fries) Quélet. The layer below the cap surface for this latter species is not composed of swollen cells either.

Tricholoma pardinum
(Persoon) Quélet

DIRTY TRICH

SYNONYM *Agaricus myomyces* var. *pardinus* Persoon

Medium to large, white with dense grayish to blackish scales and fibrils covering the dry cap; stem white or pale gray smooth with silky fibrils over apex; odor and taste farinaceous, or mealy; under conifers

CAP 4–15 cm wide, convex, becoming plane, white but densely covered with grayish or grayish-brown to black fibrillose scales, dry, opaque, with fine raised radially arranged lines over margin. **FLESH** white or pale gray, thick, unchanging. **GILLS** white, attached-notched (sinuate), close, broad. **STEM** 5–15 cm long, 10–30 mm broad, equal or swollen at base, white or pale gray, sometimes staining brown over base, smooth or coated over the apex with silky fibrils. **SPORE PRINT** white. **ODOR** and **TASTE** farinaceous.

Habit and habitat scattered or clustered on leaf litter under conifers. July to October. **RANGE** widespread.

Spores ellipsoid, smooth, colorless in KOH, not amyloid, 6–10 × 5–6.5 μm; cheilocystidia clavate, scattered, and clamp connections conspicuously present.

Comments Poisonous, causing severe gastrointestinal distress from an unknown toxin. *Tricholoma venenatum* G. F. Atkinson is very similar in appearance, but has pale tan scales and fibrils on the cap and grows under mixed hardwoods and hemlock stands. It too is quite poisonous. If the mushroom has all the features described above, but has a scaly stem surface and reddish liquid droplets forming on the gills and stem, you have *T. huronense* A. H. Smith.

Tricholoma portentosum
(Fries) Quélet

STICKY GRAY TRICH
SYNONYMS *Agaricus portentosus* Fries, *Melanoleuca portentosa* (Fries) Murrill, *Gyrophila sejuncta* var. *portentosa* (Fries) Quélet

Medium to large, grayish to blackish streaked, slippery cap with yellow undertones; gills white, but soon yellow mixed in; stem white or pale yellow over the base; odor and taste farinaceous; under conifers

CAP 5–13 cm wide, broadly conical or convex with rounded conical bump, becoming plane, gray, grayish-brown or olive-brown to black fibrillose streaked and often with undertones of yellow after expansion, slippery at first or when wet, opaque. **FLESH** white, somewhat fragile. **GILLS** white but developing distinctive yellow tones or sometimes gray mixed in, attached-notched (sinuate), close, broad. **STEM** 5–10 cm long, 10–20 mm broad, equal, white or with yellow tints especially over the base, coated with silky fibrils, solid. **SPORE PRINT** white. **ODOR** and **TASTE** farinaceous.

Habit and habitat scattered or clustered on leaf litter under conifers, especially pines. September to November or later. **RANGE** widespread.

Spores ellipsoid, smooth, colorless in KOH, not amyloid, 5–7 × 3–5 µm; cheilocystidia and clamp connections absent.

Comments This one is considered a good edible, with caution—that is, make darn sure you have made the correct identification. It fruits very late in the season. Compare with the more commonly collected and not edible *Tricholoma sejunctum* that is typically yellowish-green, contrasting sharply with the black fibrillose streaks on the cap. It has a bitter taste and occurs under hardwoods as well as conifers. *Tricholoma niveipes* Peck is somewhat similar and occurs under pitch and jack pines. It has a slippery dark gray to purplish-black cap and a white stem, but lacks any yellow colors.

Tricholoma subsejunctum
Peck

SYNONYM *Melanoleuca subsejuncta* (Peck) Murrill

Medium to large, cap slippery, yellow with black center, streaked with black fibrils to the margin; gills mostly white, some yellowish unevenly; stem white, flushed with yellow colors over middle; in mixed woods

CAP 3–10 cm wide, convex, or with a broad central bump at first, becoming plane, yellow or yellowish-green with dark gray or black over the center and radially streaked toward the margin with these dark fibrils, colors can fade with age, viscid at first, but soon moist or dry, smooth, opaque. **FLESH** white or grayish pigmented near cap skin, thick, unchanging. **GILLS** white, tinted yellow near the cap margin or where attached to the cap, edges or lower parts remaining white, mostly notched (sinuate), close, broad. **STEM** 2.5–11 cm long, 10–20 mm broad at apex, equal but sometimes swollen over base, white or flushed yellow randomly, smooth or with fine fibrils, white and narrowly hollow inside. **SPORE PRINT** white. **ODOR** farinaceous. **TASTE** farinaceous, some with bitter component.

Habit and habitat scattered or clustered on leaf litter under conifers or mixed woods with hardwoods present. July to November. **RANGE** widespread.

Spores broadly ellipsoid, smooth, colorless in KOH, not amyloid, 5–7.5 × 4–6 µm; hymenial cystidia and clamp connections absent.

Comments In older guides this would be *Tricholoma sejunctum* (Sowerby) Quélet. I find that our North American taxon is different enough from the European *T. sejunctum*, as C. H. Peck concluded, to warrant the use of Peck's name. *Tricholoma subsejunctum* is among the most commonly collected *Tricholoma* species in the Northeast.

Tricholoma subluteum Peck

SYNONYM *Melanoleuca sublutea* (Peck) Murrill

Medium to large, with conical-convex uniformly yellow cap; gills white, notched at attachment; stem white, flushed with yellow colors over middle; under conifers

CAP 4–10 cm wide, conical-convex at first, then convex, becoming plane, some with broad conical central bump, yellow or bright yellowish-green but often with olive-gray tints over the center or reddish-ochre streaked, if covered with debris it can be white, slippery at first, but soon moist or dry, smooth, with flattened radiating fibrils over margin, opaque. **FLESH** white or grayish pigmented near cap skin, thick, unchanging. **GILLS** white, sometimes pale yellow tints unevenly near the cap margin, mostly notched (sinuate), close, broad. **STEM** 5–15 cm long, 10–15 mm broad at apex, equal, some with tapered rooting base, white or flushed yellow downward, white-dotted over apex, smooth elsewhere or with fine fibrils, white and narrowly hollow inside. **SPORE PRINT** white. **ODOR** strong, unpleasant, musty cucumber-farinaceous. **TASTE** cucumber and with bitter component or farinaceous-rancid.

Habit and habitat scattered or clustered on leaf litter under conifers, frequently Norway spruce. September to November. **RANGE** widespread.

Spores subglobose or broad ellipsoid, smooth, colorless in KOH, not amyloid, 6–7 × 5–6.5 μm; hymenial cystidia and clamp connections absent.

Comments *Tricholoma subsejunctum* has a distinctive dark grayish-black cap center, with radiating dark streaks. Also see comments under *Tricholoma equestre*. I can verify *T. subluteum* smells horrible if you try to cook it in a frying pan. It is not edible.

Tricholoma subresplendens
(Murrill) Murrill

SYNONYM *Melanoleuca subresplendens* Murrill

Medium to large, white, slippery cap with tan discolorations, also sometimes blue stains; gills white; stem white, sometimes with blue stains over base; fruiting in the fall under oak and beech

CAP 3–11 cm wide, conical or convex, some with rounded central bump, becoming plane, white, but spotted with tan over the center quite often, also often turning bluish when injured, slippery at first or when wet, becoming dry, smooth or with silky fibrils, especially over the margin, opaque. **FLESH** white. **GILLS** white, attached-notched (sinuate), close. **STEM** 4–11 cm long, 10–25 mm broad, equal or enlarged downward, white, sometimes bruising bluish at base, coated with white silky fibrils, solid, white. **SPORE PRINT** white. **ODOR** and **TASTE** farinaceous.

Habit and habitat scattered or clustered on leaf litter under hardwoods, especially oak and beech. September to October. **RANGE** widespread.

Spores ellipsoid, smooth, colorless in KOH, not amyloid, 5–7 × 4–5 µm; cheilocystidia absent, but clamp connections present at base of basidia.

Comments Fruiting late in the season, usually after the leaves begin to fall in the hardwood forests. The blue staining is diagnostic but not always present on the cap and stem. However, the tan discolorations on the slippery caps are more consistent. *Tricholoma resplendens* (Fries) P. Karsten and *T. columbetta* (Fries) P. Kummer, two names sometimes used for our North American taxon, are European species that are white and viscid on the cap also. *Tricholoma resplendens* lacks blue staining and *T. columbetta* stains reddish in addition to sometimes turning blue, and lacks tan colors in the cap. These two species may not occur in North America.

Tricholoma virgatum (Fries)
P. Kummer

FIBRIL TRICH

SYNONYMS *Agaricus virgatus* Fries, *Tricholoma subacutum* Peck

Cap sharply conical, gray with darker gray radiating streaks, dry; stem white with silky fibrils; taste slowly bitter, may become acrid hot; under conifers

CAP 2–10 cm wide, conical, becoming nearly plane with pointed central bump, silvery-gray with darker gray streaks radiating from the dark gray center, dry, smooth, opaque. **FLESH** pale gray. **GILLS** white or dull gray, may discolor darker gray on edges, attached-notched (sinuate), close, broad. **STEM** 6–10 cm long, 10–20 mm broad, equal or enlarged downward, white, silky fibrillose, solid or hollow. **SPORE PRINT** white. **ODOR** not distinctive. **TASTE** bitter or slowly acrid hot, or both.

Habit and habitat scattered or clustered on leaf litter under conifers, spruce, fir, and pine. September to October. **RANGE** widespread.

Spores ellipsoid, smooth, colorless in KOH, not amyloid, 6–8.5 × 5–7 µm; cheilocystidia cylindrical or club-shaped, clamp connections absent.

Comments If the center is not sharply conic, the colors are lighter gray and not distinctly darker streaked, and the taste is instantly bitter, then *Tricholoma argenteum* Ovrebo is most likely what you have collected. If you were under mainly hardwoods, like oak and beech, and the center again is not sharply pointed but the cap is clearly dark gray streaked, and a fair number of the gill edges are dark gray discolored, you have *Tricholoma pullum* Ovrebo.

Tricholoma vaccinum
(Schaeffer) P. Kummer

FIBRIL TRICH
SYNONYM *Agaricus vaccinus* Schaeffer
Cap rusty-brown scaly, dry, margin with hanging patches of fibrils; gills white but often brown-spotted; stem with color and scales like cap; under conifers

CAP 2–7 cm wide, rounded conical or convex with the margin incurved and in youngest stages attached to the stem with a cobwebby veil, margin with hanging fibrils and patches of fibrils from that veil once expanded, rusty-brown or orange-brown becoming paler with age, dry, densely fibrillose then scaly overall, opaque. **FLESH** white or buff. **GILLS** white, often spotted yellowish-brown or cinnamon-brown, attached-notched (sinuate), close, broad. **STEM** 3–8 cm long, 5–20 mm broad, equal or enlarged downward, covered with brown fibrils and scales over a buff ground color, mostly hollow. **RING** brown cobwebby fibrils, leaving remnants on cap margin, not the stem. **SPORE PRINT** white. **ODOR** and **TASTE** not distinctive, or farinaceous.

Habit and habitat single, scattered, or clustered on leaf litter under conifers, usually spruce or pine, or both. September to October. **RANGE** widespread.

Spores ellipsoid, smooth, colorless in KOH, not amyloid, 6–7.5 × 4–5 μm; cheilocystidia and clamp connections absent.

Comments *Tricholoma imbricatum* (Fries) P. Kummer is darker dull brown with well-spaced small scales on the cap and a solid stem. Additionally, the unopened buttons lack the cobweb ring, and the mature fruit bodies lack the hanging fibrils on the cap margin. *Tricholoma pessundatum* (Fries) Quélet has chestnut-colored cap that lacks scales and is slippery. If you are under hardwoods like beech and oak, then perhaps you have *T. ustale* (Fries) P. Kummer. These latter two species cause severe gastrointestinal irritation when consumed.

Tricholoma odorum Peck

SYNONYM *Melanoleuca odora* (Peck) Murrill

Medium-sized yellowish, dry cap; gills yellow; stem yellowish-green; odor of coal tar

CAP 2–9 cm wide, broadly convex, becoming plane, some with central broad bump, pale yellowish with green hues or merely yellowish, fading quickly to buff or pinkish-buff or tan and sometimes brown over center, dry, smooth or covered with velvety matted fibrils and becoming scaly over the center, opaque. **FLESH** white or pale yellow. **GILLS** yellow or pale cream, becoming buff or pinkish-buff, attached-notched (sinuate), close, broad. **STEM** 3–12 cm long, 6–15 mm broad, equal or enlarged downward, more persistently pale yellow or greenish-yellow than cap, becoming pinkish-buff or even brownish, solid. **SPORE PRINT** white. **ODOR** of coal tar or swamp gas. **TASTE** farinaceous but disagreeable and decidedly unpleasant.

Habit and habitat single, scattered, or clustered on leaf litter under hardwoods, such as oak and beech. August to October. **RANGE** widespread.

Spores almond-shaped in side view, fusiform in face view, smooth, colorless in KOH, not amyloid, 7–11 × 5–7 µm; cheilocystidia absent, but clamp connections present at base of basidia.

Comments Two other species of *Tricholoma* can have this signature coal-tar odor, *T. sulphureum* (Bulliard) P. Kummer and *T. inamoenum* (Fries) Gillet. *Tricholoma sulphureum* retains its sulfur-yellow colors throughout development and has subdistant gills. *Tricholoma inamoenum* is white to buff in all stages of development. If the fruit bodies are white, rapidly staining yellow, and the odor is sulfurous to fruity, you most likely have *T. sulphurescens* Bresadola.

Pseudotricholoma umbrosum (A. H. Smith & M. B. Walters) Sáchez-Garcia & Matheny

AMYLOID TRICHOLOMA
SYNONYMS *Tricholoma umbrosum* A. H. Smith & M. B. Walters, *Porpoloma umbrosum* (A. H. Smith & Walters) Singer

Cap gray or dark grayish-brown, smooth then cracked; gills gray, bruising reddish; stem colored as cap, smooth becoming scaly, bruising reddish; odor pungent cucumber-farinaceous when crushed

CAP 5–10 cm wide, convex becoming plane, often centrally sunken, gray or dark grayish-brown, often with reddish-brown stains from bruising, smooth, becoming finely cracked and checked, dry. **FLESH** gray, turning reddish when exposed. **GILLS** gray, bruising reddish, attached or notched, close, broad. **STEM** 2.5–5 cm long, 15–20 mm broad, equal with tapered base, colored as cap or paler, glabrous becoming roughened and scaly with age, dry. **SPORE PRINT** white. **ODOR** pungent, cucumber-farinaceous. **TASTE** farinaceous.

Habit and habitat single, scattered, or clustered on humus and soil under conifers or hardwoods. July to October. **RANGE** widespread.

Spores ellipsoid, smooth, colorless, amyloid, 7–9 × 3–4 µm.

Comments A dark gray or grayish-brown former member of *Porpoloma* with a strong cucumber-farinaceous odor. *Porpoloma* species were recognized for their amyloid spore reaction, coupled with a fleshy stature. If you catch the spore powder on a slide, then add a drop of Melzer's reagent next to the powder and tilt the slide so the iodine liquid runs into the spore powder, you will see the blue-black amyloid reaction. Recent phylogenetic evidence using DNA molecules indicates this species belongs in its own unique genus, unrelated to *Porpoloma*, thus *Pseudotricholoma*.

Lepista glaucocana (Bresadola) Singer

SYNONYMS *Tricholoma glaucocana* Bresadola, *Clitocybe glaucocana* (Bresadola) H. E. Bigelow & A. H. Smith, *Rhodopaxillus glaucocanus* (Bresadola) Métrod

Large mushroom, faintly purple throughout when fresh, cap becoming pale buff with age; gills remaining pale purple, crowded

CAP 6–15 cm wide, convex with strongly inrolled margin, becoming plane with shallow sunken center and often a low broad bump as well, evenly watery buff with faint purple or lilac hues at first, but paler and buff or pale vinaceous-brown with expansion, glabrous, moist, opaque. FLESH white or with faint vinaceous hue. GILLS pale lilac or pale purple, bluntly attached or notched, crowded, narrow (4–7 mm). STEM 5–9 cm long, 10–25 mm broad, mostly equal or club-shaped, similar color as gills, apex cottony-scaly, with fine white longitudinal fibrils downward and often lined, white or cottony at base. SPORE PRINT vinaceous-buff (pinkish-flesh). ODOR fragrant or not distinctive. TASTE mild.

Habit and habitat in loose clusters or often in bouquetlike clusters on the ground in cedar swamps, compost beds, or mixed forests. September to October. RANGE widespread.

Spores ellipsoid, minutely bumpy overall, colorless, not amyloid, but bumps deep blue in cotton blue (cyanophilous) making the spores look blue dotted, 5.5–7.5 × 3–4.5 µm; clamp connections present.

Comments *Lepista nuda*, commonly called blewits, differs by its darker violet-purple cap, gills, and stem, with the cap becoming brown. Blewits is considered a fine edible. *Lepista irina* (Fries) H. E. Bigelow is somewhat similar in form but lacks any purple-violet hues, as does *L. subconnexa* (Murrill) Harmaja. *Lepista irina* has spores longer than 6 µm and is found in loose clusters, while *L. subconnexa* has spores less than 6 µm and is found in bouquetlike clusters. All produce pinkish-buff spore deposits.

Tricholomopsis decora (Fries) Singer

DECORATED MOP

SYNONYMS *Agaricus decorus* Fries, *Clitocybe decora* (Fries) Gillet, *Pleurotus decorus* (Fries) Saccardo, *Tricholoma decorum* (Fries) Quélet

Cap yellow with abundant dark brown or black scales; gills yellow, attached; stem yellow; found on conifer wood

CAP 3–8 cm wide, convex with margin incurved and center slightly sunken, with dark brown, grayish-brown, or nearly black scales over golden-yellow or yellowish-orange ground color, moist, densely scaly centrally, less so over the margin, opaque. FLESH yellow. GILLS yellow and paler than cap, attached, close or crowded, narrow. STEM 2–8 cm long, 3–10 mm broad, equal, yellow as the cap or paler, with minute scattered fibrillose scales or becoming glabrous. SPORE PRINT white. ODOR and TASTE not distinctive.

Habit and habitat single, scattered, or clustered on decaying logs and standing dead trees of conifers, especially hemlock. June to October. RANGE widespread.

Spores ellipsoid, smooth, colorless in KOH, not amyloid, 6–7.5 × 4–5.5 μm; cheilocystidia variously shaped but mostly swollen, abundant, clamp connections present on the hyphae.

Comments *Tricholomopsis sulphureoides* also grows on conifer wood, but the colors are much paler, the scales on the cap are paler and sparser, and it has a cobwebby ring when young. *Tricholomopsis thompsoniana* (Murrill) A. H. Smith is pale yellow-streaked and glabrous on the cap, the flesh and stem stain yellow or ochraceous, and it occurs on decaying pine wood.

Tricholomopsis sulphureoides (Peck) Singer

YELLOW OYSTER MOP

SYNONYMS *Agaricus sluphureoides* Peck, *Pleurotus sulphureoides* (Peck) Saccardo

Cap bright greenish-yellow, finely fibrillose-scaly, scales may become brownish; gills bright yellow, close, broad; stem bright yellow, mostly off-center; on conifer logs

CAP 3–7 cm wide, convex, becoming plane, with central broad bump, or sunken, sulfur-yellow, becoming paler, sparsely or more uniformly coated with similar-colored fibrillose scales that may become pale brown, especially over the center, dry, opaque. **FLESH** yellow, unchanging. **GILLS** similar yellow color as the cap, barely attached, close, broad. **STEM** 3–6 cm long, 3–7 mm broad, mostly off-center (eccentric), equal or enlarged downward, similar yellow color of cap, sparsely fibrillose over base, apparently from a collapsed thin veil, glabrous elsewhere. **SPORE PRINT** white. **ODOR** and **TASTE** not distinctive.

Habit and habitat single, scattered, or clustered, on decaying conifer logs, often hemlock. July to October. **RANGE** widespread.

Spores ellipsoid, smooth, colorless in KOH, not amyloid, 5.5–6.5 × 4.5–5 µm; cheilocystidia obvious, abundant, inflated.

Comments Peck clearly describes the cap as "subsquamulose or smooth, sulphur-yellow" in his original description (1872) and also states that "the minute scales are brown, but often wanting." *Tricholomopsis thompsoniana* is similar in color but the cap becomes variegated, streaked yellow and pallid yellow, and it is glabrous except for being "minutely tomentose on the margin." The stem of *T. thompsoniana* also stains darker yellow where handled, and the fruit bodies occur on decaying conifers. These two species occur in similar habitats and it appears *T. thompsoniana* is often mistakenly labelled *T. sulphureoides*.

Tricholomopsis rutilans
(Schaeffer) Singer

VARIEGATED MOP

SYNONYMS *Agaricus rutilans* Schaeffer, *Cortinellus rutilans* (Schaeffer) P. Karsten, *Gymnopus rutilans* (Schaeffer) Gray, *Pleurotus rutilans* (Schaeffer) Dumée, *Tricholoma rutilans* (Schaeffer) P. Kummer

Cap purplish-red or brick-red from dense fibrils and scales; gills yellow, attached (notched); stem yellow with purplish-red fibrils, central or off-center; on decaying conifer wood, or on the ground from buried wood debris

CAP 5–10 cm wide, convex, or bell-shaped, becoming plane with central broad bump, or broadly sunken, or broadly sunken with central bump; yellow ground color but the dominant color is bright purplish-red or red from a dense coating of fibrils and fibrillose scales, margin can have radial raised lines, opaque. **FLESH** yellow, unchanging. **GILLS** bright or dull yellow, attached (notched), crowded, narrow. **STEM** 5–12 cm long, 5–20 mm broad, central or off-center (eccentric), equal or enlarged downward slightly, yellow ground color but covered sparsely or sometimes densely with purplish-red fibrils, except for a yellowish glabrous zone at the very apex. **SPORE PRINT** white. **ODOR** fragrant or not distinctive. **TASTE** not distinctive.

Habit and habitat single, scattered, or clustered, sometimes in a bouquetlike fashion on decaying logs and standing dead trees of conifers, or growing from the ground and buried wood or well-decayed wood or chips. June to October. **RANGE** widespread.

Spores ellipsoid, smooth, colorless in KOH, not amyloid, 5–7 × 3–5 μm; cheilocystidia obvious, abundant, inflated.

Comments This species is very attractive with the red and yellow color combination. It apparently is inedible and to be avoided for food use.

Leucopaxillus albissimus
(Peck) Singer

WHITE LEUCOPAX
SYNONYMS *Agaricus albissimus* Peck, *Melanoleuca albissima* (Peck) Murrill
Medium to large mushroom with white or buff colored cap; gills white, crowded, narrow; stem solid, with white mycelium binding masses of needles from duff; taste typically bitter

CAP 3–20 cm wide, convex with inrolled margin, becoming plane, white or buff or pinkish-buff, sometimes pale tan or darker, especially over the center, dry, smooth, soft velvety to touch. FLESH white, thick, firm, unchanging. GILLS white or sordid cream color, bluntly attached or sub-decurrent, crowded, narrow. STEM 3–8 cm long, 10–30 mm broad, club-shaped at first, equal with age, white, smooth or with fine coating of fibrils, white and solid inside, white mycelium at base profuse and binding substrate. SPORE PRINT white. ODOR not

distinctive, or fragrant, but also noted as mealy, unpleasant, or foul. TASTE not distinctive, mealy or bitter.

Habit and habitat single, scattered, or clustered, binding large masses of needles and duff in coniferous forests, similar to how grasses bind soil into balls that you find when weeding a garden. August to October. RANGE widespread.

Spores ellipsoid, spiny or warty, colorless, amyloid, 5–8 × 4.5–5 µm.

Comments The species name *albissimus* literally means the most white, like the image provided here. However, the cap for this species seems to vary in color, and thus 10 varieties and forms have been formally described. If you are collecting under hardwoods, see *Leucopaxillus laterarius* that has a pale flesh color to the cap and a very bitter taste. Binding handfuls of substrate in a copious white mycelium is a good field character for species of *Leucopaxillus*.

Leucopaxillus laterarius
(Peck) Singer

SYNONYMS *Agaricus laterarius* Peck, *Melanoleuca laterarium* (Peck) Murrill, *Tricholoma laterarium* (Peck) Saccardo

Medium to large mushroom with pinkish-buff cap, with margin frequently lined; gills white, crowded, narrow; stem solid, with white mycelium binding masses of hardwood leaves and duff; taste typically bitter

CAP 4–20 cm wide, convex with inrolled margin, becoming plane, white or buff or pinkish-buff, especially over the center, dry, smooth, soft velvety to touch, margin often with raised lines. FLESH white, thick, firm, unchanging. GILLS white or sordid cream color, bluntly attached or subdecurrent, crowded, narrow. STEM 4–11 cm long, 10–20 mm broad, equal, white, smooth or with fine coating of fibrils, white and solid inside, white mycelium at base profuse and binding substrate. SPORE PRINT white. ODOR mealy farinaceous or very unpleasant, like coal tar. TASTE usually very bitter, or can be mild.

Habit and habitat single, scattered, or clustered, binding large masses of leaves, twigs, and duff under hardwood forests, especially beech, birch. July to October. RANGE widespread.

Spores subglobose or short-ellipsoid, spiny or warty, colorless, amyloid, 3.5–5 × 3.5–5 µm.

Comments As other field guides have also pointed out, *Leucopaxillus laterarius* is difficult to separate from *L. albissimus* based solely on macroscopic features. They can look very similar. However, if you are under hardwoods, not conifers, and the cap is flushed with pinkish-buff, then you probably have *L. laterarius*. The surest test would be to measure the spores—they are consistently less than 5.5 µm long for *L. laterarius*.

Melanoleuca alboflavida
(Peck) Murrill

SYNONYMS *Agaricus alboflavidus* Peck, *Collybia alboflavida* (Peck) Kauffman, *Tricholoma alboflavidum* (Peck) Saccardo

Medium to large, tall mushroom with yellow cap that fades and with darker brownish low broad central bump; gills white or buff, very crowded; stem rigid, lined, with subbulbous base; under hardwoods

CAP 3–12 cm wide, convex, eventually plane, broadly sunken with low central bump, dark yellow or amber-yellow, often with darker brown center, fading to pale yellow, slippery or tacky at first, then dry, smooth. FLESH white, unchanging. GILLS white or buff-cream color, attached, very crowded, broad. STEM 7–15 cm long, 5–12 mm broad, equal with subbulbous base, straight-rigid, off-white, lined, with fine coating of fibrils, white and solid inside. SPORE PRINT white. ODOR not distinctive. TASTE mild.

Habit and habitat single, scattered, or clustered, on humus under hardwood forests, on soil under shrubs in disturbed areas. July to October. RANGE widespread.

Spores ellipsoid, warty, colorless, warts amyloid, 7–9 × 4–5.5 μm; pleuro- and cheilocystidia with pointed apex.

Comments *Melanoleuca* is distinctive by its rigid stature and very crowded gills. At least 30 species are found in North America. *Melanoleuca melaleuca* (Fries) Murrill has a dark brown to black cap, the stem is streaked with dark fibrils or is dark overall, and pleurocystidia are lacking. *Melanoleuca brevipes* (Fries) Patouillard has a short stem (1–3 cm) colored white or pale buff, the cap is black to dark gray, and the gills are white.

Lyophyllum multiforme
(Peck) H. E. Bigelow

SYNONYM *Clitocybe multiformis* Peck
Cap pale brown, strongly fading to off-white or with gray tints; gills white or pale yellow, staining dull ochre; stems not fused together in large numbers, often eccentrically attached to caps; near downed logs in hardwood forests

CAP 3–10 cm wide, convex, then plane, margin often undulate with age, pale creamy-tan or orange-brown at first, hygrophanous and slowly fading to dingy gray-brown or off-white, margin paler, smooth, barely slippery, becoming dry. **FLESH** white, thick. **GILLS** white or pale yellow at first, staining dull ochre when bruised, bluntly attached, crowded, narrow. **STEM** 4–7 cm long, 5–10 mm broad, frequently off-center (eccentric), equal, off-white or pale grayish-tan, glabrous or finely fibrous-striate, dry, hollow. **SPORE PRINT** white. **ODOR** and **TASTE** not distinctive.

Habit and habitat scattered or clustered, but not multiple fruit bodies fused in large clusters at the base of the stems, on soil or humus, around fallen logs or piles of branches in hardwood forests. August to November. **RANGE** widespread.

Spores ellipsoid, smooth, colorless, not amyloid, 5–6 × 3–3.5 μm; basidia with cyanophilous bodies.

Comments This species is similar in appearance to *Lyophyllum decastes* (Fries) Singer, which differs by producing dense, bouquetlike clusters of numerous fruit bodies with stems centrally attached to the caps. The gills are not pale yellow and staining dull ochre either. The spores of *L. multiforme* are slightly longer and narrower than those of *L. decastes* whose spores are 3.5–5 × 4–5 μm. It is also possible to confuse *L. multiforme* with several different species of *Clitocybe*; therefore, you need a microscope and more advanced literature to identify many of these *Clitocybe* species.

Lyophyllum loricatum (Fries) Kühner

SYNONYM *Lyophyllum decastes* var. *loricatum* (Fries) Kühner

Large clusters of moderately large mushrooms fused at the base of the stems in grasslands or disturbed areas, with black caps; grayish gills and stem

CAP 3–10 cm wide, broadly convex with inrolled margin, then plane, some with broad central bump, black or dark grayish-black at first and with a hoary white dusting, slowly fading to brown, margin paler, smooth, dry. **FLESH** white or pale gray, thick. **GILLS** grayish at first, becoming paler, bluntly attached, crowded, moderately narrow. **STEM** 8–15 cm long, 10–20 mm broad, equal but often fused together into a large basal mass, pale grayish, becoming nearly white, glabrous, dry. **SPORE PRINT** white. **ODOR** not distinctive. **TASTE** slightly farinaceous or nutty.

Habit and habitat in massive clusters on soil or humus, among grasses, on disturbed ground, sawdust piles, hedge rows, and so on. August to September. **RANGE** widespread but not commonly collected.

Spores globose, smooth, colorless, not amyloid, 6–7 µm; basidia with cyanophilous bodies.

Comments *Lyophyllum decastes*, highly prized for the table, is a large fleshy mushroom like this species but with a much paler colored cap. *Lyophyllum loricatum* is edible, but not as highly sought after. See also *L. multiforme*.

Rugosomyces carneus
(Bulliard) Bon

PINK CALOCYBE
SYNONYMS *Calocybe carnea* (Bulliard) Donk, *Tricholoma carneum* (Bulliard) P. Kummer

Cap pink or pinkish-brown; gills white, crowded; stipe colored as cap; on lawns

CAP 1–4 cm wide, convex, becoming plane, some with a rounded bump, some shallowly depressed over center, pink or pinkish-brown, may become paler with age, smooth, dry. **FLESH** white. **GILLS** white, bluntly attached or slightly decurrent, crowded, narrow. **STEM** 1.5–5 cm long, 2–10 mm broad, equal, same colors as pileus or paler, with white hairs at base. **SPORE PRINT** white. **ODOR** and **TASTE** not distinctive.

Habit and habitat single, scattered, or clustered in grass-covered areas that are usually cultivated rather than grassy areas in natural forest settings. August to October. **RANGE** widespread.

Spores broadly elliptical, smooth, colorless, 4–6 × 2–3.5 μm; basidia with siderophilous and cyanophilous bodies.

Comments The colors of this attractive species vary in intensity, sometimes the caps are pale pink, sometimes darker pinkish-brown. It does attract attention on lawns when compared to other lawn mushrooms, like *Panaeolus foenisecii*, *Conocybe apala*, *Marasmius oreades*, and *Leucoagaricus leucothites*, that are white or some shade of brown or gray-brown.

Armillaria calvescens Bérubé & Dessureault

Cap brown, mostly glabrous, white or yellow cottony tufts on margin at first; stem club-shaped or bulbous, with white, collapsing ring, on sugar maple stumps, buried roots, or standing trees

CAP 2–10 cm wide, convex, becoming plane, some with low broad central bump, medium reddish-brown or grayish-brown or dark brown, surface glabrous or with minute black scales over the center, also with white or golden-yellow cottony tufts of universal veil scattered over margin when young, dry. **FLESH** white, solid, thick. **GILLS** cream, becoming light brown with age or injury, notched attachment, becoming short decurrent, close. **STEM** 4–9 cm long, 5–20 mm broad, club-shaped or bulbous over base, arising from cylindrical, black, unbranched rhizomorphs (look under the bark), surface pale flesh-brown or olive-brown, covered with wispy, thin, white or yellow cottony fibrils from universal veil, flattened white scales below the ring, all quickly lost. **RING** white, thin, delicate, collapsing, near the top of the stem. **SPORE PRINT** ivory white. **ODOR** and **TASTE** not distinctive.

Habit and habitat single or in small bouquetlike clusters on hardwoods, especially sugar maple, also on the ground over roots. June to September. **RANGE** known from New York, Vermont, and Michigan, but expected wherever sugar maple occurs, and is fairly common.

Spores ellipsoid, smooth, colorless, 8.5–10 × 5–7 µm; clamp connections present at base of basidia.

Comments Compare with *Armillaria gemina* that can be differentiated by the brown crustlike patches circling the edge of the persistent ring on the stipe. *Armillaria gallica* is also similar but difficult to distinguish morphologically.

Armillaria gemina Bérubé & Dessureault

Cap reddish-brown with black hairs and scales; gills white, short decurrent; stem brown with white cottony ring, rimmed with reddish-brown patches; on stumps or buried roots arising from flat, branched rhizomorphs

CAP 5–12 cm wide, convex, becoming plane, dark reddish-brown or paler and tan or yellowish-brown, surface covered with black or dark reddish-brown hairs and scales, short translucent lines over margin. **FLESH** white, solid. **GILLS** white or pale cream, becoming reddish-brown or cinnamon with age or injury, bluntly attached, becoming short decurrent, subdistant, moderately broad. **STEM** 5–20 cm long, 10–15 mm broad, club-shaped at first, becoming equal, orange-brown becoming grayish-brown to black with age or handling, covered with gauzy cream-white fibrils below the ring, arising from cylindrical, dark rhizomorphs. **RING** white, cottony with fluffy reddish-brown patches on the margin, persistent. **SPORE PRINT** pale cream. **ODOR** and **TASTE** not distinctive.

Habit and habitat usually in clusters on or around stumps, on the ground over roots or on standing trees of hardwoods. July to November. **RANGE** widespread.

Spores ellipsoid, smooth, colorless, 8–11 × 5.5–7 µm.

Comments In upstate New York this is referred to as the stumpy and is a favored edible in the fall. Several brownish *Armillaria* species in the Northeast are difficult to separate. *Armillaria solidipes* Peck [syn. *A. ostoyae* (Romagnesi) Herink] is similar to *A. gemina* and difficult to distinguish with field characters. *Armillaria gemina* is edible, but parboil then cook it thoroughly for better flavor. *Armillaria solidipes*, originally described growing under conifers in Colorado, is a severe pathogen of coniferous trees in the Rocky Mountains and western North America. It has been reported to cause gastrointestinal upset if found on hemlock and perhaps not cooked thoroughly.

Armillaria mellea (Vahl)
P. Kummer

HONEY MUSHROOM
SYNONYMS *Agaricus melleus* Vahl, *Clitocybe mellea* (Vahl) Ricken, *Lepiota mellea* (Vahl) J. E. Lange

Cap yellow with few dark scales; stem tapered to base and fused in bouquetlike clusters, with membranous ring and yellow flattened scales and patches below ring; on hardwood trees

CAP 4–12 cm wide, convex, becoming plane, variable but typically honey-yellow or yellowish-brown or olive-yellow with darker center, with scattered black hairs and scales, especially over the center, slippery when wet, becoming dry. **FLESH** white, solid. **GILLS** white or pale buff, becoming yellow-brown with age or injury, bluntly attached, becoming short decurrent with faint lines running down to the ring, close or subdistant, broad. **STEM** 5–20 cm long, 5–35 mm broad, equal but tapered to a point, clustered and fused, arising from flat, black, branching rhizomorphs (look under the bark), surface white then yellow or brown, staining darker when bruised, covered with thin white or yellow cottony fibrils and flattened scales below the ring. **RING** white, thin or thick, membranous, white above, yellowish on lower surface, persistent near the top of the stem. **SPORE PRINT** pale cream. **ODOR** and **TASTE** not distinctive.

Habit and habitat in bouquetlike clusters on hardwoods or occasionally conifers, also on the ground at the base of the trees or over roots. June to September. **RANGE** widespread.

Spores ellipsoid, smooth, colorless, 7–9 × 5.5–7 µm, clamp connections absent at base of basidia, all other *Armillaria* species have clamps.

Comments Not as frequently collected as *Armillaria gemina*, this mushroom is a serious and lethal pathogen on hardwoods in the northeastern United States.

Armillaria tabescens (Scopoli) Emel

RINGLESS HONEY MUSHROOM

SYNONYMS *Armillariella tabescens* (Scopoli) Singer, *Clitocybe monodelpha* (Morgan) Saccardo

Cap brown with minute scales and fibrils; gills off-white, becoming flesh-brown, short decurrent; stem tapered to the base, lacking a ring; in bouquetlike clusters on the ground from buried roots of living trees

CAP 2–10 cm wide, convex, becoming plane, some with low broad central bump, yellow-brown or medium reddish-brown, surface with minute erect or flattened brown or black scales, dry. **FLESH** white, solid. **GILLS** off-white, turning pinkish or brown with age or injury, short decurrent, subdistant. **STEM** 5–20 cm long, 5–15 mm broad, equal but tapering to base, surface sordid white, becoming brown, covered with scurfy fibrils. **RING** absent. **SPORE PRINT** white. **ODOR** not distinctive. **TASTE** not distinctive, or bitter.

Habit and habitat in small or large bouquetlike clusters on the ground under hardwoods, especially maple and oak, living off of the roots of those trees. July to November. **RANGE** widespread.

Spores ellipsoid, smooth, colorless, 8–10 × 5–7 µm, clamp connections present at base of basidia, sometimes difficult to demonstrate.

Comments This species often fruits in large clusters. *Armillaria tabescens* is a pathogenic mushroom on fruit trees and other hardwoods and in certain years can be fairly common in the southern and warmer regions of the Northeast.

Laccaria laccata var. *pallidifolia* (Peck) Peck

SYNONYM *Clitocybe laccata* var. *pallidifolia* Peck

Cap orange-brown or pinkish-cinnamon; gills pale pinkish or yellowish pink-colored; stem colored like cap or darker, rather firm and fibrous; spores white

CAP 1–5 cm wide, convex, becoming plane and with shallowly sunken center, orange-brown, becoming pinkish-cinnamon or apricot-buff color, dry, finely fibrillose or fibrillose-scaly, translucent lined or with low striations. FLESH same color as cap. GILLS pinkish or pinkish-flesh, bluntly attached, subdistant, broad. STEM 2–6.5 cm long, 2–4 mm broad, equal or enlarged downward, same color as cap or darker orange-brown, surface dry, fibrillose with longitudinal striations, white fuzzy mycelium over the base, inside whitish or colored as stipe surface and stuffed.

SPORE PRINT white. ODOR and. TASTE not distinctive.

Habit and habitat single, scattered, or clustered, on soil, humus, or on the ground under hardwoods or in mixed woods, with pine, birch, oak, and beech trees. June to October. RANGE widespread.

Spores globose, colorless, inamyloid, spiny, 7–10 µm excluding spines, spines 1–2 µm long.

Comments One of the most common mushrooms collected in any part of the mushroom season. It is variable in color but always ranging through the orange-brown or pinkish-cinnamon colors with pink gills. If the cap is strongly striate see *Laccaria striatula*. If the mycelium at the base of the stipe is violet, see *L. bicolor*. Other species do occur in this genus but need to be examined with a microscope, mostly for spore shape and spine dimensions on the spores.

Laccaria striatula (Peck) Peck

SYNONYM *Clitocybe striatula* Peck

Cap reddish-brown or orange-brown, obviously lined to center; gills yellowish pink-colored with ochre tints; stem darker than cap, long in relation to the cap width; spores white

CAP 0.6–3.5 cm wide, convex, becoming plane, with shallowly sunken center, translucent-lined to near center when fresh, also striate-lined over the margin, reddish-brown or orange-brown, hygrophanous, becoming pinkish-cinnamon or salmon-buff with darker brown center, dry, finely fibrillose or fibrillose-scaly, especially over the center. **FLESH** same color as cap. **GILLS** pinkish-flesh or with ochre tints, bluntly attached, subdistant or distant, broad. **STEM** 2–7 cm long, 1–4 mm broad, equal or with bulb at base, same color as cap or darker orange-brown, surface dry, glabrous but with fine longitudinal striations, white mycelium over the base. **SPORE PRINT** white. **ODOR** and **TASTE** not distinctive.

Habit and habitat single, scattered, or clustered in wet areas, often on mosses on the ground under mixed woods, with hemlock, pine, birch, oak, and beech trees. June to October. **RANGE** widespread.

Spores globose, spiny, colorless, inamyloid, 7–10 µm excluding spines, spines 1.5–3 µm long.

Comments This mushroom is smaller than *Laccaria laccata* var. *pallidifolia* in general and is rather frequently found in wet mossy areas around seepages or ephemeral waterways. *Laccaria ohiensis* (Montagne) Singer is also a smaller and similar-looking species with a striate pileus, but it can be differentiated in the field by the stem that is short, 1–2.5 cm long, and by the color of the stem that is the same color as its cap.

Laccaria bicolor (Maire)
P. D. Orton

SYNONYM *Laccaria laccata* var. *bicolor* Maire

Cap reddish-brown or orange-brown, with violet tints occasionally; gills violet, then delicate pinkish; stem colored as cap, often violet where injured; violet mycelium over base when fresh; spores white

CAP 1–7 cm wide, convex, becoming plane and with shallowly sunken center, not striate-lined, reddish-brown or orange-brown, hygrophanous, becoming pinkish-cinnamon or salmon-buff with darker brown center, dry, finely fibrillose or fibrillose-scaly, especially over the center. **FLESH** same color as cap. **GILLS** pale violet or purplish-vinaceous at first, then pinkish-flesh, bluntly attached, subdistant or distant, broad. **STEM** 2–13 cm long, 3–10 mm broad, equal or narrowly club-shaped, some with bulb at base, same color as cap or darker orange-brown, surface dry, with longitudinal striations, violet mycelium over the base when fresh, quickly losing violet color and becoming white. **SPORE PRINT** white. **ODOR** and **TASTE** not distinctive.

Habit and habitat single, scattered, or clustered on the ground in wet areas, often on mosses, including *Sphagnum*, under mixed woods with conifers and hardwoods. July to October. **RANGE** widespread and fairly common in the Adirondack region of New York.

Spores subglobose or broadly ellipsoid, spiny, colorless, inamyloid, 7–9 × 6–9 µm excluding spines, spines 1–1.8 µm long.

Comments If the collection has a long stipe, like shown here, and it is growing in sphagnum moss, but lacks any violet colors on the lamellae and at the base of the stipe, you most likely have *Laccaria longipes* G. M. Mueller which also has been collected repeatedly in the Northeast as well. If all tissues are purple-violet, see *L. amethystina*.

Laccaria ochropurpurea
(Berkeley) Peck

PURPLE-GILLED LACCARIA
SYNONYMS *Agaricus ochropurpureus* Berkeley, *Clitocybe ochropurpurea* (Berkeley) Saccardo

Large mushroom with cap violet-gray becoming silvery-white with violet tints; gills violet or purple, thick, waxy; stem colored as cap, fibrous-striate; violet mycelium over base

CAP 3–12 cm wide, convex, with shallow sunken center, becoming plane or uplifted, light violet-gray or violet-brown, becoming paler buff or off-white eventually, dry, finely fibrillose or fibrillose-scaly, especially over the center, opaque. **FLESH** violet-buff, thick. **GILLS** dark violet or purplish-vinaceous, then paler, attached, sometimes subdecurrent, subdistant or distant, broad, thick, waxy-looking. **STEM** 4–19 cm long, 6–36 mm broad, equal or swollen at base, same color as cap, surface dry, with longitudinal fibrous striations, pale violet mycelium over the base when fresh. **SPORE PRINT** white. **ODOR** and **TASTE** not distinctive.

Habit and habitat single, scattered, or clustered on the ground in grassy areas under hardwoods with oak and beech, also under white pine. July to November **RANGE** widespread.

Spores globose, spiny, colorless, inamyloid, 7–9 × 6–9 μm excluding spines, spines 1–1.4 μm long.

Comments This mushroom is very attractive and listed as edible and good. However, there are several species of *Cortinarius* that produce purple-colored gills and pale-colored caps, but the gills on *Cortinarius* eventually turn rusty-brown. If you are unsure, take a spore print; *Cortinarius* produces rusty-brown spores. Eating species of *Cortinarius*, is not recommended, unless you know for sure which species you are consuming. If you have a small violet-gilled species that produces white spores, it is most likely *Laccaria amethystina*. Also see comments under *L. trullisata*.

Laccaria trullisata (Ellis) Peck

SANDY LACCARIA

SYNONYMS *Agaricus trullisatus* Ellis, *Clitocybe trullisata* Ellis

Moderately large with grayish-purple, then ochre-brown cap; gills constantly violet or purple, thick, waxy; stem colored as cap, covered with sand; spores white; only found in sandy areas with pine

CAP 2–7.5 cm wide, convex, with shallowly sunken center, becoming plane, not striate-lined, grayish-purple, becoming reddish-brown or ochre-brown, with pale violet-gray or violet-brown colors, dry, finely fibrillose or fibrillose-scaly, especially over the center. **FLESH** pale purple, then white, thick. **GILLS** dark violet or purplish and remaining so, bluntly attached, subdistant or close, broad, thick, waxy-looking. **STEM** 2–10 cm long, 6–23 mm broad, mostly club-shaped, same color as cap, surface dry, with longitudinal fibrous striations, violet or lilac mycelium over the base when fresh, but covered with sand over lower swollen base. **SPORE PRINT** white. **ODOR** and **TASTE** not distinctive.

Habit and habitat single, scattered, or clustered in sandy areas under pines. September to November. **RANGE** widespread but restricted to sand dunes around bodies of water or inland sand eskers, and only associated with pines.

Spores fusiform or oblong-ellipsoid, not spiny but with minute irregular bumps, colorless, inamyloid, 14–22 × 5.5–8 μm.

Comments It is edible if you can get all the sand off. Compare with *Laccaria ochropurpurea* that grows with pines and also hardwoods in forested areas and has much paler colors and very different spores.

Laccaria amethystina Cooke

SYNONYM *Clitocybe amethystina* (Cooke) Peck

Small with violet-purple cap, gills, and stem, gills persistently violet or purple; spores white; in wet areas or recently dried up pools, under hardwoods

CAP 0.5–3 cm wide, convex, with shallowly sunken center, becoming plane, not striate-lined, or faintly translucent-lined, dark purple at first, hygrophanous, becoming pale vinaceous-gray or ochre-buff, dry or moist, glabrous or becoming finely fibrillose or fibrillose-scaly. **FLESH** colored as cap, thin. **GILLS** dark violet or purple and remaining so, bluntly attached, subdistant or distant, narrow, thick. **STEM** 0.5–6 cm long, 1–7 mm broad, equal or narrowly club-shaped, nearly same color as cap, developing ochre tints, surface dry, with longitudinal fibrous striations, violet mycelium over the base at first, becoming white. **SPORE PRINT** white. **ODOR** and **TASTE** not distinctive.

Habit and habitat single, scattered, or clustered in wet mossy areas or recently dried up low areas. June to November. **RANGE** widespread.

Spores globose, spiny, colorless, inamyloid, 7–10 µm excluding spines, spines 1.5–3 µm long.

Comments There are some small purple *Cortinarius* species that might be confused with this species of *Laccaria*, but *Cortinarius* produces rusty-brown spore prints and the gills turn rusty-brown from the spores.

Cystoderma amianthinum var. rugosoreticulatum
(F. Lorinser) Bon

PUNGENT CYSTODERMA

SYNONYM *Agaricus rugosoreticulatum* F. Lorinser

Small mushroom with strongly wrinkled, orange-brown, granular cap with toothlike projections on the margin; gills white attached; stem with crumbling ring; odor pungent like freshly husked green corn

CAP 3.5–5 cm wide, conical-convex, orange-brown or yellowish-brown, surface granular and radially wrinkled and grooved, margin fringed at first with triangular membranous projections. FLESH white. GILLS white, attached, close or crowded, narrow. STEM 4–8 cm long, 3–8 mm broad, equal, similar color as cap, but darker brown under the sheathing veil as it breaks up into floccose-cottony patches, paler with fine fibrillose dots above ring, white mycelioid over base. RING same color as cap, most left on cap margin, some crumbling pieces on stem. SPORE PRINT white. ODOR strong, pungent, of fresh green corn. TASTE not distinctive.

Habit and habitat scattered or in small clusters on soil, moss beds, and conifer needles. August to October. RANGE widespread.

Spores: ellipsoid, smooth, colorless, amyloid in Melzer's reagent, 5–6 × 3 µm; pileus surface composed of large inflated and globose cells.

Comments If the cap is not wrinkled but otherwise has the same features, then the species is simply *Cystoderma amianthinum* (Scopoli) Fayod. *Cystoderma amianthinum* is more commonly collected than any of the other species placed in *Cystoderma* and the related genus *Cystodermella*. Compare with *Cystoderma granosum*.

Cystoderma granosum
(Morgan) A. H. Smith & Singer

SYNONYMS *Agaricus granosus* Morgan, *Lepiota granosa* (Morgan) Saccardo

Large orange-brown granular cap; persistent flaring ring on granular-covered stem; on decaying hardwood debris

CAP 4–9 cm wide, broadly conical-convex, becoming plane with broad central conical bump, bright orange-brown, yellowish-brown, or yellowish-ochre, surface granular or somewhat scaly-granular, sometimes wrinkled. **FLESH** white. **GILLS** white, attached, crowded. **STEM** 5–10 cm long, 8–15 mm broad, equal or enlarged downward, similar color as cap below ring, white or pale tan above the ring, the sheathing veil membranous but coated with granular coating and flaring into a persistent membranous ring, white mycelioid over base. **RING** same color as cap outside, white inside (top of), persistent, flaring. **SPORE PRINT** white. **ODOR** and **TASTE** not distinctive.

Habit and habitat scattered or in small clusters on decaying wood or on the ground but growing from buried well-rotted wood of hardwoods. August to October. **RANGE** widespread.

Spores: ellipsoid, smooth, colorless, amyloid in Melzer's reagent, 4–5 × 3 μm; pileus surface composed of large inflated and globose cells.

Comments *Cystoderma fallax* A. H. Smith & Singer is similar with the flaring persistent ring and granular surfaces. However, it has a darker reddish-brown color, grows on conifer wood, and has a western distribution from the Great Lakes to the Pacific Northwest. See comments under *Cystodermella cinnabarina*.

Cystodermella cinnabarina
(Albertini & Schweinitz) Harmaja

SYNONYM *Cystoderma cinnabarinum*
(Albertini & Schweinitz) Fayod

Large bright reddish or reddish-orange granular cap, with similarly colored and granular-covered stem, ring collapsing, crumbling, often on decaying conifer wood

CAP 3–8 cm wide, convex, becoming plane, bright cinnabar-red or orange to orange-brown, surface granular and with small pyramidal granular scales, margin with white cottony fringe. **FLESH** white. **GILLS** white, attached, crowded. **STEM** 3–8 cm long, 8–15 mm broad, equal, similar color as cap from a sheathing granulose veil that breaks up into large granulose orange-red clumps separated by white cottony tissue, white above the ring, white mycelioid over base. **RING** thin, granular, evanescent, and quickly collapsing. **SPORE PRINT** white. **ODOR** and **TASTE** not distinctive.

Habit and habitat scattered or in small clusters on decaying conifer wood or conifer needles, also on moss beds under hardwoods. August to October. **RANGE** widespread.

Spores: ellipsoid, smooth, colorless, not amyloid, 3.5–5 × 2–3 µm; pileus surface composed of large inflated and globose cells; hymenial cystidia pointed and the tips encrusted with crystals.

Comments The nonamyloid spores and the pointed cystidia make *Cystodermella cinnabarina* different from others placed in *Cystoderma* that have amyloid spores and lack these distinctive cystidia. DNA analysis also reveals the nonamyloid-spored taxa are clustered together as a group, and thus the genus *Cystodermella* was proposed for *C. cinnabarina*, *C. adnatifolia* (Peck) Harmaja, *C. granulosa* (Batsch) Harmaja, and several other species formerly placed in *Cystoderma*. Only *C. cinnabarina* is covered here, as it is the most commonly collected species in this group.

Ampulloclitocybe clavipes
(Persoon) Redhead, Lutzoni, Moncalvo & Vilgalys

FAT-FOOTED CLITOCYBE
SYNONYMS *Clitocybe clavipes* (Persoon) P. Kummer, *Agaricus clavipes* Persoon, *Clavicybe clavipes* (Persoon) Harmaja, *Clitocybe squamulosoides* P. D. Orton

Medium-sized mushroom with gray-brown cap and stem, with white to cream-colored descending gills and a typically bulbous spongy stipe base; under conifers, especially white pine

CAP 2.5–7.5 cm wide, convex, becoming plane or sunken, some with a low central bump, grayish-brown or with olive tints, darker over the center, smooth, moist becoming dry. **FLESH** white. **GILLS** off-white becoming pale cream, long decurrent, subdistant. **STEM** 2–6 cm long, 5–10 mm broad, enlarging downward and characteristically bulbous at base, more or less colored as cap or paler, bulbous base with spongy texture. **SPORE PRINT** white. **ODOR** can be fragrant fruity. **TASTE** not distinctive.

Habit and habitat single or in small clusters, on the ground under conifers like white pine, occasionally in mixed conifer and hardwood stands. July to November. **RANGE** widespread.

Spores elliptical, smooth, colorless, 6–8 × 3–5 µm.

Comments Though listed as edible in many field guides, it is known to produce some moderate ill effects such as headaches and flushing with a rash when alcohol is consumed, during or after eating this mushroom. Separated from *Clitocybe* by DNA phylogenetic evidence and by the very finely ornamented basidiospores as viewed under the scanning electron microscope.

Pseudoclitocybe cyathiformis (Bulliard) Singer

SYNONYMS *Agaricus cyathiformis* Bulliard, *Clitocybe cyathiformis* (Bulliard) P. Kummer, *Cantharellula cyathiformis* (Bulliard) Singer

Cap dark brown, but becoming pale with loss of moisture, smooth and vase-shaped; gills watery or grayish-brown, decurrent; stem same color and changes as cap; on wood typically

CAP 1.5–7 cm wide, convex, becoming vase-shaped, margin inrolled at first, dark brown and translucent-lined when moist, hygrophanous then paler grayish-brown and opaque, surface smooth, dry. **FLESH** same color as cap. **GILLS** watery brown fading to grayish-brown, short decurrent becoming long decurrent as cap arches up, close or subdistant, moderately broad (up to 6 mm). **STEM** 2.5–7 cm long, 4–9 mm broad, central but sometimes eccentric, equal or enlarged downward, watery brown, fading to buff with brown-streaked fibrils over the surface, often compressed with age, hollow. **SPORE PRINT** white. **ODOR** and **TASTE** not distinctive.

Habit and habitat single, scattered, or clustered on or near logs and stumps under hardwoods or conifers. July to November. **RANGE** widespread.

Spores ellipsoid, smooth, colorless, but amyloid in Melzer's reagent, 7.5–10 × 5–6.5 µm.

Comments A somewhat similar mushroom in shape and habit that grows on decaying wood is *Pseudoarmillariella ectypoides* (Peck) Singer. For this latter species the cap is deeply vase-shaped, pale brown, and adorned with obvious scales or knots of fibrils.

Singerocybe adirondackensis (Peck) Zhu L. Yang & J. Qin

SYNONYM *Clitocybe adirondackensis* (Peck) Saccardo

A medium-sized mushroom with a funnel-shaped greasy-looking gray-brown cap that eventually loses most of its color; gills white and long decurrent; odor unpleasant

CAP 1–6 cm wide, convex or plane, soon deeply sunken and funnel-shaped, pale brown then gradually paler to gray and then buff, eventually colorless off-white from losing moisture, smooth, moist, with satiny luster, lightly slippery when wet. **FLESH** white. **GILLS** off-white or buff, short then long decurrent, close or crowded, narrow, forked. **STEM** 1–5 cm long, 2–7 mm broad, equal, colored as cap or gills, glabrous, hairy white mycelium over base, hollow. **SPORE PRINT** white or pale dingy yellow. **ODOR** not pleasant, strong fishy or rancid-farinaceous. **TASTE** unpleasant.

Habit and habitat single, scattered, or clustered on leaf litter and well-decayed wood under hardwoods, but also on duff under conifers. July to October. **RANGE** widespread.

Spores elliptical or pear-shaped, colorless, 4–6.5(–8) × 2.5–4.5 µm, cap skin with cells like inflated balloons scattered among the cylindrical cells.

Comments Yet another *Clitocybe* species that has been moved to a separate, newly described genus, because of DNA phylogenetic evidence. This species is commonly encountered. Crush a small piece of the cap flesh when you do the odor test, that usually releases more of the aromatic chemicals to detect. The swollen cells in the cap skin are also diagnostic.

Arrhenia epichysium (Persoon)
Redhead, Lutzoni, Moncalvo & Vilgalys

SYNONYMS *Clitocybe epichysium* (Persoon) H. E. Bigelow, *Omphalina epichysium* (Persoon) Quélet

Cap dark brown or almost black, often with low raised lines over strongly inrolled margin, fading to gray-brown; gills gray-brown, short decurrent; stem same color as cap

CAP 1–2.5(–5) cm wide, convex, becoming plane, eventually vase-shaped, margin inrolled at first then straight, dark grayish-brown or almost black and translucent-lined over margin often, but also becoming opaque, hygrophanous and becoming grayish or pale gray-brown, surface smooth. **FLESH** thin, watery, tough. **GILLS** gray-brown, becoming paler in age, decurrent, close or subdistant, narrow or moderately broad (up to 4 mm). **STEM** 1–3 cm long, 1–3 mm broad, mostly equal, colored as cap or darker, white cottony at base. **SPORE PRINT** white. **ODOR** and **TASTE** not distinctive.

Habit and habitat scattered or clustered on logs and stumps of conifers or hardwoods. June to November. **RANGE** widespread.

Spores ellipsoid or broadly ellipsoidal, colorless, smooth, not amyloid, 7–9 × 4–5 µm.

Comments Most field guides list this as *Omphalina*. Also see *Pseudoclitocybe cyathiformis* and comments about *Pseudoarmillariella ectypoides*, as they both grow in similar habitats and have similar shapes, but are much larger.

Arrhenia gerardiana (Peck) Elborne

SYNONYMS *Agaricus gerardianus* Peck, *Clitocybe gerardiana* (Peck) Saccardo, *Omphalina gerardiana* (Peck) Singer

Cap brown, vase-shaped, striate and with dark fibrillose dots; gills grayish-brown, long decurrent; stem darker color than cap and attached to *Sphagnum*

CAP 1–3 cm wide, convex, becoming plane, then vase-shaped, dark grayish-brown with olive tints and translucent-lined over margin when moist, hygrophanous and becoming sordid buff, surface roughened from fibrillose dark brown or black dots. **FLESH** brown, thin, fragile. **GILLS** white, soon gray-tinted, becoming dark olive-gray in age, long decurrent, subdistant, narrow. **STEM** 3–5 cm long, 1–3 mm broad, mostly equal, colored as cap or darker gray, glabrous, white cottony at base and fused onto *Sphagnum* plants. **SPORE PRINT** white. **ODOR** and **TASTE** not distinctive.

Habit and habitat scattered or clustered on *Sphagnum* in bogs and fens. June to August. **RANGE** widespread.

Spores ellipsoid or nearly cylindrical, colorless, 7–12 × 3–4.5 µm.

Comments DNA evidence from recent phylogenetic studies places the species in *Arrhenia*. However, in previous field guides, it would be found as *Clitocybe* or *Omphalina*.

Lichenomphalia umbellifera
(Linnaeus) Redhead, Lutzoni, Moncalvo & Vilgalys

SYNONYMS *Omphalia umbellifera* (Linnaeus) P. Kummer, *Omphalina umbellifera* (Linnaeus) Quélet, *Omphalina ericetorum* (Persoon) M. Lange, *Phytoconis ericetorum* (Persoon) Redhead & Kuyper

On moss and algae-covered logs, road banks; with small umbrella-like caps with scalloped margins; gills pale yellow, decurrent, distant

CAP 0.5–3.5 cm wide, convex with center shallowly sunken, becoming funnel-shaped, margin distinctly scalloped, variously colored from yellowish-brown or vinaceous-brown or ochraceous, with paler color alternating with darker color like a striped umbrella, glabrous, moist, strongly ridged from the margin to the center. **FLESH** thin. **GILLS** pale yellowish or pale clay, decurrent, distant, broad. **STEM** 1–3 cm long, 1–3 mm broad, equal, brown or reddish-brown at first, becoming slowly pale yellow or pallid grayish-tan, sparse white tufted mycelioid at base. **SPORE PRINT** white. **ODOR** and **TASTE** not distinctive.

Habit and habitat scattered on wood, mosses, soil, but associated with a green alga that is growing on these substrates. July to October. **RANGE** widespread but more common in the northern latitudes in cooler climates, common in the arctic.

Spores ellipsoid, smooth, colorless, 7–10 × 4–6 µm.

Comments As the name implies, this is a lichen, a symbiosis between a basidiomycete mushroom and a green alga. Most lichens are partnerships between algae and ascomycetes. In the image, note the bright green and white pixie cups, with their associated squamules scattered about. These are ascomycete lichens.

Rickenella fibula (Bulliard)
Raithelhuber

ORANGE MOSS AGARIC

SYNONYMS *Agaricus fibula* Bulliard, *Gerronema fibula* (Bulliard) Singer, *Mycena fibula* (Bulliard) Kühner, *Omphalia fibula* (Bulliard) P. Kummer, *Omphalina fibula* (Bulliard) Quélet

Small and delicate with orange cap with contrasting darker translucent-lined margin; gills white, decurrent and subdistant; stem yellow-orange or buff-brown, minutely hairy; on mosses

CAP 0.3–1.5 cm wide, convex with shallowly sunken center, orange or yellow-orange, fading to buff, translucent-lined, smooth, moist, glabrous. **FLESH** off-white, thin. **GILLS** off-white or cream-buff, decurrent, subdistant, broad. **STEM** 1–5 cm long, 0.5–2 mm broad, equal, yellow-orange or buff-brown or ochre-brown especially downward, minutely white fibrillose overall, hollow, very fragile. **SPORE PRINT** white. **ODOR** and **TASTE** not distinctive.

Habit and habitat single, scattered, or in small clusters on mosses, often on downed logs under hardwoods. August to November. **RANGE** widespread.

Spores ellipsoid, smooth, colorless, not amyloid, 4–5 × 2–2.5 μm; with long fusiform or cylindrical cystidia in the pileus, gills, and stem surfaces.

Comments Compare with the orange-capped *Xeromphalina* species that can be found on mosses or at least on decaying wood with mosses present. Species of *Xeromphalina* are distinguished by the dense fuzzy orange hairs at the base of the stem, the darker stem colors, the habit of producing large numbers of fruit bodies, as well as their amyloid spores.

Rickenella swartzii (Fries) Kuyper can be found in similar habitats as *R. fibula*, with similar shape and decurrent, distant, white gills, but *R. swartzii* has a dark purplish to violet-black cap center contrasting with pale brown and translucent-lined margin, and the stem apex is purplish to violet-black, pale brown below. In some manuals, this species is listed incorrectly as *R. setipes* (Fries) Raithelhuber, which is an *Omphalina* found in South America.

Rickenella swartzii

Clitocybe gibba (Persoon)
P. Kummer

FUNNEL CLITOCYBE
SYNONYMS *Agaricus gibbus* Persoon,
Clitocybe infundibuliformis f. *gibba*
(Persoon) Saccardo

Funnel-shaped pinkish-cinnamon cap;
gills white, long decurrent, crowded; stem
paler than cap

CAP 5–7.5 cm wide, convex, soon deeply
sunken becoming funnel-shaped, pinkish-
cinnamon or pinkish-tan, smooth, sometimes becoming scaly over center. **FLESH**
white. **GILLS** off-white, long decurrent,
close, narrow. **STEM** 3–7 cm long, 4–12 mm
broad, equal but often enlarged downward,
white or pale flesh, with copious white
mycelium over base, solid. **SPORE PRINT**
white. **ODOR** and **TASTE** not distinctive.

Habit and habitat single, scattered, or clustered on the ground under hardwoods,
occasionally conifers. July to October.
RANGE widespread.

Spores elliptical, smooth, colorless, 6–8 ×
3.5–5 μm.

Comments *Clitocybe gibba* var. *cernua* H. E.
Bigelow differs by the stem being the same
color as the cap, and the cap is depressed
but not funnel-shaped.

Clitocybe sinopica (Fries) P. Kummer has
an orange-brown cap that is scaly over
the center, stem colored as the cap, short
decurrent whitish gills, and a strong farinaceous, not agreeable odor and taste. It
occurs in sandy soil in disturbed areas in
June and July.

Clitocybe sinopica

Clitocybe phyllophila
(Persoon) P. Kummer

SYNONYMS *Agaricus phyllophilus* Persoon, *Clitocybe cerrusata* (Fries) P. Kummer, *Lepista phyllophila* (Persoon) Harmaja

Cap white with pale pinkish-buff or pale watery brown areas under silvery canescence; gills white, decurrent, close; odor spicy pungent, under pines

CAP 3–11 cm wide, convex, becoming plane then broadly sunken, white but when wet with pinkish-buff or pale brown spots and streaks under a thin silvery canescence, smooth, opaque, not translucent-lined. **FLESH** white. **GILLS** off-white or pale cream or pinkish-buff, bluntly attached, then short decurrent, close or crowded, narrow. **STEM** 1.5–7 cm long, 4–7 mm broad, equal or the base enlarged up to 25 mm broad, colored as cap, glabrous or with longitudinal fibrils, densely hairy white mycelium over base binding decaying conifer needles, becoming hollow. **SPORE PRINT** cream or pinkish-buff. **ODOR** strong pungent spicy. **TASTE** mild, but unpleasant in old specimens.

Habit and habitat scattered or clustered on decaying needles of white and red pine, occasionally on leaf litter under mixed woods of pine and birch. September and October. **RANGE** widespread.

Spores elliptical, colorless, 4–5 × 2.5–3.5 µm.

Comments Although not a colorful mushroom, it has a pleasant odor.

Clitocybe odora (Bulliard) P. Kummer has an odor of anise. It has a blue-green color on the cap, lacks watery spots and canescence, and is found under hardwoods. The colors and odor may fade on older fruit bodies and make it difficult to be certain of the identification, but if the flesh is crushed, a lingering scent of anise is usually present. I often smell this mushroom before I see it.

Clitocybe odora

Clitocybe martiorum J. Favre

Vinaceous or cinnamon-brown colors on all parts, and the cap with an ephemeral, silvery canescence; gills very crowded; spore deposit bright reddish-pink

CAP 3–6 cm wide, convex, with strongly inrolled margin, densely silvery canescent over dull cinnamon or vinaceous-brown or orange-brown colors, smooth, becoming glabrous. **FLESH** dingy orange-brown. **GILLS** pale vinaceous-cinnamon and more orange-brown over edges, short decurrent, very crowded. **STEM** 3–8 cm long, 8–12 mm broad, equal, more or less colored as cap, solid. **SPORE PRINT** bright reddish-pink. **ODOR** pleasant, fragrant farinaceous when crushed or cut. **TASTE** disagreeable.

Habit and habitat clustered on rich humus or well-decayed wood, under hardwoods or conifers. August to October. **RANGE** widespread.

Spores elliptical, smooth, colorless, 4.5–5 × 2–3 μm.

Comments Not a commonly collected species. Formerly it was placed in *Lepista* because of the pinkish spore deposit colors that are not typical of *Clitocybe* species but are for *Lepista*, a related genus. *Lepista* species have ornamented, bumpy spores. Compare with species of *Lepista*.

Clitocybe inornata subsp. *occidentalis* H. E. Bigelow

SYNONYMS *Agaricus inornatus* Sowerby, *Paxillus inornatus* (Sowerby) Ricken, *Atractosporocybe inornata* (Sowerby) P. Alvarado, G. Moreno & Vizzini

Cap silvery-white with watery brown spots where injured, often with short raised lines over strongly inrolled margin; gills grayish-brown, short decurrent; stem same color and bruising reaction as cap

CAP 2.5–8 cm wide, convex, becoming plane, some with broad central bump, margin strongly inrolled, pale gray at first but soon buffy-brown, with darker watery brown spots irregularly scattered, often with short raised lines over margin, covered with canescence of silvery-white fibrils, dry. **FLESH** white or ashy, thick (to 10 mm). **GILLS** gray-brown, becoming darker in age, bluntly attached, becoming short decurrent, close, moderately broad (up to 7 mm). **STEM** 3–8 cm long, 6–15 mm broad, mostly central, occasionally eccentric, mostly equal, white or silver-gray with canescence or dense matted fibrils, becoming watery brown where bruised, white or pale grayish cottony at base. **SPORE PRINT** white. **ODOR** not distinctive, or very faintly fragrant when cut. **TASTE** not distinctive, or slightly astringent-bitter.

Habit and habitat single or clustered on leaf litter and humus under hardwoods or conifers. July to October. **RANGE** widespread but rarely collected.

Spores ellipsoid, colorless, 4–6 × 2.5–3.5 (–4) μm.

Comments The European version of this species was recently placed into a separate genus, *Atractosporocybe*, based on DNA phylogenetic evidence. My collection had a distinctive fragrant odor as noted here, but that feature needs to be more fully documented for American material, as the majority of collections examined by H. E. Bigelow lack such an odor.

Clitocybe robusta Peck

Large white mushroom with strongly inrolled cap margin when young; with decurrent, close or crowded gills, a nauseating odor and a pale yellow spore print

CAP 5–17 cm wide, convex, becoming plane, with strongly inrolled margin at first, then undulating, white becoming sordid buff, with finely satiny matted or canescent surface, dry or sometimes slightly slippery when wet. **FLESH** white, firm but becoming spongy when wet. **GILLS** white becoming pale cream or pale buff, bluntly attached, then decurrent, close or crowded, broad (2–10 mm). **STEM** 4–10 cm long, 10–35 mm broad, equal or club-shaped or bulbous over base, white or buff color, with matted longitudinal fibers or sometimes scaly over apex, densely white cottony-hairy mycelium over base binding leaves and humus, white, solid, becoming hollow. **SPORE PRINT** pale yellow.

ODOR and **TASTE** rancid or nauseating and disagreeable.

Habit and habitat scattered or more frequently in dense clusters on decaying leaves and humus under hardwoods, commonly in disturbed areas like walking paths and road banks. July to October. **RANGE** widespread.

Spores elliptical or elliptical-oblong, colorless, 6–7.5 × 3–4 μm.

Comments One might confuse this species with *Clitocybe nebularis* (Batsch) P. Kummer because of the disagreeable odor and size, but *C. nebularis* has a grayish or grayish-brown cap and typically occurs under conifers from Michigan westward to the Pacific Northwest. *Clitocybe candida* Bresadola has a white spore deposit and faintly amyloid spores. Large white *Russula* and *Lactarius* species have brittle flesh, and in *Lactarius* produce latex. Also see *Gerhardtia highlandensis*.

Gerhardtia highlandensis
(Hesler & A. H. Smith) Consiglio & Contu

SYNONYM *Clitocybe highlandensis* Hesler & A. H. Smith

A moderately large stocky white mushroom with canescent cap, fragrant odor, occurring under white pine

CAP 2.5–10 cm wide, convex, becoming plane, white or with faintly pinkish-cinnamon center, smooth, watery-mottled when wet, interspersed with hoary canescent patches. **FLESH** white. **GILLS** off-white or dingy cream color, streaking paler with moisture loss, bluntly attached, becoming short decurrent, close or subdistant, quite broad (to 9 mm). **STEM** 3–9 cm long, 5–15 mm broad, equal, the base sometimes enlarged (to 25 mm), white, with hairy white mycelium over base, solid. **SPORE PRINT** white. **ODOR** fragrant, pleasant. **TASTE** not distinctive.

Habit and habitat single, scattered, or clustered on needles under white pine. July to October. **RANGE** widespread.

Spores elongate-elliptical, randomly bumpy (use oil immersion lens), colorless, 4–6 × 2.5–3 μm, bumps cyanophilic, basidia with siderophilous and cyanophilous bodies.

Comments Although it looks like a species of *Clitocybe*, this is related to *Lyophyllum* due to the basidia containing siderophilous and cyanophilous bodies, and from DNA phylogenetic analyses. Compare with *Clitocybe robusta* that is typically found under hardwoods, not conifers.

Clitocybe subditopoda Peck

Cap watery grayish-brown eventually becoming dirty white, shallowly sunken; gills grayish, short decurrent; stem colored as cap but with silvery fibrils at first; odor and taste farinaceous

CAP 1–5 cm wide, convex, becoming plane then shallowly sunken, watery grayish-brown and translucent-lined over margin, at times canescent at first, becoming paler brown or yellowing with age and eventually almost white, surface watery translucent when moist, glabrous. **FLESH** colored as cap. **GILLS** grayish with vinaceous tint, then grayish-brown, bluntly attached, becoming short decurrent, close, moderately broad (1–5 mm). **STEM** 2–6 cm long, 2.5–6 mm broad, mostly equal, with a coating of silvery fibrils over watery grayish-brown colors, otherwise glabrous, watery grayish cottony at base, hollow. **SPORE PRINT** white. **ODOR** and **TASTE** farinaceous.

Habit and habitat scattered or clustered on needle beds under pine or spruce woods, rarely hardwoods. Most frequently found on white pine needle beds. September to November. **RANGE** widespread.

Spores elliptical, colorless, 3.5–5.5 × 2.5–3.5 μm.

Comments *Clitocybe vibecina* (Fries) Quélet is nearly identical in fresh specimens and can only be separated accurately by its longer spores (5.5–7 × 3–3.5 μm). *Clitocybe ditopa* (Fries) Gillet is not as commonly collected, but also looks similar to *C. subditopoda* except for its darker smoky-brown or olive-brown cap. It can only be accurately distinguished by its globose spores (3–4 μm).

Clitocybe candicans (Persoon) P. Kummer

SYNONYMS *Agaricus candicans* Persoon, *Leucocybe candicans* (Persoon) Vizzini, P. Alvarado, G. Moreno & Consiglio, *Clitocybe alboumbilicata* Murrill

A small white- or pale-cream-capped mushroom with a canescence on the cap and upper stem when fresh; on leaves and humus under hardwoods, but also on conifer needles

CAP 1–3 cm wide, convex, becoming plane, watery pallid or cream under a white or silvery-frosted canescence when moist, smooth, becoming opaque white with loss of moisture. **FLESH** watery pallid. **GILLS** white or pale buff, becoming cream color, bluntly attached but eventually short decurrent, close. **STEM** 1–3 cm long, 1–3 mm broad, equal, more or less colored as cap or more cream color, canescent over upper half at first. **SPORE PRINT** white. **ODOR** and **TASTE** mild.

Habit and habitat scattered, clustered, or bouquetlike on leaves and humus under hardwoods or sometimes conifers. August to October. **RANGE** widespread.

Spores elliptical, smooth, colorless, 4.5–6 × 2.5–3.5 μm.

Comments There are many small species of *Clitocybe* and they can be difficult to identify accurately unless one has a microscope and H. E. Bigelow's monograph to depend upon. This species is fairly common and an easy one to recognize in the northeastern hardwood forests. If found under conifers, compare with *C. tenuissima* Romagnesi (see illustration at Mycoportal.org) which has a gray or dull light brown ground color under the shiny canescence of the cap.

Clitocybe sudorifica (Peck) Peck

SWEATING MUSHROOM

SYNONYMS *Clitocybe dealbata* var. *sudorifica* Peck, *Clitocybe dealbata* subsp. *sudorifica* (Peck) H. E. Bigelow

Medium- to small-sized white mushroom in lawns and grassy areas with moderately decurrent, closely spaced gills, often growing in arcs or rings

CAP 1.5–4 cm wide, convex, becoming plane or sunken, margin arching up with age, white or pale grayish or with pale pinkish tints when wet, smooth, opaque, dry. **FLESH** white. **GILLS** white or pale buff, somewhat darker than cap, bluntly attached but eventually decurrent, close. **STEM** 1–5 cm long, 2.5–5 mm broad, equal, mostly central, sometimes eccentric, more or less colored as cap or paler. **SPORE PRINT** white or pale cream. **ODOR** and **TASTE** mild.

Habit and habitat scattered or clustered, sometimes in arcs or rings, in grassy lawns and pastures. August to October. **RANGE** widespread.

Spores elliptical, smooth, colorless, 4–5 × 2–3 µm.

Comments A poisonous species containing muscarine, causing chills, heavy perspiration, and vision problems. Compare with *Marasmius oreades*, that also grows on lawns, looks very different, and is edible.

Clitocybe truncicola (Peck) Saccardo

SYNONYM *Agaricus truncicola* Peck

Cap white canescent, showing yellow or buff colors eventually, shallowly sunken; gills pale buff, short decurrent; stem colored as cap; odor and taste not distinctive; on hardwood logs

CAP 1–3.5 cm wide, broadly convex, becoming plane then broadly and shallowly sunken, margin inrolled at all stages, white or faintly yellowish over disc, densely white canescent at first, buff or yellow colors showing with age or as canescence is squashed down. **FLESH** white. **GILLS** pale buff, bluntly attached, becoming short decurrent, close or crowded, narrow (1–3 mm). **STEM** 1–3 cm long, 1.5–5 mm broad, mostly central, occasionally eccentric, mostly equal, white becoming pale buff, glabrous or with fine white granules over apex, white cottony at base. **SPORE PRINT** white to pale cream. **ODOR** and **TASTE** not distinctive.

Habit and habitat scattered or clustered on hardwood logs. September to November. **RANGE** widespread.

Spores subglobose or broadly ellipsoidal, colorless, 3.5–4.5(–5) × 2.5–3.5(–4) µm.

Comments Compare with other wood-inhabiting species like *Clitocybe americana* and *Pseudoclitocybe cyathiformis*, that have much darker colors. The latter species may also occur on the ground as well as decaying wood.

Clitocybe americana
H. E. Bigelow

Cap watery brown becoming pinkish-cinnamon, strongly hygrophanous turning white; gills pale buff or vinaceous-buff, short decurrent; stem paler colors than cap; odor and taste not distinctive; on hardwood stumps and logs

CAP 1–4 cm wide, convex, becoming plane with shallow sunken center, margin inrolled at first, watery brown but soon paler and pinkish-cinnamon with expansion, not translucent-lined over margin, hygrophanous and eventually nearly white. FLESH watery buff, then white. GILLS pale buff or pale vinaceous-buff or pale yellow, bluntly attached, becoming short decurrent, close or crowded, narrow (1–4 mm). STEM 1–5 cm long, 1–5 mm broad, mostly central, occasionally eccentric, mostly equal, watery pale buff or pinkish-buff, glabrous or with fine white longitudinal fibrils, white or watery buff cottony at base. SPORE PRINT pale pinkish-cream. ODOR and TASTE not distinctive.

Habit and habitat usually in clusters on hardwood logs or buried wood of birch, maple, or beech. July to September. RANGE widespread.

Spores ellipsoid or broadly ellipsoidal, colorless, 4–6 × 2.5–3.5(–4) µm.

Comments *Clitocybe diatreta* (Fries) P. Kummer looks similar but occurs on needle beds under conifers.

Cantharellula umbonata
(J. F. Gmelin) Singer

GRAYLING
SYNONYMS *Merulius umbonatus*
J. F. Gmelin, *Cantharellus umbonatus*
(J. F. Gmelin) Persoon, *Clitocybe umbonata*
(J. F. Gmelin) Konrad, *Hygrophoropsis*
umbonata (J. F. Gmelin) Kühner &
Romganesi

Cap silvery-gray to brown, often with conical bump; gills white or cream, staining vinaceous, thick and repeatedly forked; stem colored as cap and attached to hair cap mosses

CAP 2.5–5 cm wide, convex, becoming plane or shallowly sunken, most often with a conical bump at the center, smoke gray or grayish-brown or violaceous-gray, surface minutely hairy matted. **FLESH** white, bruising reddish. **GILLS** white or cream, turning wine-red where injured, long decurrent, close or crowded, narrow, thick-edged and regularly forked. **STEM** 2.5–8 cm long, 3–7 mm broad, mostly equal, white or colored as cap or darker gray, sometimes turning red over base, glabrous, white cottony at base, white turning wine-red inside at base. **SPORE PRINT** white. **ODOR** and **TASTE** not distinctive.

Habit and habitat scattered or clustered on hair cap moss plants under hardwoods. August to November. **RANGE** widespread.

Spores elongate-ellipsoidal and nearly cylindrical, colorless but amyloid in Melzer's reagent, 7–11 × 3–4 μm.

Comments A fairly common species in the fall on hair cap moss beds. Some consider this mushroom a good edible species.

Hygrophoropsis aurantiaca
(Wulfen) Maire

FALSE CHANTERELLE

SYNONYMS *Cantharellus aurantiacus*
(Wulfen) Fries, *Clitocybe aurantiaca*
(Wulfen) Studer-Steinhäuslin, *Cantharellus
ravenelii* Berkeley & Curtis

Cap orange or yellow-brown, soft, often
sunken; gills orange, decurrent and repeatedly forked; stem often eccentric, colored
as cap; on or near rotting stumps or logs

CAP 2.5–6 cm wide, convex, becoming
plane and then shallowly sunken over the
center, dark orange or yellowish-orange
or orange-brown, often darker brown
or almost black over the center, dry, soft
and felted or velvety. **FLESH** white, thin,
soft. **GILLS** pale or bright orange, decurrent, close or crowded, narrow, repeatedly
forked, soft-flimsy. **STEM** 2–10 cm long,
5–15 mm broad, equal or enlarged at base,
central but may be off-center (eccentric),
same color as cap or dark orange-brown
if cap is paler, surface glabrous or finely
velvety. **SPORE PRINT** white. **ODOR** none.
TASTE mild.

Habit and habitat scattered or clustered on
humus and well-rotted wood under conifers or in mixed woods, often associated
with well-decayed stumps. August to October. **RANGE** widespread.

Spores ellipsoid, colorless, dextrinoid
(turning orange-brown in Melzer's
reagent), smooth, 5–8 × 3–4 μm.

Comments Some say edible, some say
it can cause problems if eaten. I say it is
too soft and flimsy to bother with for the
table. There appear to be two frequently
encountered color forms, a dark one and
a pale-colored one. Note the pale colors in
the image presented and the repeatedly
forking gills. *Aphroditeola olida* (Quélet)
Redhead & Manfred Binder has a pink cap
and stem with white decurrent forking
gills, a candylike odor and occurs under
conifers as well. The spores are not dextrinoid and the DNA phylogeny places this
species with the wax caps, not with the
bolete-related *Hygrophoropsis*.

Gerronema strombodes
(Berkeley & Montagne) Singer

SYNONYMS *Chrysomphalina strombodes*
(Berkeley & Montagne) Clémençon,
Omphalina strombodes (Berkeley &
Montagne) Murrill, *Clitocybe strombodes*
(Berkeley & Montagne) Singer
Vase-shaped, grayish-brown flattened
fibrils masking pale yellow tints on cap;
gills decurrent, pale yellow; stem pale yel-
low or nearly white; in clusters on conifer
or hardwood logs

CAP 2–10 cm wide, plane and shallowly or
deeply sunken over the center, becoming
vase-shaped, brown or grayish-brown from
flattened radiating fibrils that completely
overlay a pale yellow ground color that only
shows after cap expands, colors fading
with age. **FLESH** pale yellow, thin. **GILLS**
pale yellow, decurrent, broad, subdistant,
thin. **STEM** 2–6 cm long, 3–8 mm broad,
equal, pale yellow or pale grayish or almost
white, glabrous or with silvery-white matted fibrils. **SPORE PRINT** white. **ODOR** not
distinctive, or slight fragrant. **TASTE** not
distinctive, or slightly bitter.

Habit and habitat single, scattered, or most
often in clusters on hardwood and conifer logs. June to September. **RANGE** widespread in southeastern North America, but
does occasionally occur in the Northeast.

Spores elliptical or nearly cylindrical,
colorless in water, inamyloid, smooth,
7.5–9 × 4–5 µm; stipe tissues sarcodimitic,
with inflated, fusoid, thick-walled cells
bound by or intermingled with narrow,
branched, thin-walled filaments.

Comments *Gerronema strombodes* was
originally described from Ohio. A more
commonly collected yellow-gilled species on conifer logs in the Northeast
is *Chrysomphalina chrysophylla* (Fries)
Clémençon. The gills are much darker
golden-yellow.

Clitocybula oculus (Peck)
Singer

SYNONYMS *Agaricus oculus* Peck, *Gymnopus oculus* (Peck) Murrill, *Omphalia oculus* (Peck) Saccardo

Clustered bouquetlike mushrooms on hardwood logs; cap pale ashy-white with gray-brown center; gills white, subdecurrent; stipe colored as cap, coated with gray-brown or blackish scales and dots (use a lens)

CAP 1–4 cm wide, convex, becoming plane and often shallowly sunken over center, off-white or pale ashy-white with gray-brown or dark brown center and gray-brown radially arranged fibrils to margin or similar-colored scales over center, surface smooth, moist. **FLESH** whitish, thin. **GILLS** white, bluntly attached becoming subdecurrent, close. **STEM** 2–8 cm long, 2–5 mm broad, equal, similar color as cap, surface roughened with fine gray-brown or blackish scales and dots, hollow, easily splitting lengthwise. **SPORE PRINT** white. **ODOR** and **TASTE** not distinctive.

Habit and habitat densely packed into bouquetlike clusters on hardwood logs and woody debris on the forest floor. July to September. **RANGE** widespread.

Spores subglobose or broadly elliptical, smooth, colorless, amyloid in Melzer's reagent, 5–6.5 × 4–5.5 µm, cheilocystidia clavate.

Comments A similar bouquetlike species often featured in field guides is *Clitocybula familia* (Peck) Singer that grows on decaying conifer logs, has a bald stipe, and the cap lacks a central dark depression. In addition, its spores are globose and it lacks cheilocystidia. *Clitocybula familia* is considered edible, but the edibility of *C. oculus* is not known. *Clitocybula abundans* (Peck) Singer is similar in colors and the bouquetlike growth, and is distinguished by the glabrous stem, that may have a fine white pubescence at the apex but lacks the darkly colored scales and dots.

Cyptotrama asprata (Berkeley) Redhead & Ginns

GOLDEN-SCRUFFY COLLYBIA

SYNONYMS *Agaricus aspratus* Berkeley, *Armillaria aspratus* (Berkeley) Petch, *Xerula asprata* (Berkeley) Aberdeen, *Xerulina asprata* (Berkeley) Pegler

Bright golden-yellow cap and stem with granules or roughly scaly on stem; gills white, thick, distant; bluntly attached to decaying wood

CAP 0.5–2 cm wide, convex, becoming plane, bright yellow or dark golden-yellow, surface granular or becoming wrinkled or pitted, especially over the center, and the margin may be radially lined with age, dry. **FLESH** white or pale yellow, thin. **GILLS** white, thick, bluntly attached, eventually subdecurrent, distant, may become interveined. **STEM** 1–5 cm long, 1–3 mm broad, equal but subbulbous at base, tough-pliant, similar color as cap or paler, surface roughened with dark golden scales and granules. **SPORE PRINT** white. **ODOR** and **TASTE** not distinctive.

Habit and habitat scattered or clustered on decaying hardwood logs and debris. June to October. **RANGE** widespread.

Spores broadly elliptical or somewhat lemon-shaped, smooth, colorless, not amyloid, 8–12 × 6–7 µm; cheilocystidia narrowly fusiform with rounded tip.

Comments This species can be found in older North American field guides in the genus *Xerulina* or as *Cyptotrama chrysopepla* (Berkeley & M. A. Curtis) Singer. The species is not uncommon in the Northeast. Most of its relatives are tropical.

Asterophora lycoperdoides
(Bulliard) Ditmar

POWDER CAP

SYNONYMS *Agaricus lycoperdoides* Bulliard, *Nyctalis lycoperdoides* (Bulliard) J. Schröter, *Asterophora agaricoides* Fries

Cap small white, turning into a brown powdery mass; gills typically nonexistent or poorly formed; fruiting in clusters on the blackened or infected caps of larger mushrooms

CAP 1–2 cm wide, convex, white and cottony, surface soon cracked then becoming brown and powdery from asexual spores, dry. **FLESH** white. **GILLS** not well-developed, narrow, thick when present, white becoming brown. **STEM** 2–5 cm long, 3–10 mm broad, equal, colored as cap, smooth or minutely hairy. **SPORE PRINT** when formed, white. **ODOR** and **TASTE** farinaceous or mealy.

Habit and habitat clustered on infected and often rotting, blackened mushrooms, typically *Russula* or perhaps *Lactarius* species. June to November. **RANGE** widespread.

Spores of two types, the basidiospores when present oval, smooth, colorless 5–6 × 3.5–4 µm, the asexual spores on the pileus, the chlamydospores, are thick-walled, brown, globose, with rounded spines, 12–18 µm.

Comments Compare with the even smaller-capped *Collybia tuberosa* and the other small collybias that are found on decaying and blackened mushrooms, but produce well-formed gills and brown or yellow tuberlike structures at the base of their stems. They also never produce brown powdery asexual spores on the cap surfaces.

Asterophora parasitica (Bulliard ex Persoon) Singer is not as commonly collected as *A. lycoperdoides,* but has a conical-convex, silvery-white, silky cap that becomes gray-brown. It fruits in clusters on the blackened or infected caps of larger mushrooms, typically *Russula* or perhaps *Lactarius* species.

Asterophora parasitica

Collybia tuberosa (Bulliard) P. Kummer

TUBEROUS COLLYBIA

SYNONYMS *Agaricus tuberosus* Bulliard, *Gymnopus tuberosus* Gray, *Microcollybia tuberosa* (Bulliard) Lennox

Very small white mushrooms on blackened decaying larger mushrooms or rich humus, the stems attached to a reddish-brown apple-seed-shaped tuber, or sclerotium

CAP 0.3–1 cm wide, convex, becoming plane, often shallowly sunken or with a small central bump, or sunken and with a bump, white with a buff or brown center, smooth or finely hairy, opaque or faintly translucent-lined over margin. **FLESH** white, thin. **GILLS** white, attached, close or subdistant, narrow. **STEM** 1–5 cm long, 0.5–1 mm broad, equal but often flexuous, white, surface glabrous but finely white-dotted at apex and white-velvety at base, mostly attached to a reddish-brown sclerotium (asexual reproductive structure) shaped and sized like an apple seed. **SPORE PRINT** white. **ODOR** and **TASTE** not distinctive.

Habit and habitat clustered on blackened remains of decaying mushrooms or occasionally humus or decaying wood on the forest floor. August to November. **RANGE** widespread.

Spores elliptical or teardrop-shaped, smooth, colorless, 4–6 × 2–3.5 μm.

Comments *Collybia cookei* is similar but arises from smaller, round, yellowish-orange sclerotia. *Collybia cirrhata*, also similar in looks, lacks any type of sclerotium at the base of the stipe. Note the reddish-brown seed-shaped sclerotia in the image of this species.

Gymnopus dryophilus
(Bulliard) Murrill

OAK-LOVING COLLYBIA

SYNONYMS *Agaricus dryophilus* Bulliard, *Collybia dryophila* (Bulliard) P. Kummer, *Marasmius dryophilus* (Bulliard) P. Karsten

Cap reddish-brown becoming orange-brown; gills white, very crowded; stem white at first then colored as cap, stringy-pliant, easily splitting; odor and taste not distinctive

CAP 1–5 cm wide, convex, becoming plane, some shallowly sunken, dark red-brown at first, soon orange-brown, eventually quite pale, pinkish-buff or pale orange-yellow overall, glabrous, moist and somewhat slippery at first, soon dry. **FLESH** white. **GILLS** white or pinkish-buff becoming pale buff, attached or almost free, crowded, narrow or moderately broad. **STEM** 1–5 cm long, 2–5 mm broad, mostly equal but some club-shaped or bulbous at the base, stringy-pliant and easily splitting, white at first, soon colored as cap, glabrous, becoming faintly lined, hollow. **SPORE PRINT** white with tint of cream. **ODOR** and **TASTE** not distinctive.

Habit and habitat scattered or clustered, occasionally in bouquetlike clusters on humus, leaf litter, or well-decayed wood under hardwoods or mixed conifers and hardwoods, oak is not always present. June to September. **RANGE** widespread.

Spores teardrop-shaped or elliptical, smooth, colorless, 5.6–6.4 × 2.8–3.5 μm; cheilocystidia narrowly club-shaped, flexuous, contorted, some irregularly lobed, hyphae of cap cylindrical, many ends bifurcate.

Comments A jelly fungus known as *Syzygospora mycetophila*, formerly called *Christiansenia*, occasionally attacks this mushroom, producing white or pale buff gall-like structures on all parts of the fruit bodies. See that species for another view of *Gymnopus dryophilus*. In the early spring a species somewhat similar to *G. dryophilus* and found in the same habitats is *G. subsulphureus* (Peck) Murrill. *Gymnopus subsulphureus* has a pale yellow-buff cap, yellow gills and stem, and pinkish-ochre rhizoids at the base of the stem.

Gymnopus confluens
(Persoon) Antonín, Halling & Noordeloos

TUFTED COLLYBIA

SYNONYMS *Agaricus confluens* Persoon, *Collybia confluens* (Persoon) P. Kummer, *Marasmius confluens* (Persoon) P. Karsten

Cap thin-fleshed, cinnamon becoming white; gills pinkish-buff, crowded; stem long thin, pinkish-cinnamon with white dense fuzzy covering; odor and taste not distinctive; occurring in bouquetlike clusters on humus

CAP 1–3.5 cm wide, convex, becoming plane, some with low broad central bump, reddish-brown at first, soon cinnamon-brown, then yellow-brown, eventually quite pale, pinkish-buff or nearly white overall, surface glabrous or finely matted-hairy, dry. FLESH whitish, very thin, pliant. GILLS pinkish-buff becoming cream, attached or almost free, crowded, narrow. STEM 2.5–9 cm long, 1.5–4 mm broad, equal, pliant-tough, pale pinkish-cinnamon with a dense white or grayish fuzz overall, hollow. SPORE PRINT white with tint of cream. ODOR not distinctive. TASTE not distinctive, or weakly of garlic.

Habit and habitat loosely clustered but most often in bouquetlike clusters on humus and leaf litter under hardwoods or mixed conifers and hardwoods. June to October. RANGE widespread.

Spores teardrop-shaped or elliptical, smooth, colorless, 7–9 × 3.5–4.2 µm; cheilocystidia narrowly club-shaped, flexuous, contorted, some irregularly lobed, hyphae of cap cylindrical, branched.

Comments *Gymnopus polyphyllus* (Peck) Halling is somewhat similar, but is larger, darker colored at first, has a strong garlic odor and does not produce bouquetlike clusters. The spores are longer in *G. confluens* and the cap hyphae are very different as well.

Gymnopus polyphyllus

Gymnopus biformis (Peck) Halling

SYNONYMS *Collybia biformis* (Peck) Singer, *Marasmius biformis* Peck

Cap cinnamon-brown with furrowed surface and depressed center; gills white close; stem densely velvety-hairy, colored as cap, tough

CAP 1–2.5 cm wide, convex, becoming plane, shallowly sunken, also sometimes with a small central bump, reddish-brown fading to cinnamon-brown or paler, surface furrowed, bumpy-lined from near center to margin, dry. **FLESH** whitish, very thin, pliant. **GILLS** white, attached, close, narrow. **STEM** 1.5–5 cm long, 1–4 mm broad, equal, pliant-tough, similar to cap colors or orange-brown, surface densely velvety-hairy overall, whitish hairy above, more orange-cinnamon over the base, hollow. **SPORE PRINT** white or cream. **ODOR** and **TASTE** not distinctive.

Habit and habitat scattered or clustered, typically on soil, occasionally on leaf or needle litter on the forest floor under hardwoods or mixed conifer-hardwood forests. Also frequently along unimproved roadways in forested land. July to September. **RANGE** widespread.

Spores teardrop-shaped or elliptical, smooth, colorless, 6.4–8.6 × 3.2–4.4 µm; cheilocystidia cylindrical, contorted, or lobed, hyphae of cap also similar in form.

Comments *Gymnopus subnudus* (Ellis ex Peck) Halling is similar in cap color and texture, but differs by the more highly colored and widely spaced lamellae, the lack of pubescence on the upper portions of the two-toned stem, the bitter taste, and the growth on sticks, leaves, and humus.

Gymnopus subnudus

Gymnopus spongiosus
(Berkeley & M. A. Curtis) Halling

HAIRY-STALKED COLLYBIA
SYNONYMS *Marasmius spongiosus* Berkeley & M. A. Curtis, *Collybia spongiosa* (Berkeley & M. A. Curtis) Singer

Small with rusty-brown cap and swollen stem; gills white; cap becoming much paler than the dense, spongy, rusty-brown matted hairs over the base of the stem; ammonia turns tissues green

CAP 1–3 cm wide, convex, becoming plane, some with low broad central bump, reddish-brown, fading to grayish-orange then white along margin, surface glabrous, smooth or bumpy over margin, dry but may be slippery when wet. **FLESH** white, to 3 mm thick. **GILLS** white or pale cream, attached, close or crowded, narrow. **STEM** 2–7 cm long, 1–4 mm broad at apex, wider below, equal or narrowly club-shaped, pallid and mostly bald at apex, reddish-brown below or overall, base swollen, spongy from densely matted reddish-brown hairs, tough, hollow. **SPORE PRINT** white. **ODOR** and **TASTE** not distinctive.

Habit and habitat single, scattered, or clustered on leaf and needle litter under hardwoods like oaks, or in mixed woods with pine. July to October. **RANGE** widespread.

Spores teardrop-shaped or elliptical, smooth, colorless, 6–8.4 × 3.5–4.2 µm. A drop of household ammonia will turn stem surface green.

Comments Compare with *Gymnopus biformis* and *G. subnudus*.

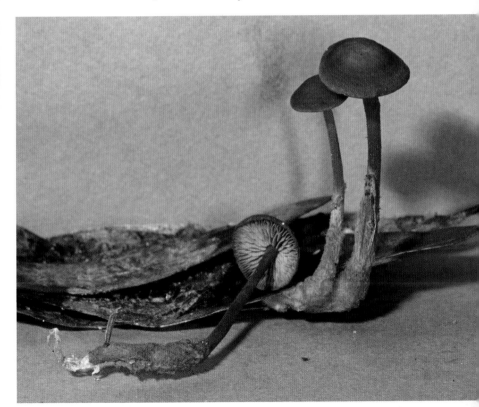

Gymnopus androsaceus
(Linnaeus) Della Maggiora & Trassinelli

SYNONYMS *Agaricus androsaceus* Linnaeus, *Marasmius androsaceus* (Linnaeus) Fries, *Setulipes androsaceus* (Linnaeus) Antonin

Very small with black horse-hairlike stems, dark brown then pinkish-brown wrinkled caps; gills distant, brownish-pink, broad; black threads (rhizomorphs) at the stem base

CAP 0.2–1.5 cm wide, convex, then plane, mostly with center sunken, sometimes with small central bump, margin corrugate-lined, dark brown with pink tint at first, then clay-pinkish with reddish-brown center, glabrous, smooth, dry. **FLESH** thin. **GILLS** grayish-cream, soon brownish-pink and tinged with gray, attached, distant, broad. **STEM** 1.2–6 cm long, 0.2–1 mm broad, equal, round or compressed, black or blackish-brown, shiny, dry, horse-hairlike, tough, with black threads (rhizomorphs) branching out from the base. **SPORE PRINT** white. **ODOR** not distinctive. **TASTE** not distinctive, or somewhat bitter.

Habit and habitat scattered or clustered and often in large numbers on conifer needles, twigs, cone, and cone scales, but also, not as often though, on hardwood twigs and stems. July to November. **RANGE** widespread.

Spores apple-seed-shaped or ovate, smooth, colorless, not amyloid, 7–10 × 3–4.5 µm; cheilocystidia are broom cells (or miniature mutant cow udders as Michael Kuo accurately describes their look).

Comments This mushroom can be hard to find until your eyes accommodate to the low light in the forests. It is small and usually fruits in large numbers, so staring at the ground when under conifers will increase your chances of finding them. The wiry black stems are characteristic of many species of *Marasmius*, where this species was formerly placed.

Connopus acervatus
(Fries) K. W. Hughes, Mather & R. H. Petersen

CLUSTERED COLLYBIA

SYNONYMS *Agaricus acervatus* Fries, *Collybia acervata* (Fries) P. Kummer, *Gymnopus acervatus* (Fries) Murrill, *Marasmius acervatus* (Fries) P. Karsten

Densely clustered bouquetlike, with dark purplish-brown cap eventually turning yellow-brown; gills white, crowded; stipe dark purplish-brown or paler, coated over the base with white velvety hairs; in moist rich woods, bogs

CAP 2–5 cm wide, convex, becoming plane, radically changing color, dark purplish-brown or reddish or chestnut-brown when moist, hygrophanous and becoming paler yellow-brown or buff from margin inward with center retaining darker colors, surface smooth, moist. FLESH white, thin. GILLS white or with pink tint in age, barely attached, crowded. STEM 4–10 cm long, 2–6 mm broad, equal, dark grayish-black, becoming purplish-brown, base coated with white or ashy-white velvety coating, surface smooth, hollow. SPORE PRINT white. ODOR and TASTE not distinctive.

Habit and habitat in bouquetlike clusters on the ground, on decaying wood, on moss beds in bogs, under conifers or mixed hardwoods and conifers in moist forests. July to October. RANGE widespread.

Spores elliptical, smooth, colorless, 5.5–7 × 2.5–3 μm.

Comments The stems are fused together at the base, so it is easy to pick the entire bouquet of mushrooms to resituate for taking pictures. Edibility unknown.

Rhodocollybia lentinoides
(Peck) Halling

SYNONYMS *Agaricus lentinoides* Peck,
Collybia lentinoides (Peck) Saccardo,
Gymnopus lentinoides (Peck) Murrill
Cap cinnamon with darker center; gills
white with jagged, saw-toothed edges;
stem club-shaped, fragile, watery grayish-
white, not distinctly lined or deeply ridged;
found under spruce

CAP 2–4.5 cm wide, convex, becoming
plane, some with low broad central bump,
dull cinnamon-brown with darker brown
center, fading to pinkish-buff over margin,
surface glabrous, moist, translucent-lined
over margin. FLESH white, up to 4 mm
thick. GILLS white or pale buff, attached
or separating from stipe, close, broad
(up to 5 mm), edges distinctly jagged,
saw-toothed. STEM 2–6 cm long, 3–6 mm
broad, equal or club-shaped, fragile, easily
splitting, watery grayish-white and translucent, glabrous, faintly lined or dusted

with granular dots, or both, hollow. SPORE
PRINT pale pinkish-buff. ODOR and TASTE
not distinctive.

Habit and habitat scattered or clustered,
on humus and needle litter under spruce,
especially *Picea abies*. June to October.
RANGE widespread where spruce occurs,
but rarely collected.

Spores teardrop-shaped or elliptical,
smooth, colorless, many dextrinoid in
Melzer's reagent, 5.5–7.5 × 3.2–4.4 µm;
cheilocystidia absent.

Comments Very similar to *Rhodocollybia
butyracea* but distinguished by the saw-
toothed edges of the gills, the paler colors
of the cap, and the shorter spores. If it is
a small mushroom from under conifers,
with a dark brown, moist but textured
cap surface under a lens, and the gills are
white or ash-gray with smooth edges, the
stem is equal and translucent and white,
the spore print is pale pinkish-buff, then
you have *R. unakensis* (Murrill) Halling.

Rhodocollybia maculata
(Albertini & Schweinitz) Singer

SPOTTED COLLYBIA

SYNONYMS *Agaricus maculatus* Albertini & Schweinitz, *Collybia maculata* (Albertini & Schweinitz) P. Kummer

Cap often large, white with rusty-red stains; gills white, crowded; stem white, rooting, developing rusty-red stains; taste bitter

CAP 5–17 cm wide, convex, becoming plane, some with low broad central bump, white or pinkish-buff over center, often spotted or streaked with rusty-red, surface glabrous, dry. **FLESH** white, 8–15 mm thick. **GILLS** white or pale cream or pinkish-buff, also often spotted and streaked with rusty-red, attached, crowded, moderately broad (4–8 mm). **STEM** 5–15 cm long, 8–15 mm broad, equal and often with short root at the base, firm and fibrous, white developing rusty stains like cap and gills, matted hairy becoming glabrous, hollow. **SPORE PRINT** pinkish-buff or ochraceous-buff. **ODOR** not distinctive but often pungent. **TASTE** bitter.

Habit and habitat single, scattered, or clustered on buried wood humus and needle litter under conifers or in mixed woods with conifers such as spruce, pine, and hemlock. July to October. **RANGE** widespread.

Spores globose or subglobose, smooth, colorless, dextrinoid inner wall in Melzer's reagent, 5.6–6.4 × 4.8–5.6 μm.

Comments *Rhodocollybia maculata* var. *scorzonerea* (Batsch) Lennox differs by its yellow lamellae and a stipe that also may become yellow.

Hymenopellis furfuracea
(Peck) R. H. Petersen

SYNONYMS *Collybia radicata* var. *furfuracea* Peck, *Oudemansiella furfuracea* (Peck) Zhu L. Yang, G. M. Mueller, G. Kost & Rexer, *Xerula furfuracea* (Peck) Redhead, Ginns & Shoemaker

Large brown wrinkled caps; gills white, attached, broad; stem long, hairy and scruffy dotted, long taprootlike pseudorhiza; hardwood forests or on lawns from buried wood

CAP 2–12 cm wide, broadly convex or conical-convex or bell-shaped, becoming plane, often with central broad bump, margin becoming uplifted, dark brown or grayish-brown or yellowish-brown, surface usually wrinkled radially, more so with age, glabrous, feeling rubbery under the skin when fresh and wet, becoming dry. **FLESH** white, thick. **GILLS** white, attached or notched with decurrent tooth, close or subdistant, broad. **STEM** 5–20 cm long, 5–20 mm broad, equal but often swollen just above the base then tapering into a long deeply penetrating pseudorhiza, white over apex and taproot, brown or grayish-brown elsewhere, longitudinally fibrous-striate, densely hairy and dotted with clumps of fibrils, dry, snapping easily. **SPORE PRINT** white. **ODOR** and **TASTE** not distinctive.

Habit and habitat scattered or clustered in humus and leaf litter under hardwoods, especially with beech, also grassy areas near dead trees, arising from buried decaying wood and roots. July to November. **RANGE** widespread.

Spores broadly ovoid or broadly ellipsoid, smooth, colorless, not amyloid, 12–18 × 9–13 µm.

Comments *Hymenopellis rubrobrunnescens* is typically larger, has a naked stem, and the gills and stem stain dark reddish-brown when bruised, albeit sometimes slowly. *Hymenopellis megalospora* (Clements) R. H. Petersen has a slippery or slimy pale brown or even buff-colored cap and a white glabrous stem. The spores are larger in this last species as well, reaching 22 µm in length.

Hymenopellis rubrobrunnescens (Redhead, Ginns & Shoemaker) R. H. Petersen

SYNONYMS *Xerula rubrobrunnescens* Redhead, Ginns & Shoemaker, *Oudemansiella rubrobrunnescens* (Redhead, Ginns & Shoemaker) Zhu L. Yang, G. M. Mueller, G. Kost & Rexer

Large rusty-brown wrinkled caps; gills white but staining rusty-brown; stem long, white staining rusty-brown and with long taprootlike pseudorhiza; hardwood forests from buried roots

CAP 2–9 cm wide, convex, becoming plane with central broad bump, rust-brown or cinnamon-brown or with honey-brown or hazel, smooth on the center but surface strongly wrinkled elsewhere, more so with age, glabrous, feeling gelatinous-rubbery under the skin when fresh and moist, becoming dry. **FLESH** white, turning rusty-brown. **GILLS** white, but edges often rusty-brown and turning rusty-brown from injury, attached or notched with decurrent tooth, close or subdistant, broad. **STEM** 8–20(–40) cm long, 2–20 mm broad, equal but often swollen just above the base then tapering into a long deeply penetrating pseudorhiza, white but staining intensely rusty-brown where injured or handled, longitudinally ridged, glabrous, dry, snapping easily. **SPORE PRINT** white. **ODOR** and **TASTE** not distinctive.

Habit and habitat single or scattered in humus and leaf litter under hardwoods, arising from buried decaying roots. July to November. **RANGE** widespread.

Spores lemon-shaped or almond-shaped, smooth, colorless, not amyloid, 13.5–16 × 8–9 µm; pleurocystida bowling-pin-shaped.

Comments See comments under *Hymenopellis furfuracea*, which is much more commonly encountered than *H. rubrobrunnescens*. Note the tapering rootlike pseudorhiza in this image that is characteristic for *Hymenopellis* and also the brown-spored genus *Phaeocollybia*.

Baeospora myosura (Fries) Singer

CONIFER-CONE BAEOSPORA

SYNONYMS *Agaricus myosurus* Fries, *Collybia myosura* (Fries) Quélet, *Mycena myosura* (Fries) Kühner

Cap pale brown with paler margin; gills white, crowded; in clusters on fallen, decaying cones of Norway spruce or white pine most often

CAP 0.5–2 cm wide, convex, becoming plane, tan or cinnamon, paler with age as they fade from the margin inward, smooth, moist. **FLESH** white, very thin. **GILLS** white, finely attached, crowded, narrow. **STEM** 1–5.5 cm long, 1–2 mm broad, equal, pale at first, becoming grayish-brown, with minute hairs over surface, coarsely long white hairy at base. **SPORE PRINT** white. **ODOR** and **TASTE** not distinctive.

Habit and habitat clustered on decaying conifer cones of Norway spruce, white pine most frequently, but also reported from Douglas fir cones, and even the nonconifer magnolia "cones." September to November. **RANGE** widespread.

Spores elliptical-elongate, smooth, colorless, amyloid, 3–4.5 × 1–2.5 µm.

Comments Compare with *Strobilurus* species. If it is the early spring, and you are under Norway spruce, you most likely have *Strobilurus esculentus*, which lacks amyloid spores and has thick-walled cystidia. Examination of the spores and tissues with a microscope is the best way to make accurate identifications for these small cone-inhabiting species, since they look very similar. However, the difference in fruiting period seems to hold for *S. esculentus* vs. *Baeospora myosura*.

Baeospora myriadophylla
(Peck) Singer

LAVENDER BAEOSPORA

SYNONYMS *Agaricus myriadophyllus* Peck, *Collybia myriadophylla* (Peck) Saccardo, *Mycena myriadophylla* (Peck) Kühner

Cap lavender developing ochre-brown colors over center, then overall; gills lavender, crowded; stipe colored as pileus; on decaying wood of conifers, especially hemlock, but also hardwoods

CAP 1–4 cm wide, convex, becoming plane or sunken, lavender, hygrophanous and quickly changing color over the center to pale lavender, then developing brown or ochre-brown colors from the center outward with age, smooth, moist. **FLESH** color as cap surface, very thin. **GILLS** lavender, finely attached or nearly free, crowded, narrow. **STEM** 2–5 cm long, 1–3 mm broad, equal, same colors as cap or slightly paler, with long white hairs at base. **SPORE PRINT** white. **ODOR** and **TASTE** not distinctive.

Habit and habitat single, scattered, or clustered on decaying conifer or hardwood logs, especially hemlock, but also reported from poplar. June to October. **RANGE** widespread.

Spores broadly elliptical, smooth, colorless, amyloid, 3.5–4.5 × 2–3 μm.

Comments The fruit bodies can become dark dull brown overall with age, but the small stature, crowded lamellae, and habit of fruiting from decaying woody substrates are helpful in identifying the species if you do not have a microscope and the lavender colors have faded.

Strobilurus esculentus
(Wulfen) Singer

SYNONYMS *Agaricus esculentus* Wulfen, *Collybia esculenta* (Wulfen) P. Kummer, *Marasmius conigenus* subsp. *esculentus* (Wulfen) J. Favre, *Pseudohiatula esculenta* (Wulfen) Singer, *Pseudohiatula conigena* var. *esculenta* (Wulfen) M. M. Moser

Small, gray-brown-capped mushroom, growing on Norway spruce decaying cones, usually in wet areas, after snowmelt

CAP 1–2 cm wide, convex, becoming plane, center sunken or with a low bump, dark gray-brown or yellow-brown with gray tints, but when very young almost colorless, dry, glabrous, smooth or obscure radiately raised lines over margin, opaque. **FLESH** white or pale brown, thin. **GILLS** white or pale gray, barely attached or becoming free, crowded to close, broad. **STEM** 3.5–14 cm long, 1–3 mm broad, equal, white over apex, sordid pale yellow or darkening to ochraceous-yellow downward, smooth, but finely powdered (use a lens), tough and rigid. **SPORE PRINT** white. **ODOR** not distinctive. **TASTE** mild.

Habit and habitat clustered on fallen, decaying cones of Norway spruce (*Picea abies*). April. **RANGE** New York, apparently very rare.

Spores ellipsoid, smooth, colorless, not amyloid, 4.5–7 × 2.5–4 µm; pleurocystidia thick-walled (up to 2.5 µm thick) and with large colorless crystals at apex.

Comments A microscope is necessary to confirm the diagnostic thick-walled and crystal-capped pleurocystidia of *Strobilurus esculentus*. *Strobilurus albipilatus* (Peck) Wells & Kempton occurs on decaying cones of Douglas fir and various pine species across the northern coniferous forests of North America, from New York to British Columbia to Alaska from September to November. The cap colors vary from vinaceous-buff to grayish-sepia-brown and the cap margin is translucent-lined. The pleurocystidia are thin-walled and resin-encrusted, not crystal-encrusted. See also *Baeospora myosura*.

Strobilurus conigenoides
(Ellis) Singer

MAGNOLIA-CONE MUSHROOM
SYNONYMS *Agaricus conigenoides* Ellis, *Collybia conigenoides* (Ellis) Saccardo, *Pseudohiatula coniginoides* (Ellis) Singer

Small, white, hairy-capped mushroom with lined margins when expanded; growing on Magnolia fruits only

CAP 0.5–1 cm wide, convex, becoming plane, white or pale ochre-tan overall at first, but soon mostly white and then center may show tan or orange-tan, dry, minutely hairy, radiately translucent-lined to near center. **FLESH** white or pale brown, thin. **GILLS** white, barely attached or nearly free, subdistant. **STEM** 2–3 cm long, 0.5–1 mm broad, equal with tapered base, white over apex at first, but can be tan or darkening to orange-brown downward, finely downy overall, white mycelioid at base. **SPORE PRINT** white. **ODOR** not distinctive. **TASTE** mild.

Habit and habitat clustered on fallen, decaying fruits of *Magnolia*. August to November. **RANGE** widespread with its host.

Spores ellipsoid, smooth, colorless, not amyloid, 4–6 × 2–3.5 µm; pleurocystidia and cheilocystidia thick-walled, broadly ventricose with scanty encrustations at apex.

Comments The stem of these tiny mushrooms can be quite variable in color, ranging from white to tan to orange-brown.

Crinipellis setipes (Peck) Singer

SYNONYMS *Agaricus stipitarius* var. *setipes* Peck, *Collybia stipitaria* var. *setipes* (Peck) Peck, *Marasmius setipes* (Peck) G. F. Atkinson & House

Small hairy cap with a dark brown center; gills white; stem long, thin, dark brown, finely hairy overall, tough, on hardwood debris

CAP 0.5–1.5 cm wide, convex, becoming plane, sunken with a small conical central bump, yellowish-brown or tan, becoming paler buff or almost white in radiating lines from the center, surface shaggy with brown matted hairs, margin grooved and fringed shaggy with tufted hairs. **FLESH** white, very thin. **GILLS** white, free or nearly free, close, narrow. **STEM** 2–13 cm long, 0.5–3 mm broad, very slender, almost hair-like but tough, inserted into the substrate, equal, dark gray-brown or red-brown, finely hairy overall, hollow. **SPORE PRINT** white. **ODOR** and **TASTE** not distinctive.

Habit and habitat solitary but more often in small clusters on decaying hardwood debris, usually not leaves. July to October. **RANGE** widespread.

Spores: ellipsoid, smooth, colorless, not amyloid, 7–9 × 4–5 µm; cap hairs dextrinoid (turning reddish-brown) in Melzer's reagent.

Comments *Crinipellis piceae* Singer is very similar in form and color, but differs by occurring on needles of spruce with the stem attached to a distinct circular pad.

Crinipellis campanella (Peck) Singer

SYNONYM *Collybia campanella* Peck
Small rusty-brown, conical or bell-shaped cap, densely matted with rusty-brown hairs, central bump black, naked; gills white, crowded, almost free; stem dark rusty-brown, covered with tufted hairs; on conifer twigs

CAP 0.6–1.8 cm wide, conical bell-shaped, the very top sunken with a small conical bump on the center, reddish-brown or rusty-brown with pale yellow ground color, with the central bump almost black, surface shaggy with reddish-brown radially arranged and matted hairs, margin fringed with tufted hairs. **FLESH** white, very thin. **GILLS** white, free or nearly free, crowded, narrow. **STEM** 2–4 cm long, 1–1.5 mm broad, equal, tough, dark reddish-brown, short, tufted-hairy overall, inserted into the substrate. **SPORE PRINT** white. **ODOR** and **TASTE** not distinctive.

Habit and habitat solitary but more often in small clusters on decaying twigs and branches of conifers, especially arborvitae, but also on hardwood debris. August to September. **RANGE** widespread but not commonly collected.

Spores: ellipsoid, smooth, colorless, not amyloid, 7–9.5 × 4–5.5 μm; cap hairs dextrinoid (turning reddish-brown) in Melzer's reagent.

Comments This colorful species is somewhat wiry-pliant like species of *Marasmius*, but the matted hairs on the cap are distinctive. *Marasmius* species do not have hairs on the cap or stem. The red-brown reaction of the cap hairs to Melzer's reagent place the species in *Crinipellis*.

Resinomycena rhododendri
(Peck) Redhead & Singer

SYNONYMS *Agaricus rhododendri* Peck, *Omphalia rhododendri* (Peck) Saccardo, *Omphaliopsis rhododendri* (Peck) Murrill, *Marasmius resinosus* Peck

Small, slender, white, sticky-resinous on the stem, cap, and gill edges; on decaying hardwood leaves and twigs or decaying rhododendron leaves

CAP 0.5–2 cm wide, convex, becoming plane, usually sunken centrally, white or creamy-white, often corrugate-lined over margin, glabrous or finely dusted when dry, but tacky-sticky when fresh. **FLESH** white, thin. **GILLS** white, bluntly attached to subdecurrent, close, narrow, edges beaded with resin, sticky. **STEM** 1–5 cm long, 0.5–1 mm broad, equal, white, silky white mycelioid pad at base, with a mucous-like covering or, as drying out, spots with glistening resinous drops appear, very tacky. **SPORE PRINT** white. **ODOR** and **TASTE** not distinctive.

Habit and habitat scattered or clustered and sometimes in bouquetlike fashion on decaying leaves and twigs of hardwoods like oaks, beech, hickory or on rhododendron leaves. August to November. **RANGE** widespread.

Spores ellipsoid or broadly cylindrical, smooth, colorless, 5.5–8.5 × 2.5–4 µm.

Comments A small white mushroom on leaves that looks like a species of *Mycena* or perhaps *Omphalina* because of the sunken cap center. The tenacious stickiness of the stem is a dead giveaway for *Resinomycena*. If the cap is brown instead of white, and on rhododendron leaves, and has sticky stems, then it is *Resinomycena brunnescens* Redhead & Singer. If growing under conifers like larch, spruce, fir, or pine, and the cap is about 10 mm wide, the lamellae are widely spaced, and it is white and sticky, then it could be *R. acadiensis* Redhead & Singer that also has larger spores (9–13 × 4–5 µm).

Marasmius cohaerens
(Persoon) Cooke & Quélet

FUSED MARASMIUS
SYNONYMS *Agaricus cohaerens* Persoon, *Mycena cohaerens* (Persoon) Gillet

Bouquetlike clusters of medium-sized mushrooms on leaf litter; cap reddish-brown with pale bloom at first; gills are pinkish-tan, becoming brown; stem pale above and black and shiny below, with cottony-covered base

CAP 1–6.5 cm wide, conical, convex, or bell-shaped, then plane with uplifted margin, dark red-brown or flesh-pink or yellow-brown, when fresh with a bloom that is easily rubbed off, glabrous, smooth or somewhat wrinkled, dry. **FLESH** brown, turning pallid, thin. **GILLS** pale pinkish-tan becoming brown, attached, subdistant, broad. **STEM** 2–8 cm long, 1–3 mm broad, equal and often curved, buff or yellowish-tan at first, but turning black or blackish-brown from the base upward, shiny, dry, glabrous, tough, with copious cottony white or tan-colored mycelium covering the base. **SPORE PRINT** white. **ODOR** not distinctive, or unpleasant for some. **TASTE** not distinctive, or slightly bitter or acidic.

Habit and habitat mostly clustered in bouquetlike clumps on leaf litter and woody debris in hardwood forests. July to October. **RANGE** widespread.

Spores elongate ellipsoid, smooth, colorless, not amyloid, 6–11 × 3–5.5 μm; with abundant thick-walled, dark brown setae on the gills.

Comments If you find a variant with the gills narrow and crowded together, it can be referred to as *Marasmius cohaerens* var. *lachnophyllus* (Berkeley) Gilliam. Several other varieties have also been described. Compare with *Connopus acervatus* that forms bouquetlike clusters and can be found in similar habitats, but it has white then pinkish-colored gills, and the stem is reddish or purplish-brown and not becoming black.

Marasmius fulvoferrugineus
Gilliam

Small horse-hairlike black-stemmed mushrooms with dark reddish or orange-brown scalloped caps; gills pale cream and widely spaced

CAP 2–4.5 cm wide, bell-shaped becoming conical or convex, often with barely sunken center with small rounded bump, deeply corrugated or scalloped to center and often also wrinkled with age, dark reddish-brown or cinnamon-brown or cinnamon, glabrous, dry. **FLESH** white, thin. **GILLS** pale creamy-white, free, distant, broad. **STEM** 5–8 cm long, 0.5–1 mm broad, equal, very straight, pale yellow at apex, black or blackish-brown elsewhere, shiny, dry, glabrous, horse-hairlike, hollow. **SPORE PRINT** white. **ODOR** and **TASTE** farinaceous.

Habit and habitat scattered or clustered on leaf litter and woody debris in hardwood forests. July to October. **RANGE** more common in southeastern North America, but occurs in the Northeast as well.

Spores elongate and lance-shaped, smooth, colorless, not amyloid, 15–18 × 3–4.5 µm; cystidia rare or lacking.

Comments *Marasmius siccus* (Schweinitz) Fries with its orange or yellowish-orange caps is generally found in the northern areas, while *M. fulvoferrugineus* has more darkly reddish-brown or orange-brown cap colors and is mainly in the southeastern area. *Marasmius siccus* has pleurocystidia, *M. fulvoferrugineus* lacks these sterile cells or they are very rare.

If the cap and stem have delicate pinkish or purplish-pink colors, then you have *Marasmius pulcherripes* Peck.

Marasmius siccus

Marasmius bellipes Morgan

Cap small, deep purple-red, corrugate-lined; gills white or pale pinkish and widely spaced; stems horse-hairlike, smooth, deep purplish-red

CAP 0.4–1.5 cm wide, conical bell-shaped at first and strongly corrugate-lined, becoming convex, center with small bump and strongly wrinkled, deep purplish-red, eventually fading to reddish-brown or light brown, glabrous but minutely velvety, dry. **FLESH** white, thin. **GILLS** white or pinkish, eventually pale yellow, barely attached or free, distant, broad. **STEM** 2–4 cm long, 0.3–0.75 mm broad, equal, deep purplish-red overall at first, turning yellowish-brown or grayish-brown with age, shiny, dry, glabrous, horse-hairlike. **SPORE PRINT** white. **ODOR** and **TASTE** not distinctive.

Habit and habitat scattered or clustered on leaf litter in hardwood forests. June to September. **RANGE** widespread.

Spores elongate and narrowly club-shaped or canoe-shaped, often curved, smooth, colorless, not amyloid, 8.5–12 × 3–3.5 µm; pleurocystidia hyaline, thin-walled, variously shaped but often with rounded subcapitate apex, scattered.

Comments *Marasmius pulcherripes* Peck is more commonly collected and somewhat similar in appearance, except the colors are pinkish or reddish-brown on the cap, fading to yellow hues, and pink or grayish-red at first on the stem, not deep purple-red overall as seen in *M. bellipes.* The spores are also longer, 11–16 µm for *M. pulcherripes.*

Marasmius pallidocephalus
Gilliam

Small, cap yellow-brown or pinkish-brown, bumpy; gills buff, widely spaced; stem red-brown, horse-hairlike, with black rhizomorphs radiating from base; on conifer needles

CAP 0.2–1.4 cm wide, convex, becoming plane, center shallowly sunken, brown at first but soon yellow-brown or pale pinkish-brown or pale orange-brown, darker on the center, glabrous, somewhat bumpy, slightly slippery when wet or dry. **FLESH** white, thin. **GILLS** light buff, eventually pale yellow-brown, barely attached or free, subdistant, broad. **STEM** 1–4 cm long, 0.2–0.8 mm broad, equal, mostly red-brown, shiny, dry, glabrous, horse-hairlike, with black rhizomorphs (tough shoe-stringlike structures). **SPORE PRINT** white. **ODOR** and **TASTE** not distinctive.

Habit and habitat in small or large clusters on needles of conifers, especially spruce. June to October. **RANGE** widespread.

Spores narrowly elliptical or apple-seed-shaped, smooth, colorless, not amyloid, 6–10 × 2.5–4 µm; hymenial cystidia absent.

Comments *Gymnopus androsaceus*, formerly *Marasmius androsaceus*, can look very similar to *M. pallidocephalus* once the cap colors of *G. androsaceus* have begun to fade. However, for *G. androsaceus*, the center of the cap usually remains reddish-brown, the gills are pinkish-brown and distant, the stems are black, and the black rhizomorphs at the base of the stems are prolific.

Marasmius oreades (Bolton) Fries

FAIRY-RING MUSHROOM

SYNONYMS *Agaricus oreades* Bolton, *Collybia oreades* (Bolton) P. Kummer

Cap yellowish or grayish-brown or tan, often with broad bump; gills off-white, subdistant; stem covered with yellowish to reddish-brown fibrils, straight, rubbery-tough; in grassy areas

CAP 1–5 cm wide, bell-shaped, becoming convex then plane, often with low rounded central bump, margin uplifted and wavy with age, dark brown in button stages, becoming pale tan or yellowish-brown or grayish-brown, hygrophanous and then pale yellowish-cream or almost white, glabrous, smooth or pitted, dry. **FLESH** white or pale cream, thick and firm. **GILLS** off-white or pale tan, barely attached, close or subdistant, broad. **STEM** 2–8 cm long, 2–6 mm broad, equal, straight and tough, pale cream with yellowish-brown or reddish-brown colors over lower portions, dry, covered with matted fibrils, solid or stuffed, basal mycelium inconspicuous. **SPORE PRINT** white. **ODOR** not distinctive, or pungent. **TASTE** not distinctive.

Habit and habitat clustered in circles, fairy rings, or in scattered troops, on lawns and grassy areas in parks, playgrounds, meadows, and so on. May to October. **RANGE** widespread.

Spores canoe-shaped or narrowly ellipsoid, smooth, colorless, not amyloid, 7–10 × 4–6 μm.

Comments This mushroom does not have a lot of flesh on it, but the sweet taste of the cap is apparently very good when used as a flavoring and thus sought after by mushroom hunters for food. Since it occurs on lawns, be sure not to collect and eat the poisonous *Clitocybe dealbata* or any of the lawn-loving, brown-spored *Inocybe* species.

Marasmius strictipes (Peck)
Singer

ORANGE-YELLOW MARASMIUS
SYNONYMS *Collybia strictipes* Peck, *Gymnopus strictipes* (Peck) Murrill

Cap mostly two-toned orange-yellow and cream, with wrinkled center; gills pale yellow crowded; stem white, straight, tough

CAP 2–6.5 cm wide, convex becoming plane, often with low rounded central bump and uplifted margin, yellow or orange-yellow at first, hygrophanous and becoming pale yellowish-cream over the margin, but remaining darker and often orange or ochre-yellow over the center, glabrous but minutely wrinkled over the center at least, dry. FLESH pale yellow or white, thin. GILLS pale pinkish-buff or pale yellow tending to white, attached, crowded, broad. STEM 3–9 cm long, 3–9 mm broad, equal, straight and tough, white or pale yellow, dry, finely dusted or with matted fibrils, hollow, with abundant pale yellow or white basal mycelium binding the substrate. SPORE PRINT white. ODOR not distinctive. TASTE not distinctive, or radishlike.

Habit and habitat single, scattered, or clustered on leaf litter and woody debris in hardwood forests. June to October. RANGE widespread.

Spores apple-seed-shaped or narrowly ellipsoid, smooth, colorless, not amyloid, 6–10.5 × 3–4.5 μm.

Comments Compare with *Marasmius oreades* that tends to have a similar stiff-spined look but grows on lawns mainly and the colors are darker or at least more muted than the yellow or orange-yellow caps of *M. strictipes*.

Marasmius rotula (Scoparius)
Fries

PINWHEEL MARASMIUS

SYNONYM *Agaricus rotula* Scoparius

In troops on wiry black stems; cap white, radially corrugated with dark brown bump centrally; gills creamy-white, attached to a collar, distant; on decaying hardwood stumps, logs, and woody debris

CAP 0.2–1.5 cm wide, broadly convex or shaped like a foot cushion, sunken centrally and with low rounded central bump, deeply radially corrugate to center, pale yellowish or grayish-brown in very young stages, but quickly white except for brown on the central bump, appearing glabrous but minutely granular under a lens, dry. FLESH white, thin. GILLS off-white or pale cream, attached to a collar that is free from the stem, distant, broad. STEM 1.5–8.5 cm long, 0.5–1 mm broad, equal, straight and tough, pale cream over apex, brown or black from base upward, dry, glabrous, shiny, with black threads (rhizomorphs) sometimes present. SPORE PRINT white. ODOR not distinctive. TASTE not distinctive, or bitterish.

Habit and habitat scattered or often in dense clusters on decaying hardwood logs, stumps, branches. May to October. RANGE widespread.

Spores narrowly ellipsoid or apple-seed-shaped, smooth, colorless, not amyloid, 6–10 × 3–4.5 μm.

Comments These small, rather tough *Marasmius* species are able to shrivel in dry weather and then revive when it is wet and produce more spores. They seem to bloom overnight, like little white flowers, after a rain or when the humidity is up, enabling them to produce more sexual spores. *Marasmius capillaris* Morgan is a smaller version with a thinner stem that only grows on decaying oak leaves.

Mycetinis scorodonius (Fries)
A. W. Wilson & Desjardin

GARLIC MARASMIUS

SYNONYMS *Agaricus scorodonius* Fries, *Gymnopus scorodonius* (Fries) J. L. Mata & R. H. Petersen, *Marasmius scorodonius* (Fries) Fries

Cap small, reddish-brown but soon yellowish-brown to pale buff or pinkish-buff; gills white or pale pinkish-cream; stem off-white above, reddish-brown elsewhere; strong odor of garlic

CAP 0.5–3 cm wide, convex or broadly bell-shaped, then plane with upturned wavy margin, often with barely sunken center with a small rounded bump, reddish-brown then pale brown at first, soon pinkish-yellow or pale buff, glabrous, dry, may be slightly wrinkled. **FLESH** white, thin. **GILLS** white or pale pinkish-cream, bluntly attached or free, subdistant, broad. **STEM** 2–6 cm long, 0.5–3 mm broad, equal, white or pale yellow over apex, reddish-brown elsewhere, progressively darker with age, dry, glabrous, shiny like a polished cow horn. **SPORE PRINT** white. **ODOR** and **TASTE** of garlic or onion.

Habit and habitat scattered or clustered on needles and woody debris under conifers, also occurring on dead grass stems. July to October. **RANGE** widespread.

Spores ellipsoid or apple-seed-shaped, smooth, colorless, not amyloid, 6–10 × 3–5 µm; cheilocystidia with finger-like projections.

Comments Based on phylogenetic studies using DNA molecules, this species was removed from *Marasmius* along with a few other garlic-emitting and related species and placed in a separate genus, *Mycetinis*. *Marasmius olidus* Gilliam also produces an odor of garlic or onion, but differs by occurring on oak leaves and the gills and cap are yellowish-brown.

Marasmius sullivantii
Montagne

Cap bright reddish-orange; gills white with pinkish edges when young, barely attached, close; stem hornlike, white becoming reddish-orange or darker downward; on hardwood leaf litter

CAP 0.5–2.5 cm wide, convex, becoming plane, bright reddish-orange or rusty-brown, fading to tawny with age and expansion of cap, glabrous, dry, may be faintly lined over margin. **FLESH** white, thin. **GILLS** white, barely attached or becoming free, close, narrow. **STEM** 0.5–4 cm long, 1–1.5 mm broad, equal, rigid and hornlike, white or pale cream at first, eventually reddish-orange and black in the lower parts, dry, glabrous, covered with fine hairs at first, white cottony at base. **SPORE PRINT** white. **ODOR** not distinctive. **TASTE** mild or slightly bitter.

Habit and habitat single, scattered, or clustered on leaf litter in hardwood forests. July to September. **RANGE** widespread but not common.

Spores ellipsoid, smooth, colorless, not amyloid, 7–9 × 3–4 μm; cheilocystidia with fingerlike projections, pleurocystidia variously shaped.

Comments A beautiful small species that can be found occasionally in hardwood forests of the Northeast. It would not be confused with any other species because of its size, colors, and hornlike stem.

Mycena acicula (Schaeffer) P. Kummer

SYNONYMS *Agaricus aciculus* Schaeffer, *Hemimycena acicula* (Schaeffer) Singer, *Marasmiellus acicula* (Schaeffer) Singer, *Trogia acicula* (Schaeffer) Corner

Small, spindly, fragile mushroom with bright coral-red cap and lemon-yellow stem; stem at first white powdery pubescent-covered; on leaves and twigs under hardwoods

CAP 0.3–1 cm wide, ovate, becoming bell-shaped, margin flaring with age, white-frosted at first over the dominant dark red color, sometimes yellowish at margin with age and slowly fading to bright orange-yellow, translucent-lined when expanded, glabrous, dry. **FLESH** yellow, thin. **GILLS** pale orange or white or tinged yellowish near the cap, bluntly attached, close or subdistant, moderately broad. **STEM** 1–6 cm long, up to 1 mm broad, equal, densely white powder coated over pale lemon-yellow or orange-yellow, becoming glabrous, dry, with white hairy mycelioid base. **SPORE PRINT** white. **ODOR** and **TASTE** not distinctive.

Habit and habitat single, scattered, or clustered on decaying leaves in wet areas under hardwoods. May to September. **RANGE** widespread.

Spores narrowly canoe-shaped or spindle-shaped (tapered at both ends), smooth, colorless, not amyloid, 9–11 × 3.5–4.5 µm.

Comments They look like tiny red torches on the dark forest floor. One might confuse them with small red hygrocybes, but the spindly stem that is covered with a fine powdery fuzz readily separates this tiny fragile mushroom from any wax cap.

Mycena sanguinolenta
(Albertini & Schweinitz) P. Kummer

SYNONYM *Agaricus sanguinolentus* Albertini & Schweinitz

Dark reddish-brown, very delicate and fragile, producing reddish or reddish-brown juice when injured; gills with dark reddish-brown edge; on conifer needles, moss beds

CAP 0.3–1 cm wide, convex or conical-convex, occasionally bell-shaped, faintly translucent-lined at first, but soon radially ridged from margin to near center, pale grayish hoary-covered at first then glabrous, reddish-brown or vinaceous with a darker center, moist or dry. **FLESH** red, exuding a red juice when cut. **GILLS** red or grayish-red, edges dark red-brown, bluntly attached and some with decurrent tooth, subdistant or distant, moderately broad. **STEM** 2–6 cm long, 1–1.5 mm broad, equal and tubular, very fragile, colored as the cap, white fine hairs over the base, elsewhere with fine white dots, or pruina, eventually glabrous, dry, hollow. **LATEX** produced from cap, gills, and stem as red or red-brown juice when cut. **SPORE PRINT** white. **ODOR** and **TASTE** not distinctive.

Habit and habitat mostly scattered or clustered on conifer needles, moss beds, and decaying leaves. July to September. **RANGE** widespread.

Spores elongate ellipsoid, smooth, colorless, weakly amyloid, $8–10 \times 4–6$ µm; cheilocystidia abundant, fusiform or ventricose with a short tapered neck, hyaline.

Comments *Mycena sanguinolenta* is one of at least three "bleeding" *Mycena* species. *Mycena atkinsoniana* A. H. Smith is larger and occurs on beech or oak leaves. *Mycena haematopus*, whose latex is more purplish-red and bloodlike, grows on dead hardwood debris and is the most common of the three.

Mycena atkinsoniana

Mycena rutilantiformis
(Murrill) Murrill

SYNONYM *Prunulus rutilantiformis* Murrill

Medium-sized mushrooms with purplish-brown hygrophanous caps; gills dark purplish-red, the edges darker than the faces; stem yellow with scattered purplish fibrils over apex; odor and taste of radish

CAP 2–7 cm wide, convex, becoming broadly convex, often with low broad central bump, translucent-lined over margin, dark reddish-brown or purplish-brown, hygrophanous and becoming paler yellowish-brown from center outward, glabrous, moist or slight slippery, surface uneven. **FLESH** yellowish or white, moderately thick. **GILLS** purplish-red, edges more darkly colored than the faces, bluntly attached, close or subdistant, often cross-veined, broad. **STEM** 3–8 cm long, 5–10 mm broad, equal, dark golden-yellow, with scattered purplish fibrils over apex, otherwise glabrous, but may be longitudinally lined or ridged, dry, hollow and somewhat fragile. **SPORE PRINT** white. **ODOR** of radish. **TASTE** radishlike or bitter.

Habit and habitat single, scattered, or clustered on duff and decaying debris under hardwoods and conifers. June to September. **RANGE** widespread.

Spores ellipsoid or ovoid, smooth, colorless, amyloid, 8–10 × 4–5 µm; cheilocystidia and pleurocystidia obvious, ovoid or fusoid with reddish-brown content.

Comments *Mycena pelianthina* (Fries) Quélet is only vaguely similar in its radish odor and its appearance with purplish-brown or pale lilac-gray-brown lamella with dark purple edges. However, the cap is pale lilac-brown fading to ochre or beige and the stem is mostly white or pale lilac-brown with darker purplish flattened fibrils, lacking any hint of golden-yellow color. The spores are also smaller (5.5–7 × 3–3.5 µm). *Mycena pelianthina* is apparently a rare species in North America.

Mycena pura (Persoon)
P. Kummer

SYNONYMS *Agaricus purus* Persoon, *Mycena ianthina* (Fries) P. Kummer

Most often lilac or purplish, but variable in color from reddish to white; gills often developing cross veins; odor and taste of radish

CAP 2–6 cm wide, oval, then convex or bell-shaped, becoming plane, often with low broad central bump, translucent-lined and also radially lined or irregularly ridged from margin to near the center, color variable, often lilac or purplish, but reddish, lilac-gray, yellowish or white with faint bluish tints also develop, hygrophanous and becoming paler, glabrous, moist or dry, surface uneven. **FLESH** watery gray or white, thin. **GILLS** pale grayish-lilac, but can be pinkish or pale purple or white, bluntly attached, subdistant, developing cross veins with age, broad. **STEM** 4–10 cm long, 2–6 mm broad, equal, same color as cap or nearly white, glabrous, dry, hollow, sparse white mycelioid at base. **SPORE PRINT** white. **ODOR** and **TASTE** radishlike.

Habit and habitat single, scattered, or clustered on duff and decaying debris under hardwoods and conifers. June to September. **RANGE** widespread.

Spores elongate ellipsoid or cylindrical, smooth, colorless, amyloid, 6–10 × 3–3.5 μm; cheilocystidia and pleurocystidia large, ventricose with long broad tapered tips or inflated saccate, hyaline.

Comments Highly variable in color, but the odor and taste will help in the determination. If the cross veins have developed between the lamellae, this is also a good feature to help confirm the identification.

Mycena rosella (Fries)
P. Kummer

SYNONYMS *Agaricus rosellus* Fries, *Mycena rosea* Saccardo

On conifer needles with flesh-pink or salmon-pink caps with darker central bump; gills pale rose with darker rose-red edges; stem pale rose with brown tints

CAP 0.5–2 cm wide, oval becoming broadly conical or bell-shaped or convex, often with a central broad, low bump, moderately lined or radially ridged to near center with age, also translucent-lined when fresh, flesh-pink or salmon-pink or bright pink with ochre-brown on the center, fading with age to yellowish hues, glabrous, moist or slightly slippery when fresh. **FLESH** sordid pink or white, thin. **GILLS** pale rose, edges dark rose-red, bluntly attached, close or subdistant, moderately broad. **STEM** 2–7(–8) cm long, 1–2.5 mm broad, equal, pale rose or grayish-rose or with brown tints, somewhat translucent, slightly slippery but not slimy, glabrous. **SPORE PRINT** white. **ODOR** and **TASTE** not distinctive.

Habit and habitat usually clustered on needle duff under conifers, pines, spruce in particular. August to October. **RANGE** widespread.

Spores ellipsoid, smooth, colorless, amyloid, 7–9 × 4–5 μm; cheilocystidia club-shaped or fusoid and covered with short rodlike projections.

Comments This mushroom can fruit in large numbers on conifer needles in the fall. The dingy pink caps are easy to spot on the forest floor.

Sphagnurus paluster (Peck)
Redhead & V. Hofstetter

SYNONYMS *Agaricus paluster* Peck, *Collybia palustris* (Peck) A. H. Smith, *Tephrocybe palustris* (Peck) Donk, *Tephrophana palustris* (Peck) Kühner

On *Sphagnum* beds, cap grayish with translucent lines; gills white or pale grayish; stem long, hollow and very fragile

CAP 1–3 cm wide, conical or bell-shaped, then plane with small central bump, margin translucent-lined, grayish or brownish-gray or olive-brown, often with fine white pruina (very small granules) over the margin, hygrophanous and becoming pale gray, smooth, moist, becoming dry, glabrous. **FLESH**, gray, thin, watery soft. **GILLS** whitish or pale gray, bluntly attached with short decurrent tooth, close, broad. **STEM** 2–10 cm long, 1–5 mm broad, equal, pale grayish-brown or gray, with fine white pruina or fibrils overall, becoming glabrous, dry, hollow, very fragile. **SPORE PRINT** white or pale cream. **ODOR** and **TASTE** farinaceous.

Habit and habitat scattered or clustered on *Sphagnum* in bogs, fens and anywhere this moss grows in large beds, parasitizing the moss. July to November. **RANGE** widespread.

Spores ovate or ellipsoid, smooth, colorless, not amyloid, 5–6 × 3–3.5 µm.

Comments If you wish to see this species, you will have to find a good-sized patch of *Sphagnum*, either in a fen or a bog would be your best chances. Because it is parasitizing the sphagnum mosses, occasionally, if the infection is well established, circular "bleached" patches on large sphagnum mats will be visible. Often *Galerina paludosa* is also present on these same moss mats, at least in the Adirondack Park area.

Tetrapyrgos nigripes (Fries)
E. Horak

BLACK-FOOTED MARASMIUS

SYNONYMS *Marasmius nigripes* Fries, *Heliomyces nigripes* (Fries) Morgan, *Marasmiellus nigripes* (Fries) Singer, *Pterospora nigripes* (Fries) E. Horak

Small white, wrinkled cap; gills white distant; stem mostly black but frosted with tiny white hairs; on twigs

CAP 0.5–2 cm wide, convex, becoming plane, some with a broad central bump, often wrinkled-pleated, white, dry, glabrous or faintly powdery dusted, opaque. **FLESH** white, thin but rubbery. **GILLS** white, sometimes staining red, attached or subdecurrent, distant or subdistant, broad, forking and with cross veins. **STEM** 2.5–5 cm long, 1–1.5 mm broad at apex, equal, naked at insertion point on the substrate, white overall at first or only at very apex with age, otherwise black, fine white fibrillose coated overall, dry, flexible-tough. **SPORE PRINT** white. **ODOR** and **TASTE** not distinctive.

Habit and habitat scattered or clustered on leaf litter and twigs in hardwood forests. August to October. **RANGE** widespread.

Spores star or jax-shaped (the game) with 3–5 arms, smooth, colorless in KOH, 8–9 µm; cheilocystidia obvious and with numerous fingerlike projections over apex.

Comments The white caps might catch your eye first, since these are small inconspicuous mushrooms. The cap is slightly rubbery and the stem is flexible, not breaking easily. The genus is more common in the tropics.

Xeromphalina campanella

(Batsch) Kühner & Maire

FUZZY FOOT

SYNONYMS *Agaricus campanellus* Batsch, *Omphalia campanella* (Batsch) P. Kummer, *Omphalina campanella* (Batsch) Quélet, *Omphaliopsis campanella* (Batsch) O. K. Miller

Small orange-brown cap with central depression and radial lines over margin; gills yellow and decurrent; stem on older ones dark brown with fuzzy orange tufted hairs at base; on decaying conifer wood

CAP 0.5–2 cm wide, convex, center shallowly sunken, orange-brown, yellowish-brown, rusty, or cinnamon-brown, darker on the center, dry, glabrous, margin radially lined with age. **FLESH** thin. **GILLS** pale yellow or orange-yellow, decurrent, distant, with cross veins common, broad. **STEM** 1–5 cm long, 0.5–3 mm broad, equal but with bulb at base, pale watery yellow overall, soon with the lower stem dark brown or reddish-brown, progressively dark reddish-brown to black upward, glabrous, wiry tough, often curved, base with dense orange-brown or yellow long tufted hairs. **SPORE PRINT** pale buff. **ODOR** and **TASTE** not distinctive.

Habit and habitat typically densely clustered on decaying logs, stumps and standing dead trees of conifers, favoring hemlock. July to October. **RANGE** widespread.

Spores ellipsoid, smooth, colorless in KOH, amyloid (faintly so because the spore walls are so thin), 5.5–7 × 3–4.5 µm; cheilocystidia obvious, abundant, inflated.

Comments This is the most commonly collected species of *Xeromphalina* and often occurs in large numbers on well-decayed hemlock stumps. *Xeromphalina kauffmanii* A. H. Smith is similar but on hardwood stumps with bright rusty-colored caps. *Xeromphalina cauticinalis* (Fries) Kühner & Maire has long stems and is found scattered on the ground attached to needles and woody debris under conifers. Both have the characteristic orange tufted hairs at the base of the stem. *Xeromphalina tenuipes* (Schweinitz) A. H. Smith produces larger fruit bodies (cap 2–5 cm), with tough, densely hairy, golden orange-brown stems and grows scattered on decaying hardwood logs and debris in late spring or early summer.

Mycena haematopus
(Persoon) P. Kummer

BLEEDING MYCENA

SYNONYM *Agaricus haematopus* Persoon

Cap red-brown with paler, clearly toothed or scalloped margin; gills white, staining or spotting reddish-brown; stem when nicked producing purplish-red juice; mostly in bouquetlike clusters on wood

CAP 1–5 cm wide, oval becoming convex or bell-shaped, then convex or plane, some with broad central bump, margin a sterile toothed or scalloped band of tissue, faintly translucent-lined at first, soon radially ridged from margin to center, white hoary, then glabrous, red-brown or vinaceous-brown, becoming paler toward the margin, moist or dry. **FLESH** pallid or of cap color, exuding a purplish-red juice when cut. **GILLS** white, staining reddish-brown, edges white cottony, barely attached, sub-distant, broad. **STEM** 2–10 cm long, 1–2 mm broad, equal, colored as the cap or with purplish hues, at first densely coated with tiny hairs, eventually glabrous, dry, hollow.

LATEX produced from cap, gills and stem as purplish-red juice when cut. **SPORE PRINT** white. **ODOR** not distinctive. **TASTE** not distinctive, or slightly bitter.

Habit and habitat mostly clustered in bouquetlike tufts or scattered on fairly well-decayed, hardwood logs, these often covered with moss. July to September. **RANGE** widespread.

Spores broadly ellipsoid, smooth, colorless, amyloid, 8–11 × 5–7 μm; cheilocystidia abundant, fusiform or ventricose with a long tapered neck, hyaline.

Comments There are two other commonly collected reddish latex-producing *Mycena* species found growing on the ground. *Marasmius atkinsoniana* is found on decaying hardwood leaves, especially beech, while *M. sanguinolenta*, a much smaller species, grows on needle beds of conifers or mosses. Also *M. haematopus* var. *marginata* J. E. Lange that also occurs on wood has reddish-edged (marginate) gills, but it is infrequently encountered.

Mycena subcaerulea (Peck) Saccardo

SYNONYMS *Agaricus subcaeruleus* Peck, *Mycena cyaneobasis* Peck, *Mycena cyanothrix* G. F. Atkinson

Small mushrooms on decaying hardwood logs and debris with pale blue or greenish-blue caps and bright blue mycelioid bases when fresh and moist

CAP 0.5–2.5 cm wide, oval then broadly conical, bell-shaped, becoming plane, variable in color, when young with pale blue or greenish-blue tints, or mixed with brown or grayish-brown, becoming sordid yellowish with age, but blue tints lingering on the margin, translucent-lined when fresh and expanded, glabrous, slippery at first. **FLESH** pallid, thin. **GILLS** white or tinged gray, barely attached, close or crowded, moderately broad. **STEM** 3–8 cm long, 1–2 mm broad, equal, pale blue or greenish-blue over the apex, otherwise translucent gray or pale brown, densely coated with minute pubescence, dry, hollow, elastic, with bright blue mycelioid base at first, this color soon fading to white. **SPORE PRINT** white. **ODOR** and **TASTE** mild.

Habit and habitat single, scattered, or clustered on duff and decaying debris under hardwoods, especially beech, oak, basswood and elm. June to September. **RANGE** widespread.

Spores globose or subglobose, smooth, colorless, amyloid, 6–8 × 6–7 μm.

Comments The delicate blue is beautiful, but disappears quickly as the mushrooms loses moisture. *Mycena amicta* (Fries) Quélet is similar-colored species but found on conifer needles, debris and wood, and the spores are ellipsoid, not globose.

Mycena galericulata (Scopoli) Gray

COMMON MYCENA

SYNONYMS *Agaricus galericulatus* Scopoli, *Collybia rugulosiceps* Kauffman

Clustered on hardwood logs and stumps, cap broadly conical, brown then nearly white with brown center; gills white, becoming pink; stem white above, brown below, glabrous

CAP 1–6 cm wide, conical or bell-shaped with broad central bump, brown at first, fading to pale tan or tan on disc nearly white elsewhere, glabrous, surface uneven, radially lined or bumpy, margin becoming frayed, dry or slight tacky when wet. **FLESH** white, thin. **GILLS** white becoming uniformly pink, barley attached, subdistant, broad. **STEM** 5–9 cm long, 2–5 mm broad, equal and sometimes with radicating base, white above, tan or brown below, dry, glabrous, hollow. **SPORE PRINT** white. **ODOR** and **TASTE** not distinctive, or slightly farinaceous.

Habit and habitat scattered or clustered on well-decayed hardwood logs, stumps, or from buried wood. May to October. **RANGE** widespread.

Spores ellipsoid, smooth, colorless, amyloid, 8–10 × 5.5–7 µm; cheilocystidia abundant, club-shaped or swollen on the top and covered with rodlike projections, hyaline.

Comments *Mycena inclinata* is similar but the cap margin has distinct rounded "teeth" when young and typically develops yellow stains or discolorations on the stem and cap. Also, the stem base of *M. inclinata* is covered with white clumps of fibrils when young, while *M. galericulata* has a bald or glabrous stem.

Mycena epipterygia var. lignicola A. H. Smith

YELLOW-STALKED MYCENA

SYNONYMS *Agaricus epipterygius* Scopoli, *Mycena citrinella* (Persoon) P. Kummer, *Prunulus epipterygius* (Scopoli) Murrill

On conifer wood with bright yellow-green cap and stem; stem slippery or slimy when fresh

CAP 1–2 cm wide, oval becoming broadly conical or bell-shaped, subtly lined or radially ridged to near center with age, bright greenish-yellow or grayish-green with darker olive center, fading with age to almost white, glabrous, slippery when fresh. **FLESH** yellowish, thin. **GILLS** white or faintly yellowish, barely attached, subdistant, narrow. **STEM** 6–8 cm long, 1–2 mm broad, equal, bright lemon-yellow, fading to white slowly when exposed, slimy, glabrous. **SPORE PRINT** white. **ODOR** and **TASTE** farinaceous.

Habit and habitat scattered or clustered on decaying conifer wood. July to October. **RANGE** widespread.

Spores ellipsoid, smooth, colorless, amyloid, 9–13 × 5.5–8 μm; cheilocystidia club-shaped and covered with rod-like projections.

Comments If you find a similar-looking species but it is on the ground in coniferous forests, it would be *Mycena epipterygia* var. *epipterygia*. This variety lacks an odor and generally has some brown color in the cap.

Mycena leaiana (Berkeley) Saccardo

ORANGE MYCENA

SYNONYMS *Agaricus leaianus* Berkeley, *Collybia leaiana* (Berkeley) Fairman, *Prunulus leaianus* (Berkeley) Murrill

Small or moderately sized bright orange mushrooms in bouquetlike clusters on downed logs of hardwoods; gills orange-salmon color with darker reddish-orange edges

CAP 1–5 cm wide, convex or becoming bell-shaped, typically translucent-lined to near center, rich orange or reddish-orange or yellowish-orange, slowly fading to pale orange or nearly white, glabrous, slippery at first, drying shiny. **FLESH** white or watery yellow. **GILLS** light orange-salmon, staining orange-yellow when bruised, edges bright reddish-orange, bluntly attached or notched, close or crowded, broad. **STEM** 3–7 cm long, 2–4 mm broad, equal, evenly pale lemon-yellow but covered moderately with orange or orange yellow pubescent dots or powder, slippery at first, with orange pubescent mycelioid base, exuding a thin orange juice when cut. **SPORE PRINT** white (but staining white paper orange from pigments in the gill tissues). **ODOR** and **TASTE** not distinctive, or slightly farinaceous.

Habit and habitat in typically dense bouquetlike clusters on decaying hardwood logs, especially beech in our area, but can be found on many different species of hardwoods. June to October. **RANGE** widespread.

Spores ellipsoid, smooth, colorless, amyloid, 7–10 × 5–6 µm.

Comments This species fruits very early in the season and continues to fruit into the fall. Because of the bright orange colors, it is easily seen in the dimly lit forests. Not recommended for eating, it produces bioactive compounds, leainafulvenes, that are cytotoxic, antibacterial, and mutagenic.

Omphalotus illudens
(Schweinitz) Bresinsky & Basl

JACK O'LANTERN MUSHROOM

SYNONYMS *Agaricus illudens* Schweinitz, *Clitocybe illudens* (Schweinitz) Saccardo, *Lentinus illudens* (Schweinitz) Hennings, *Panus illudens* (Schweinitz) Fries

Large bright orange mushroom growing in bouquetlike clusters from fused stem bases; gills orange, crowded, decurrent; found on hardwoods stumps, at the base of standing trees, or from buried wood, especially oak

CAP 3–20 cm wide, broadly convex with low broad central bump, becoming plane and eventually somewhat funnel-shaped with age, bright orange or yellowish-orange, becoming dull orange with some brown hues in age, glabrous, moist or slightly greasy or dry, smooth and opaque. **FLESH** orange, thick. **GILLS** bright orange or becoming paler orange, sometimes beaded with orange liquid drops, decurrent, close or crowded. **STEM** 9–20 cm long, 10–40 mm broad, equal or often tapered downward, orange or pale orange, longitudinally fibrous-striate, dry. **SPORE PRINT** creamy-white. **ODOR** and **TASTE** not distinctive.

Habit and habitat densely clustered in bouquetlike fashion with the stem bases fused, on various hardwood trees, especially oak, right at the base of standing trees (usually dead ones or stumps), also from buried wood. July to November. **RANGE** widespread.

Spores globose or subglobose, smooth, colorless, not amyloid, 3–5 μm.

Comments If you find a bouquet of these poisonous mushrooms that cause nasty gastrointestinal distress, do not eat them, but do take them to the bedroom. Put them on aluminum foil on your dresser or where you can see them when you wake up in the middle of the night. Point the gills in your direction—they glow a soft greenish light called bioluminescence, basically the same light put out by fireflies. Compare with the American golden chanterelle, *Cantharellus flavus*, which is a delicacy, but do not make the mistake many have by confusing the two.

Flammulina velutipes (Curtis) Singer

VELVET FOOT

SYNONYMS *Agaricus velutipes* Curtis, *Collybia velutipes* (Curtis) P. Kummer, *Gymnopus velutipes* (Curtis) Gray, *Pleurotus velutipes* (Curtis) Quélet

Cap orange-brown or yellow-brown, slimy; stem yellow above, orange-brown below and densely velvety, becoming almost black; often on dead standing hardwoods; late in the season and fruiting in winter

CAP 2–5(–7) cm wide, convex, becoming plane with or without a broad central bump, dark reddish-brown or orange-brown or yellow-brown, paler yellowish around the margin, colors fading with age, surface slippery or slimy becoming tacky, glabrous, the skin is tough and somewhat elastic-rubbery. FLESH white or pale yellow, thin. GILLS white or pale peachy-yellow, attached or barely attached, close, broad. STEM 2–8(–11) cm long, 3–5(–10) mm broad, equal, tough-pliant, pale yellow over upper areas, darker orange-brown over lower areas, surface covered with a fine velvety coating that becomes heavier and rusty-brown to black over lower areas. SPORE PRINT white. ODOR and TASTE not distinctive.

Habit and habitat single but more often in small clusters on decaying hardwood logs, especially dead standing elms, poplars, and willows. October to May. RANGE widespread.

Spores long elliptical, smooth, colorless, not amyloid, 7–9(–11) × 3–6 μm; cheilocystidia present, variously shaped.

Comments Also called the winter mushroom since it fruits late in the fall, usually once the weather turns cold at night. It can also be found on a winter's day during a brief thaw, usually on dead elms. It is an edible species that is commercially grown and sold in stores as enoki, enokitake, or enokidake. The commercially grown form is white or cream-colored entirely, with long stems and tiny caps, resembling noodles. The flavor of this commercially produced mushroom is very good. If you wish to use this wild mushroom for food, please learn to identify *Galerina marginata*, that also has a brown cap, grows on wood, can fruit later in the season but not the winter typically, and is deadly poisonous.

Leucopholiota decorosa
(Peck) O. K. Miller, T. J. Volk & Bessette

SYNONYMS *Agaricus decorosus* Peck, *Armillaria decorosa* (Peck) A. H. Smith, *Tricholomopsis decorosa* (Peck) Singer, *Tricholoma decorosum* (Peck) Saccardo

Medium-sized mushroom with brown scaly cap and stem; gills white, notched at attachment; stem white above sheathing scales, ring zone present; on hardwood logs; spore print white

CAP 3–6 cm wide, broadly convex, becoming plane, margin incurved, dark brown or rusty-brown, dry, covered in thick, dense pointed fibrillose scales, margin shaggy scaly. FLESH white, unchanging. GILLS white, notched, close, broad. STEM 2–8 cm long, 6–11 mm broad, equal, white over apex but sheathed entirely with dark brown or rusty-brown scales and fibrils similar to cap. RING a zone of brown fibrils at the upper edge of the sheath. SPORE PRINT white. ODOR not distinctive. TASTE not distinctive, or bitter.

Habit and habitat single, scattered, or clustered, on downed logs of hardwoods in forest with beech, hemlock and maples. August to October. RANGE widespread.

Spores ellipsoid, smooth, colorless, amyloid, 5.5–6 × 3.5–4 µm.

Comments Found in rich moist woods, this species used to be rarely encountered, but seems to show up with regular frequency now on collecting tables at regional forays. Edibility is unknown. Compare with *Pholiota squarrosa* that looks somewhat similar, also grows on wood, but produces brown spores.

Neolentinus lepideus (Fries)
Redhead & Ginns

SCALY LENTINUS

SYNONYMS *Lentinus lepideus* (Fries) Fries, *Clitocybe lepidea* (Fries) P. Kummer, *Panus lepideus* (Fries) Corner

Large white mushroom on decaying logs and wood, with large flat brown scales on the cap; gills decurrent, subdistant, sawtooth-edged; stem with ring, brown scales below the ring

CAP 5–20 cm wide, convex then plane and with low broad bump over the center, margin incurved and often beaded with clear droplets when young, white with crust-brown flattened scales, tacky or dry. **FLESH** white, then slowly yellowing, thick, tough. **GILLS** off-white or buff, slowly yellowing, bruising to brown, decurrent, subdistant, easily splitting transversely, edges even becoming irregularly saw-toothed with age, broad. **STEM** 4–15 cm long, 10–30 mm broad, central mostly, equal but tapered to a point at the base, off-white, colored as pileus, with scaly or submembranous white ring near the apex, brown scaly below the ring, very firm, white, solid inside. **RING** white, often missing with age. **SPORE PRINT** white. **ODOR** pungent or disagreeable. **TASTE** mild.

Habit and habitat single or scattered on decaying conifer logs, also sometimes on oak, fence posts and railroad ties. June to October. **RANGE** widespread.

Spores cylindrical or nearly so, smooth, colorless, not amyloid, 9–14 × 4–5 µm.

Comments A very hard, heavy, firm mushroom. It is also listed as edible, but not considered choice and really not collected in large numbers.

Megacollybia rodmanii
R. H. Petersen, K. W. Hughes & Lickey

PLATTERFUL MUSHROOM

Large grayish-brown caps with white lens-shaped fissures; stem white, stocky, with radiating white cords; on well-decayed hardwood logs and woody debris

CAP 3–20 cm wide, convex becoming plane and often shallowly sunken, dark gray-brown, becoming pale gray-brown, glabrous but radially streaked with dark fibrils, dry, with tiny to large white lens-shaped fissures radiating out from the center (the skin splits open showing the flesh). FLESH white, thick. GILLS white, bluntly attached, close, broad. STEM 5–12 cm long, 10–25 mm broad, equal, straight and stout, white, dry, glabrous but fibrous, with thick white cords radiating out from the base. SPORE PRINT white. ODOR not distinctive. TASTE mild.

Habit and habitat single, scattered, or clustered in rows on decaying moss covered, hardwood logs, stumps, branches, from buried wood. May to October. RANGE widespread.

Spores ellipsoid, smooth, colorless, not amyloid, 6–10 × 5–7.5 µm; cheilocystidia abundant, club-shaped, hyaline.

Comments Formerly referred to as a species of *Tricholomopsis* and more recently as *Megacollybia platyphylla*, it is among the first of the large mushrooms to appear in late spring. The white lens-shaped splits showing the white flesh on the cap are diagnostic. There seem to be two distinctive forms in the Northeast, *Megacollybia rodmanii* f. *murina*, just described, and *M. rodmanii* f. *rodmanii*, which has a brown or olive-brown pileus that occasionally produces white cords that are often inconspicuous. We now know that *M. platyphylla* is genetically different and only found in Europe.

Hypsizygus ulmarius (Bulliard)
Redhead

ELM OYSTER

SYNONYMS *Agaricus ulmarius* Bulliard, *Pleurotus ulmarius* (Bulliard) Gray, *Lyophyllum ulmarium* (Bulliard) Kühner

Cap large, white, often with tan over center, cracking eventually; gills bluntly attached or notched, subdistant, white or cream; stem well-developed, central or off-center; coming out of old wounds on living or dead hardwood trees

CAP 5–15 cm wide, convex with inrolled margin at first, becoming nearly plane, white turning creamy-buff or tan over center, dry, smooth, often becoming cracked like a dry mudflat. **FLESH** white, thick. **GILLS** white becoming cream, bluntly attached or notched, close or subdistant, broad. **STEM** 5–10 cm long, 10–25 mm broad, equal or enlarged at base, off-center (eccentrically attached) or central, off-white, developing tan discolorations with age, dry, fibrous, cracking when dry, firm and solid inside. **SPORE PRINT** white. **ODOR** and **TASTE** not distinctive.

Habit and habitat single, or clustered in twos or threes, occasionally more and fruiting from old branch scars or wounds on living hardwoods, especially elm and box elder. August to December. **RANGE** widespread.

Spores subglobose or broadly ellipsoid, smooth, colorless, not amyloid but walls cyanophilic, 5.5–6 × 5–5.5 µm.

Comments *Hypsizygus ulmarius* is not causing the death of elms. The real parasite is an ascomycete. Elm oyster is an edible saprotroph on already dead tissues of the plant. Compare with *Pleurotus ostreatus* and *P. pulmonarius* that have decurrent gills, much better flavor, and are more commonly found on other kinds of hardwoods, like beech, poplar, and sugar maple. If you find a bouquetlike cluster with watery-spotted caps on sugar maple or poplar, and if the spores are globose and 4–5 µm, then you have *Hypsizygus tessulatus* (Bulliard) Singer [syn. *Hypsizygus elongatipes* (Peck) H. E. Bigelow and *H. marmoreus* (Peck) H. E. Bigelow]. See also *Ossicaulis lignatilis*.

Ossicaulis lignatilis (Persoon) Redhead & Ginns

SYNONYMS *Agaricus lignatilis* Persoon, *Hypsizygus circinatus* (Gillet) Kuntze, *Nothopanus lignatilis* (Persoon) Bon, *Pleurocybella lignatilis* (Persoon) Singer, *Pleurotus lignatilis* (Persoon) P. Kummer Cap medium-sized, chalk-white, with powdery surface; gills bluntly to barely attached, close or subcrowded, white; stem well-developed, off-center; on decaying wood of living or dead, hollowed-out, large-diameter maples

CAP 5–6 cm wide, convex, margin inrolled at first, becoming plane, chalk-white occasionally with pale rose or vinaceous tints over center, dry, smooth, but with fine powdered or granular or sub-woolly covering. **FLESH** white, thick, tough. **GILLS** white but drying a pale creamy-white, bluntly attached or becoming barely attached but with a decurrent tooth, close or subcrowded, broad. **STEM** 1.5–4 cm long, 1.5–6 mm broad, equal, often strongly curved, off-center (eccentrically attached), rarely central, chalk-white, some with faint rose tints, dry, surface suedelike or roughened above, firm and solid inside, white. **SPORE PRINT** white. **ODOR** fleetingly farinaceous when tissues damaged. **TASTE** not distinctive.

Habit and habitat single or in small clusters on well-decayed wood in hollowed out, large diameter trunks of living or recently downed sugar maple trees, but also known from beech and poplar. August to October. **RANGE** widespread but restricted to climax forests.

Spores broadly ellipsoid, smooth, colorless, walls not amyloid, not cyanophilic in Cotton Blue, 4–6 × 3–3.5 µm; with coralloid branching cells in the cap skin.

Comments *Hypsizygus ulmarius* might be confused with this species, but *H. ulmarius* is larger, has a cracked cap with age, and the cap is not powdery-granular. *Hypsizygus ulmarius* is found on dead elms or box elder most frequently, as well. The spores of the two are quite different in shape and in reaction to cotton blue reagent, and *H. ulmarius* lacks coralloid elements in its cap surface.

Pleurotus pulmonarius (Fries)
Quélet

SYNONYMS *Agaricus pulmonarius* Fries, *Pleurotus ostreatus* f. *pulmonarius* (Fries) Pilát

White or beige or tan, fan-shaped or liver-shaped caps, often with lined margins; gills white, close; stem typically very short or absent; on hardwoods

CAP 2–12 cm wide, convex becoming plane, from the top fan-shaped or liver-shaped and attached to substrate at the side of the cap or from a lateral stem (pleurotoid), sometimes nearly circular if on top of the woody substrate, margin inrolled at first, becoming flat and undulate and often finely lined, white or beige of pale tan, slightly greasy to the touch when fresh, then dry, smooth. **FLESH** white, thick. **GILLS** white, decurrent, close or subdistant, radiating out from attachment point. **STEM** absent or short and then 0.5–1 cm long, 5–10 mm broad, equal, eccentrically or laterally attached, whitish, hairy, white and solid inside. **SPORE PRINT** lilac-gray. **ODOR** fragrant like candle wax. **TASTE** not distinctive.

Habit and habitat scattered or clustered and usually shelving on hardwood logs, stumps, especially beech. July to September. **RANGE** widespread.

Spores cylindrical or elongate-ellipsoid, smooth, colorless, not amyloid, 7–10 × 2.5–5 µm.

Comments A fine edible species. Compare with *Pleurocybella porrigens*. *Pleurotus ostreatus* usually fruits in cooler weather months and the cap color is decidedly brown. *Pleurotus populinus* O. Hilber & O. K. Miller is similar to *P. pulmonarius* in colors, but grows on poplar or aspen trees, standing or downed, tends to be fleshier like *P. ostreatus*, and has a white, not lilac, spore deposit. If the fruit bodies are mostly white with a stem that is only slightly off-center and adorned with a delicate but obvious cottony ring that may also hang from the cap margin, you have *Pleurotus dryinus* (Persoon) P. Kummer. This latter species also produces a white spore deposit and is edible, but it usually only occurs in small numbers on hardwoods like oak and beech.

Pleurotus ostreatus (Jacquin)
P. Kummer

OYSTER MUSHROOM

SYNONYMS *Agaricus ostreatus* Jacquin, *Agaricus fuligineus* Persoon, *Crepidotus ostreatus* (Jacquin) Gray

Tan or ash-gray or brown, fan-shaped or semicircular, opaque caps; gills white, close; stem typically very short, eccentric or lateral; on hardwoods

CAP 5–20 cm wide, convex becoming plane or margin arching upward, from the top fan-shaped or semicircular or elongate-oyster-shell-shaped and attached to substrate at the side of the cap or from a lateral stem (pleurotoid), sometimes nearly circular if on top of the woody substrate, margin inrolled at first, quite variable ranging from tan, ash-grayish, brown or even with bluish-black tinges, dry, smooth, opaque. **FLESH** white, thick. **GILLS** white, decurrent, close or subdistant, radiating out from attachment point. **STEM** absent or short and then 0.5–1 cm long, 5–10 mm broad, equal, eccentrically or laterally attached, whitish, hairy, white and solid inside. **SPORE PRINT** lilac-gray. **ODOR** pleasant. **TASTE** not distinctive.

Habit and habitat scattered or clustered and usually shelving on hardwood logs, stumps. July to September. **RANGE** widespread.

Spores cylindrical or elongate-ellipsoid, smooth, colorless, not amyloid, 7–10 × 2.5–5 μm.

Comments A fine edible species. The *Pleurotus ostreatus* "complex" consists of three distinct species in northeastern North America: *P. ostreatus*, *P. pulmonarius*, and *P. populinus*. All are equally worthy of a chef's attention. In the summer and fall, one typically finds *P. pulmonarius* on various hardwoods and *P. populinus* restricted to aspen or poplar trees. *Pleurotus ostreatus* tends to fruit in the cooler months in the fall and even winter. See comments under *P. pulmonarius* as well. If you are finding one or just a few fruit bodies on the tree and it is an elm or box elder, see *Hypsizygus ulmarius*.

Pleurocybella porrigens
(Persoon) Singer

ANGEL'S WINGS

SYNONYMS *Agaricus porrigens* Persoon, *Nothopanus porrigens* (Persoon) Singer, *Pleurotus albolanatus* Peck

Milk-white, fan-shaped or spatula-shaped caps, often arching up; gills creamy-white, close; on conifer wood

CAP 3–10 cm in the longest dimension, convex or broadly convex or arching up and out (like a snake's head), from the top fan-shaped or elongate spatula-shaped and attached to substrate at narrow end (pleurotoid), margin incurved at first, milk white, dry, smooth or covered with fine hairs, opaque. **FLESH** white, thin. **GILLS** white becoming cream-yellow, decurrent, close, narrow, radiating out from stublike attachment point. **STEM** absent or barley present as a stublike tissue. **SPORE PRINT** white. **ODOR** and **TASTE** not distinctive.

Habit and habitat scattered or clustered and loosely shelving on conifer logs, stumps, especially hemlock. September to October. **RANGE** widespread.

Spores globose or subglobose, smooth, colorless, not amyloid, 6–7 × 5–6 µm.

Comments Compare with the much more robust and edible species of *Pleurotus*. Although *P. porrigens* is listed as edible in many field guides, a deadly poisoning account from Japan in the early 2000s should make mycophagists cautious.

Phyllotopsis nidulans
(Persoon) Singer

ORANGE MOCK OYSTER

SYNONYMS *Agaricus nidulans* Persoon, *Claudopus nidulans* (Persoon) Peck, *Crepidotus nidulans* (Persoon) Quélet, *Pleurotus nidulans* (Persoon) P. Kummer

Medium-sized, orange, woolly, fan-shaped caps, gills bright orange, flesh soft, odor strong of natural gas or propane

CAP 2–8 cm wide, fan-shaped, shell-like, kidney-shaped or spathulate and lobed around margin, margin inrolled at first, bright or dull orange or becoming orange-buff, dry, densely woolly (tomentose), opaque. **FLESH** orange, soft. **GILLS** bright or pale orange, attached to a common point and radiating out, close or subdistant. **STEM** absent or very reduced and not well-developed and then at edge of cap (lateral), fuzzy orange. **SPORE PRINT** pale pinkish. **ODOR** strong and disagreeable, sulfurous, somewhat like the additives put into natural gas or propane, thiols or mercaptans. **TASTE** not distinctive, or unpleasant.

Habit and habitat single or clustered and overlapping on dead logs of hardwoods or conifers. September. **RANGE** widespread.

Spores sausage-shaped or cylindrical, smooth, colorless in KOH, not amyloid, 6–8 × 3–4 µm; clamp connections present on the hyphae.

Comments This fungus is worth collecting just to experience the odor. Just because the common name is mock oyster, does not mean it is edible. The smell should keep such thoughts out of your head.

Panus neostrigosus Drechsler-Santos & Wartchow

SYNONYMS *Agaricus strigosus* Schweinitz, *Lentinus strigosus* Fries, *Panus rudis* Fries

Reddish-purple, tough, fuzzy caps; gills white or tan, decurrent, crowded, narrow; stem short, often off-center on cap, fuzzy, tough; on recently dead hardwood stumps and logs

CAP 2–10 cm wide, convex with margin strongly inrolled, becoming sunken or vaselike, purple or reddish-purple at first, soon fading to reddish-brown or pinkish-brown or tan with a hint of violet or not, dry, densely woolly-hairy overall. **FLESH** white, tough. **GILLS** white or tan, sometimes with violet tones when young, decurrent, crowded, narrow. **STEM** 1–4 cm long, 5–10 mm wide, central or off-center and may become lateral, equal, or tapered to attachment point, colored like cap, densely hairy, tough. **SPORE PRINT** white. **ODOR** not distinctive. **TASTE** not distinctive, or bitter.

Habit and habitat single, scattered or more often in dense clusters on recent stumps and downed hardwood logs in early stages of decay. May to November. **RANGE** widespread.

Spores elliptical, smooth, colorless, not amyloid, 4.5–6 × 2.5–4 µm.

Comments The purple colors can be lost rather rapidly, but usually a hint of violet remains somewhere on the fruit body. Not a true mushroom, but related to the polypores, based on anatomical structure and DNA phylogeny. This fungus is an early colonizer of recently dead hardwoods and can be found in urban settings were stumps have been left after tree removal.

Panellus serotinus (Persoon) Kühner

LATE FALL OYSTER

SYNONYMS *Agaricus serotinus* Persoon, *Panus serotinus* (Persoon) Kühner

Large olive-green or violet-brown caps with green tints on downed logs; gills yellowish, crowded; stem pluglike, yellowish, lateral; usually appearing after the first frost

CAP 3–10 cm wide, laterally attached, convex with margin strongly inrolled, becoming lobed, from the top kidney-bean-shaped or fan-shaped, dark olive-green or green mixed with yellows or violet or browns, slimy or slippery when wet, becoming dry, somewhat woolly at attachment point. **FLESH** white, thick. **GILLS** ochre or yellowish at first, becoming paler yellowish or white, bluntly attached or subdecurrent, crowded, edges may turn violet. **STEM** 0.5–2 cm long, 5–10 mm wide at the flared apex, attached to the side of the cap, equal, or tapered to attachment point, yellow or ochre-brown, hairy fuzzy covering, tough. **SPORE PRINT** white or pale yellow. **ODOR** not distinctive. **TASTE** not distinctive, or bitter.

Habit and habitat single or more often in shelving clusters on downed hardwood and conifer logs. October to December. **RANGE** widespread.

Spores sausage-shaped, smooth, colorless, amyloid or not amyloid when fresh, but usually amyloid after storing as dried collections, 4–6 × 1.5–2 µm.

Comments The harbinger of the end of the mushroom season. The leaves are beginning to fall and the weather has turned much colder. This species is listed as edible, but must be cooked slow and long. In Japan, the mukitake as it is known, seems to be a favored edible.

Panellus stipticus (Bulliard)
P. Karsten

LUMINESCENT PANELLUS

SYNONYMS *Agaricus stipticus* Bulliard, *Panellus farinaceus* (Schumacher) P. Karsten, *Panus farinaceus* (Schumacher) P. Karsten, *Panus stipticus* (Bulliard) Fries

Shelving clusters on downed hardwood logs with rather tough, pale tan, kidney-bean-shaped hairy caps and pinkish-brown gills with sticky edges

CAP 1–3 cm wide, laterally attached, convex with strongly inrolled margin, then plano-convex, from the top kidney-bean-shaped or shell-shaped, tan or pale yellowish-brown or buff, fine velvety or woolly, dry and opaque. **FLESH** colored as cap, thin, tough. **GILLS** colored as cap or pinkish-brown, bluntly attached, close, narrow, edges sticky-resinous. **STEM** 0.5–1 cm long, 3–10 mm wide at the flared apex, attached to the side of the cap, equal, or tapered to attachment point, pale buffy-brown or whitish from fuzzy covering, tough. **SPORE PRINT** white. **ODOR** not distinctive. **TASTE** usually bitter.

Habit and habitat in small or large shelving clusters on downed hardwood logs, on fence posts or wood in wood piles. May to September. **RANGE** widespread.

Spores sausage-shaped, smooth, colorless, amyloid, 3–6 × 2–3 μm.

Comments This species is very common. It is strongly bioluminescent. I suggest, as I did for *Omphalotus illudens*, take this collection to bed, place it on your dresser and when you wake up in the middle of the dark night with your eyes accommodated to the darkness, this little mushroom puts out quite a glow. Why do they have this bioluminescence? The low-level light emitted probably attracts nocturnal insects to feast on the mushroom. When the insect moves on, it might then accidentally take with it pieces of mushroom tissue that can asexually propagate clones of the mushroom if they are dropped onto suitable substrates.

Plicaturopsis crispa (Persoon) D. A. Reid

SYNONYMS *Cantharellus crispa* Persoon, *Plicatura crispa* (Persoon) Rea, *Trogia crispa* (Persoon) Fries

Small overlapping fan-shaped and lobed fruit bodies with fuzzy-zoned tops and white, crinkled, veinlike gills; on downed sticks and branches of hardwoods

CAP 1–2.5 cm wide, fan-shaped, shell-like, or spathulate and lobed around margin, with a false stem and concentric ridges and color bands forming zones, margin inrolled, orange-brown or yellowish-brown or tan with a paler almost white margin, dry, covered with densely woolly fibrils (tomentose), opaque. **FLESH** thin, flabby-pliant. **GILLS** off-white, attached to a common point and radiating out, mostly wavy-crinkled, thin, some forking, narrow. **STEM** absent. **SPORE PRINT** white. **ODOR** and **TASTE** not distinctive.

Habit and habitat clustered and overlapping on decaying sticks and limbs in hardwood forests. September. **RANGE** widespread.

Spores sausage-shaped, smooth, colorless in KOH, faintly amyloid, 3–4.5 × 1 μm; clamp connections present on the hyphae.

Comments An easily identified saprobe found on decaying branches because of the colors, the rumpled gills and the flabby-pliant texture.

Cheimonophyllum candidissimum (Berkeley & M. A. Curtis) Singer

SYNONYMS *Agaricus candidissimus* Berkeley & M. A. Curtis, *Pleurotus candidissimus* (Berkeley & M. A. Curtis) Saccardo, *Geopetalum candidissimum* (Berkeley & M. A. Curtis) Murrill, *Pleurotellus candidissimus* (Berkeley & M. A. Curtis) Konrad & Maublanc

Small, white, shell-shaped with inrolled and scalloped margin; gills white or pale cream, with granular fringed edges; stem stublike or absent; on branches often lacking bark

CAP 0.1–2 cm wide, semicircular or kidney or shell-shaped, white, minutely hairy, faintly lined over margin, margin inrolled and scalloped, dry. **FLESH** white. **GILLS** white or pale cream, bluntly attached or slightly descending stem when present, subdistant, broad, edges fringed or granulate. **STEM** when present at the cap edge, very short, equal, white. **SPORE PRINT** white. **ODOR** and **TASTE** not distinctive.

Habit and habitat usually clustered downed branches of hardwoods, these often lacking their bark. July to October. **RANGE** widespread.

Spores globose, smooth, colorless, 5–6 × 4.5–5.5 µm.

Comments This species can easily be mistaken for small white shell-shaped species of *Crepidotus*, *Clitopilus*, or *Claudopus*, that often share similar habitats. However, the spores are brown for *Crepidotus* and fleshy-brown or pinkish-brown for the other two genera.

Schizophyllum commune
Fries

COMMON SPLIT GILL
Small, white or grayish, fuzzy and fan or shell-shaped with scalloped margin; gills pale gray, separating their entire length; stem absent; on decaying hardwoods, and milled hardwood lumber

CAP 1–5 cm wide, semicircular or fan-shaped or shell-shaped, margin scalloped and lobed, often splitting and becoming frayed with age, white or pale gray or tan, densely hairy, dry. **FLESH** gray, thin, but flexible-tough. **GILLS** pale gray or whitish, subdistant, thick and separating their length, fibrous between the folds. **STEM** absent. **SPORE PRINT** white. **ODOR** and **TASTE** not distinctive.

Habit and habitat usually clustered on downed limbs and smaller branches of hardwoods, also on lumber made from hardwoods. All months. **RANGE** widespread.

Spores cylindrical, smooth, colorless, 3–4 × 1–1.5 μm.

Comments *Schizophyllum* is an odd member of the true mushrooms, even though it produces the typical-looking gills. It is not fleshy-soft like most mushrooms, but tough like a polypore. DNA analysis shows that *Schizophyllum* is a member of the true mushrooms, not the polypores. The tough flesh does not seem to be a deterrent though, since the fruit bodies are used as food in Mexico and in Central America, much like other wild mushrooms harvested for the dinner table.

Resupinatus striatulus
(Persoon) Murrill

SYNONYMS *Agaricus striatulus* Persoon, *Geopetalum striatulum* (Persoon) Kühner & Romagnesi, *Pleurotus striatulus* (Persoon) P. Kummer

Very small, dark grayish-brown inverted cups with whitish corrugate-grooved margins; gills pale grayish, distant, broad; densely clustered on well-decayed wood

CAP 0.4–0.8 cm wide, convex or becoming nearly plane, semicircular or fan-shaped and attached directly to the substrate, pale gray or gray-brown, corrugate-lined over margin and paler than the rest of the cap, surface glabrous, dry. **FLESH** thin, soft. **GILLS** white or pale gray, distant, radiating from attachment point, broad. **STEM** absent. **SPORE PRINT** white. **ODOR** and **TASTE** not distinctive.

Habit and habitat densely clustered on well-decayed pieces of hardwood logs, stumps, branches, also on conifers. June to November. **RANGE** widespread.

Spores: globose or subglobose, smooth, colorless in KOH, 4–6 × 3.5–5 μm.

Comments One generally does not easily see this fungus; one has to deliberately go looking for it because of its small size and dark colors. *Resupinatus applicatus* (Batsch) Gray, the black jelly oyster, is somewhat similar in size and colors, but it arises from a humplike pseudostipe at the point of attachment, and the surface of the cap nearest the substrate is covered with dark fuzz. It also is somewhat rubbery, often has a dark bluish tint, and lacks the corrugate-lined margin. *Resupinatus alboniger* (Patouillard) Singer is also somewhat similar, but the spores are elongate curved, like a sausage, while *R. applicatus* has globose spores like *R. striatulus*.

Gilled Mushrooms with Pink or Pinkish-Brown Spore Prints

The spore deposit color can vary somewhat in the pink-spored mushrooms. The color is always dark or bright pinkish, often with reddish, salmon, ochre, or brownish hues as well. If the spore print is pale pinkish-buff, see *Lepista* and *Phyllotopsis* in the white-spored gilled mushroom group. The paler pink color is indicative of very different spore wall architecture, indicating they are not related to the truly darker pink-spored mushrooms, which is borne out by DNA phylogenetic evidence.

There are two families of pink-spored mushrooms—those with free gills, the Pluteaceae, and the more diverse group with attached gills, the Entolomataceae. The Entolomataceae has more than 200 species documented from eastern North America alone so far. For the majority of species, an examination of the microscopic details is necessary to make accurate identification, until you learn the species. In addition, for those pink-spored mushrooms with attached gills, mainly in the genus *Entoloma*, the bulk of species in North America are still not completely documented.

Gills free, spores smooth, growing mostly on wood (Pluteaceae)

Volvariella and **Volvopluteus** Volva at base of stem
Pluteus Lacking a volva

Gills attached, decurrent, bluntly attached, or barely attached, spores angular in end view, growing mostly on the ground, some on wood (Entolomataceae)

SPORES NOT OBVIOUSLY ANGULAR IN SIDE VIEW
GILLS DECURRENT, OR STEM LATERAL AND GROWING ON WOOD
Clitopilus Either medium to large fleshy fruit body, with white or pale gray caps with central stem and farinaceous taste, *or* fruit bodies small, thin-fleshed, white, lacking a stem or stem highly reduced, taste mild, on wood; in both cases, spores ellipsoid or almond-shaped with longitudinal ridges, clamps absent
Clitocella Medium to large fleshy fruit body, with crowded narrow gills, bitter taste on soil or humus; spores subglobose, thin-walled, with obscure low bumps, clamps absent
GILLS NOT DECURRENT
Clitopilopsis Gills bluntly attached, small, gray, growing on sandy or mineral soils, spores globose, thick-walled, barely bumpy and then only on some spores, clamps absent
Rhodocybe Gills variously attached, variously colored, spores variously shaped, thin-walled, clearly bumpy, 6- to 12-angular in end view, clamps absent
Rhodophana Gills barely attached, small, red-brown colors, spores almond-shaped with large bumps, 6- to 12-angular in end view, clamps present

SPORES ANGULAR IN SIDE VIEW WITH 4–7 OR MORE ANGLES
CAPS MOSTLY CONICAL IN SHAPE
Inocephalus Conical cap with flattened radiating hairs, gills barely attached small to medium, fragile, spores 4- or 5-angled

Nolanea Conical or egg-shaped caps glabrous, lacking flattened hairs, gills bluntly or barely attached, spores 5- to 7-angled, usually not elongate, pileus with cylindrical end cells

Pouzarella Conical cap with dense uplifted hairs, stem hairy, base of stem with long tufted hairs, gills barely attached, mostly small, spores 6- to 9-angled (with two or more constrictions), aborted basidia with internal dark pigment

Trichopilus Conical cap with granular or fuzzy coating, gills barely attached, medium to large, cheilocystidia often bowling-pin-shaped

Stem absent, growing on wood or other macrofungi

Claudopus Stem highly reduced or absent, cap white, gray, or brown-colored

Small

Alboleptonia White or pale pinkish or with cream colors overall, cap silky over the center and sometimes to the margin, gills often bluntly attached or decurrent, small to medium, spores 5- to 7-angled and elongate, cap skin with inflated clavate end cells

Leptonia Cap often sunken on the center, with abundant minute scales on the center at least

Medium-sized to large with glabrous caps

Entocybe Cap convex or broadly conical, medium to large, spores globose or subglobose, 6- to 12-angled in end view (like *Rhodocybe*), weakly angled in side view, irregularly bumpy, clamps present

Entoloma Cap convex or plane, medium or mostly large, gills notched or bluntly attached or short decurrent, spores globose, (4- to) 5- to 6-angled in all views, cap skin of cylindrical, horizontally arranged cells

Calliderma Cap convex or plane, glabrous, becoming wrinkled, large, spores subglobose, (4- to) 5- to 6-angled in end and side views, cap skin composed of erect, inflated cells

Volvariella and *Pluteus*

Volvariella bombycina
(Schaeffer) Singer

TREE VOLVARIELLA
SYNONYMS *Agaricus bombycinus* Schaeffer, *Pluteus bombycinus* (Schaeffer) Fries, *Volvaria bombycinus* (Schaeffer) P. Kummer, *Volvariopsis bombycina* (Schaeffer) Murrill, *Volvaria flaviceps* Murrill

Large white, silky cap; gills white then pink, free; stem white, the base encompassed by a thick white or brown cup (volva); on dead or living hardwood trees, often above head height

CAP 5–20 cm wide, oval then broadly conical-convex or bell-shaped, may become nearly plane, white or tinged with yellow over the center, with a dense coating of radiating silky fibrils, opaque. FLESH white. GILLS white becoming pink, free, crowded, broad. STEM 6–20 cm long, 10–20 mm broad, equal with an enlarged base, white, glabrous. UNIVERSAL VEIL white, but sometimes brownish, thick saclike tissue encompassing the stem base (a volva). SPORE PRINT pinkish. ODOR and TASTE not distinctive.

Habit and habitat single, but also scattered or clustered on decaying logs, stumps and standing dead trees or on wounds in living trees of hardwoods such as maple, elm, beech and oak. July to October. RANGE widespread.

Spores ellipsoid, smooth, pale straw yellow in KOH, not amyloid, 6.5–10 × 4.5–7 μm; cheilocystidia obvious, abundant, inflated.

Comments This species seems to have a good deal of color variation of the cap and volva. Unfortunately, it is not commonly collected, so studying the variations—or harvesting it for the table, as it is a fairly good edible—is not reliable. Future studies may help sort out why there are color variants of this species and most likely will uncover cryptic species. *Volvopluteus gloiocephalus* (de Candolle) Vizzini, Contu & Justo is entirely white and on the ground. *Volvariella volvacea* (Bulliard) Singer is a large species that grows clustered in gardens or compost piles and on wood chips, and the cap is gray and streaked with hairs. Several other, mostly smaller species exist.

Pluteus cervinus (Schaeffer)
P. Kummer

FAWN MUSHROOM, DEER MUSHROOM

SYNONYMS *Agaricus cervinus* Schaeffer, *Pluteus atricapillus* (Batsch) Fayod, *Agaricus pluteus* Batsch

Medium to large gray-brown cap; gills free, white then pink; stem white or with gray-brown colors mixed in; taste of radish; on hardwood logs and stumps

CAP 3–12 cm wide, convex, becoming plane, with or without a low broad central bump, dark brown or pale brown or grayish-brown, sometimes radially streaked, tacky or slightly slippery when wet, then dry, glabrous or finely scaly-fibrillose over the center. **FLESH** white. **GILLS** white becoming pink, free, close or crowded, broad. **STEM** 5–13 cm long, 5–10 mm broad, equal, white but often with fine brown or grayish-brown fibrillose, dry, glabrous, basal mycelium white. **SPORE PRINT** salmon-pink. **ODOR** not distinctive, or radishlike. **TASTE** radishlike.

Habit and habitat single, scattered, or clustered on decaying logs and woody debris of hardwoods, occasionally conifers, occasionally on the ground but from buried wood. June to October. **RANGE** widespread.

Spores broadly ellipsoid, smooth, colorless, not amyloid, 6–8 × 4.5–6 μm; pleurocystidia fusiform, thick-walled, with the apex producing 2–5 apical horns or hooks.

Comments A variable species and recent studies suggest many cryptic species may exist that need to be identified and described using molecular techniques since morphology is not completely helpful. If the colors are very dark brown and the stipe is covered with very dark gray-brown scales and fibrils, and you are in a boreal forest, you most likely have *Pluteus rangifer* Justo, E. F. Malysheva & Bulyondova recently described and occurring widely in Eurasia and North America. Also compare *P. petasatus* with white cap and brown scales on the center.

Pluteus tomentosulus Peck

Entirely white at first, cap broadly conical at first with short pubescence; gills free, white then flesh-pink; stem white; odor and taste not distinctive; on decaying conifer logs and woody debris

CAP 3–10 cm wide, conical-convex becoming bell-shaped, with or without a low broad central bump that becomes wrinkled with age, chalk white, dry, felted with short pubescence, margin opaque. **FLESH** white. **GILLS** white becoming flesh-pink, free, close or crowded, broad, edges white fringed. **STEM** 5–10 cm long, 4–10 mm broad, equal with a bulb at the base, white, fine granular pubescent overall, sometimes longitudinally lined, dry, basal mycelium white, solid and white within. **SPORE PRINT** flesh-pink. **ODOR** and **TASTE** not distinctive.

Habit and habitat single, scattered, or clustered on decaying logs and woody debris of conifers such as eastern hemlock and spruce, but also on other conifers. June to September. **RANGE** widespread.

Spores broadly ellipsoid, smooth, colorless, not amyloid, 5–7 × 4.5–6 μm; pleurocystidia and cheilocystidia similar, thin-walled, swollen with a tapering, often cylindrical apex.

Comments *Pluteus petasatus* has a convex cap that is white with brown fibrils and scales on the center and *P. nothopellitus* Justo & M. L. Castro produces convex caps that are white and smooth. Both of these species have thick-walled cystidia with horns at the apex and are commonly collected in the Adirondack Park, while *P. tomentosulus* is less frequently encountered.

Pluteus petasatus (Fries) Gillet

SYNONYMS *Agaricus petasatus* Fries, *Plureus cervinus* var. *petasatus* (Fries) Massee

Entirely off-white at first, cap developing brown scales; gills free, white then flesh-pink; stem white but becoming brown over base; odor and taste not distinctive; on decaying hardwood logs and woody debris

CAP 4–14 cm wide, convex becoming plane, with or without a low broad central bump, pale white or very washed out grayish-brown, with pale brown flattened fibrils radially arranged and center with brown flattened scales with age, tacky at first, then dry, glabrous, margin opaque. **FLESH** white. **GILLS** white becoming flesh-pink, free, close or crowded, broad. **STEM** 4–12 cm long, 4–20 mm broad, equal, white becoming brown below, glabrous or near the base flattened fibrillose, dry, basal mycelium white, solid and white within. **SPORE PRINT** flesh-pink. **ODOR** and **TASTE** not distinctive.

Habit and habitat single, scattered or in small clusters on decaying hardwood logs and woody debris such as wood chips, sawdust piles, mulch, often in urban areas. June to September. **RANGE** widespread.

Spores ellipsoid, smooth, colorless, not amyloid, 5–7 × 4.5–5 μm; pleurocystidia thick-walled, with several apical horns or hooks.

Comments *Pluteus petasatus* can be confused with pale forms of *P. cervinus*, but that species has a decidedly radish odor and taste. See also the comments under *P. tomentosulus*.

Pluteus aurantiorugosus
(Trog) Saccardo

SYNONYMS *Agaricus aurantiorugosus* Trog, *Pluteus caloceps* G. F. Atkinson, *Pluteus leoninus* var. *coccineus* Massee, *Pluteus coccineus* (Massee) J. E. Lange

Cap brilliant red or orange-red, surface granular or velvety, center usually wrinkled; gills free, white then pink; stem white or dull yellow, finely fibrillose; on decaying hardwood debris

CAP 2–6 cm wide, conical-convex becoming convex, then plane, with a low broad central bump, bright red or orange-red when young, progressively more orange-yellow with age, glabrous or slightly granular, usually center wrinkled, moist becoming dry. **FLESH** pale yellow, thin. **GILLS** white becoming delicate pink, free, close or crowded, broad. **STEM** 3–6 cm long, 3–8 mm broad, equal, white or dull yellow especially over apex, becoming reddish over the base, finely fibrillose-hairy, dry. **SPORE PRINT** salmon-pink. **ODOR** and **TASTE** not distinctive.

Habit and habitat single or scattered on decaying hardwood logs and stumps or rotting woody debris. July to October. **RANGE** widespread, but rare.

Spores broadly ellipsoid, smooth, colorless, not amyloid, 5–8 × 3.5–6 μm; pleurocystidia inflated saccate or clavate or fusiform, thin-walled, lacking horns or hooked projections; pileus surface composed of round, inflated cells.

Comments This brightly colored mushroom is one you probably would not pass up; it is just too colorful. If you find a collection, please take a picture (or several) and post them on Facebook, or even better MushroomObserver.org. You will be very popular, at least with mycophiles. In fact, please make a collection and dry it carefully (after taking some notes and images) and then send the collection to a mycological herbarium of your choice to document the find. If the cap is brown and the stem is bright orange-red, it is *Pluteus aurantipes* Minnis, Sundberg & Nelsen.

Pluteus chrysophlebius
(Berkeley & M. A. Curtis) Saccardo

YELLOW PLUTEUS
SYNONYMS *Pluteus admirabilis* (Peck) Peck, *Agaricus admirabilis* Peck, *Pluteus chrysophaeus* (Schaeffer) Quélet, *Pluteus aurantiacus* Murrill, *Pluteus melleus* Murrill

Small fragile, cap yellow, often wrinkled; gills white then soon pink, free; stem pale yellow, translucent; on hardwood logs and stumps

CAP 1–3 cm wide, conical or convex then plane, with or without a central bump, often wrinkled over center, dull yellow or ochre-yellow or with greenish or olive tints, moist, glabrous, finely lined over margin. **FLESH** yellow, thin. **GILLS** white becoming pink, free, close or crowded, broad. **STEM** 2–5 cm long, 1–3 mm broad, equal, pale yellow and translucent, dry, glabrous, fragile, basal mycelium white. **SPORE PRINT** salmon-pink. **ODOR** and **TASTE** not distinctive.

Habit and habitat single, or clustered on decaying hardwood logs and woody debris. July to October. **RANGE** widespread.

Spores subglobose or broadly ellipsoid, smooth, colorless, not amyloid, 5–7 × 4.5–6 μm.

Comments Previously listed as *Pluteus admirabilis* in most field guides, but recent studies have revealed the presence of this older name that must be used. Several other not so well known species are also considered synonyms of *P. chrysophlebius*. *Pluteus chrysophlebius* is rather variable in ranges of the yellow coloration and development of wrinkling on the cap. *Pluteus rugosidiscus* Murrill seems to be separate from this species based on the green pigments mixed in with the yellow colors on the cap, the clearly wrinkled cap center and initial comparison of DNA profiles. Otherwise, the two are very similar.

Pluteus flavofuligineus
G. F. Atkinson

SYNONYM possibly *Pluteus leoninus* (Schaeffer) P. Kummer

Cap dark brown becoming yellowish-brown or yellowish-ochre, surface granular or velvety; gills free, white then pink; stem pale pinkish then dull yellow, glabrous; on decaying hardwood debris

CAP 2–7 cm wide, conical-convex becoming bell-shaped, then plane, with or without a low broad central bump, very dark brown when young, progressively more yellow as cap expands and then yellowish-ochre or dull brownish-yellow with darker disc, granular or velvety, dry. **FLESH** white, thin. **GILLS** white becoming pink, free, close, broad. **STEM** 4–10 cm long, 4–8 mm broad, equal, pink becoming dull yellow, glabrous, dry. **SPORE PRINT** salmon-pink. **ODOR** and **TASTE** not distinctive.

Habit and habitat single or scattered on decaying hardwood debris. June to October. **RANGE** widespread.

Spores ellipsoid, smooth, colorless, not amyloid, 6–7 × 4.5–5.5 µm; pleurocystidia fusiform, thin-walled, lacking hooked projections.

Comments A recent limited molecular study suggested our North American species might be the same as the European *Pluteus leoninus*, but there was some ambiguity to the results. *Pluteus flavofuligineus* is rather common and distinctive, but can be variable in colors. Maybe there are cryptic species to yet be uncovered, but we prefer the use of Atkinson's name at this time. If the colors are uniformly dark brown and the stipe is also granulose covered, then you most likely have *Pluteus granularis* Peck.

Pluteus longistriatus (Peck)
Peck

PLEATED PLUTEUS

SYNONYM *Agaricus longistriatus* Peck

Cap gray-brown, conical-convex, strongly pleated-lined to near the center, surface granular; flesh soft and flabby; stem white, longitudinally fibrillose lined; on decaying hardwood logs and debris

CAP 1–5 cm wide, conical-convex or bell-shaped, eventually plane with or without a low broad central bump, strongly pleated-lined or furrowed to near center, gray or gray-brown, dry, granular. **FLESH** very thin, very soft-flabby, collapsing readily. **GILLS** white becoming flesh-pink, free, close or crowded, eventually broad, very soft and collapsing. **STEM** 2–8 cm long, 1.5–4 mm broad, equal often enlarged toward base, white, longitudinally fibrillose-lined, dry, flabby and fragile. **SPORE PRINT** flesh-pink. **ODOR** and **TASTE** not distinctive.

Habit and habitat single, scattered, or clustered on decaying hardwood logs and woody debris. July to September. **RANGE** widespread.

Spores ellipsoid, smooth, colorless, not amyloid, 6–7.5 × 5–5.5 μm; pleurocystidia and cheilocystidia, thin-walled, mostly swollen.

Comments The cap and gills are quite soft and easily collapsing, especially if the weather is hot, reminding one of a *Bolbitius* species.

Entoloma and Allies

Clitopilus prunulus (Scopoli) P. Kummer

SWEETBREAD MUSHROOM
SYNONYMS *Paxillopsis prunulus* (Scopoli) J. E. Lange, *Pleuropus prunulus* (Scopoli) Murrill, *Clitopilus orcellus* (Bulliard) P. Kummer

Cap white or pale gray, soft; gills soon fleshy-pink, decurrent; stem colored as cap; spore print dark salmon-pink

CAP 5–10 cm, convex, eventually plane with wavy margin, white or buff or pale ashy-gray, surface soft, suedelike, glabrous, dry but when wet out may be slightly slippery, opaque. **FLESH** white. **GILLS** white, becoming pinkish or pale flesh-brown, decurrent, close, narrow, soft. **STEM** 4–8 cm long, 3–15 mm broad, occasionally off-center (eccentric), equal or enlarged toward the base, white or pale gray, glabrous, except for white mycelioid base, white and solid inside. **SPORE PRINT** salmon-pink. **ODOR** and **TASTE** farinaceous or of fresh unbaked bread dough.

Habit and habitat single, scattered, or clustered on humus or in grassy areas under hardwood and mixed coniferous forests, also under planted Norway spruce stands in the fall. July to October. **RANGE** widespread.

Spores: fusiform or football-shaped, 8–12 × 5–7 μm, with (5–)6 longitudinal ridges running from end to end; mostly 6-angled in polar view.

Comments A good edible species. Compare with the toxic *Clitocybe dealbata* that produces white or pale buff-pinkish spores and grows on open lawns. The cap can be variable in color for *Clitopilus prunulus* and might be confused with the gray-colored *Entoloma abortivum* if the aborted fruit bodies of the attacked *Armillaria* are not present. Fortunately, *E. abortivum* is considered an edible species, but *Entoloma* species in general should be avoided, since several in this genus do cause severe gastrointestinal irritation.

Entoloma abortivum (Berkeley & M. A. Curtis) Donk

ABORTED ENTOLOMA

SYNONYMS *Clitopilus abortivus* (Berkeley & M. A. Curtis) Saccardo, *Rhodophyllus abortivus* (Berkeley & M. A. Curtis) Singer

Fruit bodies in two forms: a moderately large gray mushroom with short decurrent pinkish-gray gills, usually associated with one or more white or pale pinkish globose or irregular-shaped balls (carpophoroids)

CAP 4–10 cm wide, convex, becoming plane and some with low broad bump over center, gray or gray-brown, surface smooth, matted fibrous or radially silky, becoming glabrous. **FLESH** white. **GILLS** pale gray, then with a flesh-pink cast, short decurrent, close. **STEM** 3–10 cm long, 5–15 mm broad, equal or enlarged downward, sometimes club-shaped, similar color as cap, glabrous or with small dots over apex, white mycelioid over base, with irregularly shaped, white or pale pinkish balls attached to the base or nearby. **SPORE PRINT** reddish-cinnamon or salmon-pink. **ODOR** of cucumber. **TASTE** faintly farinaceous.

Habit and habitat scattered or abundant in clusters on the ground, leaf litter or on decaying stumps and logs, usually under hardwoods. August to November. **RANGE** widespread.

Spores: angular, pale straw color, 8–10 × 5–6 μm; 5- to 6-angled in side view.

Comments The irregular formed balls are actually parasitized fruit bodies of *Armillaria* species. So the *Entoloma* is not aborted as the name implies, but it actually is the aggressor in this case of parasitism. The mushrooms are edible and of good flavor, cook them thoroughly. The aborted, attacked *Armillaria* is also edible.

Entoloma subsinuatum
Murrill

> Cap white with pale yellow or gray tints, thick, opaque; gills yellow; stem white, often with bulbous base, covered with coarse white fibrils and patches, odor strongly farinaceous

CAP 5–15 cm wide, convex, becoming plane, white or very pale cream, sometimes with pale gray tints, not hygrophanous, surface smooth, glabrous, becoming matted fibrous or minutely scaly with age, opaque, dry. **FLESH** white, thick. **GILLS** yellow, soon with a flesh-pink cast, attached, subdistant. **STEM** 5–12 cm long, 10–25 mm broad, equal or club-shaped or base bulbous swollen, white, sordid yellow in places, coarsely fibrillose or flocked with white patches overall, white mycelioid over base, solid and white inside, becoming spongy. **SPORE PRINT** reddish-cinnamon or salmon-pink. **ODOR** strongly farinaceous. **TASTE** rancid, very unpleasant.

Habit and habitat scattered or in small clusters on the humus and leaf litter or on grassy areas under conifers and hardwoods, including pine, spruce, fir, oak and birches present. July to September. **RANGE** Known from New York and Maine where it was originally described, but infrequent.

Spores: angular, pale straw color, 7–8.5 × 6.5–8 µm; 5- to 7-angled in side view.

Comments Poisonous. This is a toxic species like its counterpart in Europe, *Entoloma sinuatum* (Bulliard) P. Kummer, that has been known to cause death in Europe. Poisoning cases by this, or similar yellow-gilled species of *Entoloma*, have been documented as causing 1–2 days of severe vomiting and diarrhea. If the cap is large (12 cm broad) and yellowish-brown with pale yellow gills, and the stipe is stout (20–30 mm thick) and white, then one has *Entoloma whiteae* Murrill. There is another species found in the Northeast that has yellow gills, a gray-brown cap, and an off-white stem; it has been called *E. sinuatum* (also *E. lividum*) in some field guides. That species may not yet have a name. See also *E. luridum*.

Entoloma luridum Hesler

Cap pale cream becoming white, conical bell-shaped becoming convex with
broad conical bump; gills dark yellow;
stem white, sparsely fibrillose; odor not
distinctive

CAP 5–9 cm wide, conical bell-shaped,
becoming convex with well-defined
broadly conical broad central bump, yellow or very pale cream, hygrophanous
and then almost white, surface smooth,
glabrous or slightly silky, moist and then
faintly translucent-lined over margin.
FLESH white, thick. **GILLS** dark yellow,
becoming paler buff-yellow and then with
a flesh-pink cast, attached, close, very
broad. **STEM** 5–9 cm long, 6–12 mm broad,
equal or narrowly club-shaped, white, glabrous or lightly fibrillose, white mycelioid
over base, hollow. **SPORE PRINT** pinkish-
cinnamon. **ODOR** not distinctive. **TASTE** not
distinctive, or of green grass.

Habit and habitat scattered or in small
clusters on the humus or soil under mixed
conifer and hardwoods. July to October.
RANGE widespread.

Spores: angular, pale straw color, 8–10 ×
7–8 μm; 5- to 6-angled in side view.

Comments Differing from *Entoloma
subsinuatum* by the darker yellow gills,
broad conical bump on the cap, and the
lack of an odor. Edibility for this species
is unknown, but because *E. subsinuatum*
is toxic, *E. luridum* should be avoided. Do
not confuse this species with *Tricholoma
equestre* that has white spore deposit and
different cap colors.

Entoloma melleicolor Murrill

Cap moderately large, dull yellow with a yellowish-brown raised center at first, becoming uniformly champagne yellow; gills white, attached-notched; stem white; odor and taste farinaceous; fruit bodies very fragile; under hardwoods with oak and beech

CAP 4–9 cm wide, conical-convex, then convex, becoming plane and with low broad central bump, lobed and undulate around the margin with expansion, dull olive-yellow with dull yellowish-brown (clay-color) and faintly canescent center at first, becoming uniformly pale muted yellow overall (champagne), turning brown when dried, surface smooth, moist or slightly slippery, glabrous, opaque or very faintly translucent-lined at margin with age. FLESH white becoming watery olive-yellow, thick (to 6–10 mm), very fragile. GILLS white, soon pale pink, attached-notched, close, broad (8–10 mm), edges becoming eroded, very fragile. STEM 4–10 cm long, 10–20 mm broad, equal,

white or pale cream over base, fibrillose lined, mostly glabrous, matted pubescent over apex, white mycelioid over base, stuffed becoming hollow. SPORE PRINT pinkish-brown. ODOR cucumber-farinaceous. TASTE nutty-farinaceous.

Habit and habitat scattered or clustered on the humus under hardwoods with beech and red oak. June to July. RANGE Known so far from Iowa and New York.

Spores: angular, pale straw color, 10–12 × 8–10 µm; 5- to 7-angled in side view; lacking cystidia, clamps present.

Comments The champagne yellow cap, white gills turning pink, and white stocky stem, coupled with the cap turning uniformly brown when dried and the large spores, make this species different from all others in the genus. *Entoloma burlinghamiae* Murrill has a yellow, slippery cap with a brown center that reaches 4 cm, and the gills and stem are white at first like *E. melleicolor*. However, the spores of *E. burlinghamiae* are more elongate and smaller, 7–9 × 5–6(–7) µm.

Entoloma rhodopolium f. *nidorosum* (Fries) Noordeloos

SYNONYM *Agaricus nidorosus* Fries
Cap medium-sized, yellowish-brown and distinctly translucent-lined to near center when fresh, strongly hygrophanous and changing colors; gills white; stem white; odor alkaline-nitrous

CAP 2–7 cm wide, convex, becoming plane and often slightly sunken over the center, pale yellow-brown and distinctly translucent-lined to near the center, strongly hygrophanous and becoming pale grayish-yellow or sordid white in streaks outward from the center, surface smooth, moist, glabrous. **FLESH** pallid or colored as cap surface, thin. **GILLS** white, soon pink cast, attached, close or subdistant. **STEM** 3–12 cm long, 3–9 mm broad, equal, white, glabrous or with small white granular dots over apex, white mycelioid over base, brittle, hollow. **SPORE PRINT** reddish-brown. **ODOR** weakly or intensely alkaline (nitrous or chlorinelike). **TASTE** unpleasant, rancid.

Habit and habitat scattered or clustered in moist areas on the humus under hardwoods, including beech, oak, birch, ash. July to September. **RANGE** widespread.

Spores: angular, pale straw color, 7–10 × 6–8 μm; 5- to 7-angled in side view.

Comments *Entoloma rhodopolium* lacks the alkaline odor, but is very similar in all other aspects. This species group is highly variable and in need of study in all temperate zones.

Entoloma lividoalbum (Kühner & Romagnesi) Kubička is somewhat similar to *E. rhodopolium* but has a moderately large, dark yellowish-brown cap, with broad conical bump; it also has a farinaceous smell and taste. *Entoloma rhodopolium* has a paler yellow-brown cap, plane and distinctly translucent-lined to near the center, with a slender stem, and lacks the farinaceous smell and taste.

Entoloma lividoalbum

Calliderma indigofera (Ellis)
Largent

SYNONYMS *Agaricus indigoferus* Ellis, *Entoloma indigoferum* (Ellis) Saccardo

Cap large, deep indigo blue, radially wrinkled; gills white, attached-notched, broad; stem blue

CAP 7–10 cm wide, convex, becoming plane, deep indigo blue, fading to violet-blue, surface slightly wrinkled radially, moist, glabrous, opaque. **FLESH** white. **GILLS** white or creamy-white, soon with a flesh-pink cast, attached-notched, close, broad. **STEM** 5–7.5 cm long, 6–12 mm broad, equal, pale blue, glabrous but fibrous splitting showing white through narrow splits, brittle and splitting, white and solid inside. **SPORE PRINT** dull flesh-pink. **ODOR** and **TASTE** not distinctive.

Habit and habitat scattered or clustered on the humus under Atlantic white cedar in swampy areas or under pitch pine in pine barrens. July to September. **RANGE** Known only from the New Jersey Pine Barrens, rare.

Spores: angular, pale straw color, 7–8 × 6.5–7.5 µm; 5- to 6-angled in side view; cap surface composed of variously shaped, inflated, erect cells; clamps present.

Comments A very unusual species of entolomatoid mushrooms, looking more like a russula until the gills turn pink from the maturing spores. The cap is smooth and wrinkles because of the erect swollen cells making up the cap skin (an hymeniform layer). This wrinkling of the cap skin occurs frequently in other mushrooms with an hymeniform cap skin, such as *Psathyrella* and *Bolbitius*, which are brown-spored mushrooms. Other species of *Calliderma* are found in Europe and in the neotropics, but they also are not commonly encountered.

Entoloma strictius (Peck) Saccardo

STRAIGHT-STALKED ENTOLOMA

SYNONYM *Nolanea strictior* (Peck) Pomerleau

Cap yellow-brown or caramel-brown, conical becoming plane with a prominent rounded bump, translucent-lined when fresh, strongly hygrophanous, opaque and yellowish; stem straight, silvery-white streaked

CAP 3–7 cm wide, broadly conical-convex, becoming plane with a prominent rounded central bump, yellow-brown or reddish-caramel-brown when moist and distinctly translucent-lined to near the center, strongly hygrophanous and becoming pale straw-yellow or cream-buff in streaks outward from the center, surface smooth, moist, glabrous. **FLESH** white, thin. **GILLS** white, soon pink, attached, close, broad. **STEM** 5–15 cm long, 4–10 mm broad, equal or somewhat enlarged downward, similar colors to cap or paler when fresh, becoming silvery-white or streaked with pale gray-brown, glabrous but finely fibrillose-striate, often twisted, white mycelioid over base, brittle, hollow. **SPORE PRINT** rose-salmon. **ODOR** weakly grasslike or not distinctive. **TASTE** not distinct or slightly farinaceous.

Habit and habitat single, scattered, or clustered in moist areas, especially bogs but also on drier sites, on the humus or mosses under conifers or in mixed hardwoods and conifers. June to September. **RANGE** widespread.

Spores: angular, pale straw color, 9.5–12 × 7.5–9 µm; 5-angled in side view.

Comments The more common variety of this species, *Entoloma strictius* var. *isabellinum* Peck, is the species described and imaged here because it is more frequently collected in the Northeast. *Entoloma strictius* var. *strictius* has a "grayish brown" cap that is generally "paler when dry" (from Peck's original description) and in my experience, this grayish-brown-capped variety is less commonly encountered. They both should be known as *Nolanea*, but the taxonomic rearrangements have not been completed as yet. Compare also with the much larger and stout *E. lividoalbum* that occurs with oak.

Nolanea verna (S. Lundell) Kotlaba & Pouzar

EARLY SPRING ENTOLOMA

SYNONYM *Entoloma vernum* S. Lundell

Cap dark blackish-brown, conical-convex; gills grayish, almost free; stem similar color to cap, but also silvery fibrillose-striate; early to late spring

CAP 2.5–5 cm wide, conical or conical-convex, with a rounded central bump, very dark blackish-brown, dark brown or dark grayish-brown with glistening silvery-white embedded fibrils when moist and only faintly translucent-lined at very margin, strongly hygrophanous and becoming gray-brown or pale brown, surface smooth, moist, glabrous. **FLESH** colored as cap surface, thin. **GILLS** gray-brown, soon pink cast, barely attached, close or subdistant, broad. **STEM** 3–9 cm long, 2–7 mm broad, equal or somewhat enlarged downward, sometimes compressed and with longitudinal grooves, similar colors to cap or paler, glabrous but finely fibrillose-striate, often twisted, white mycelioid over base, brittle, hollow. **SPORE PRINT** rose-salmon. **ODOR** and **TASTE** weakly farinaceous or not distinctive.

Habit and habitat scattered or clustered often in open areas like unmown old grass fields, waste areas, road sides, or on humus or mosses under conifers or in mixed hardwoods and conifers. March to May. **RANGE** widespread.

Spores: angular, pale straw color, 9–11 × 7–9 µm, Q = 1.1–1.4; 5- to 7-angled in side view; hyphae of cap skin with fine brown encrustations as well as brown internal pigments.

Comments I must agree with Michael Kuo (MushroomExpert.com) about there being several "verna-like" species with conical caps and brownish colors, that can be found in the spring, at least in the Northeast of North America. One may well be *Nolanea papillata* Bresadola. Other species seem to exist as well, and they do not appear to have names. If you are finding a very dark-colored, conical-capped species in March through May, it may or may not be *N. verna*.

Nolanea sericea (Quélet)
P. D. Orton

SYNONYMS *Entoloma sericeum* Quélet, *Rhodophyllus sericeus* (Quélet) Quélet, *Acurtis sericeus* (Quélet) Singer

Cap dark reddish to grayish-brown, conical; gills gray-brown, barely attached; stem similar color to cap, silver fibrillose-striate; on grass; in the summer and fall

CAP 2–4(–7) cm wide, conical, becoming plane, sometimes with a small central conical bump, dark red or gray or sepia-brown, hygrophanous and silvery radial streaked, smooth, moist, glabrous, silky-shining as drying. **FLESH** colored as cap or paler, thin. **GILLS** pale gray-brown, soon with pink cast, eventually reddish-brown, barely attached, close, broad. **STEM** 2–8 cm long, 2–7 mm broad, equal, similar colors to cap, some grayish-silvery, fibrillose-lined, apex white-dotted, white mycelioid over base, brittle, hollow. **SPORE PRINT** brownish-pink. **ODOR** and **TASTE** rancid-farinaceous.

Habit and habitat scattered or clustered, in open, disturbed areas like old fields, waste areas, or road sides in mixed hardwood and coniferous forests. July to September. **RANGE** widespread.

Spores: angular, pale straw color, 7–10.5 × 6.5–9.5 µm, Q = 1.0–1.2; 5–6 rounded angles in side view; hyphae of cap skin with brown encrustations; clamps present in hymenium.

Comments Similar to *Nolanea verna* in form, colors, and habitat preferences, but differing by the time of fruiting, the lack of internal brown pigments in hyphae of the cap skin, and by rounded angular spores with Q = 1.0–1.2 (length of spore divided by the width). *Nolanea papillata* Bresadola is another species with dark brown colors that occurs in the summer and fall, that has a papillate bump on the cap center and a Q value of the spores greater than 1.2. A microscope is needed to make accurate identifications with these similar-looking fungi, and several other similar-looking species also occur.

Nolanea conica Saccardo

SYNONYMS *Entoloma conicum* (Saccardo) Hesler, *Entoloma subquadratum* Hesler, *Entoloma alboumbonatum* Hesler

Cap pale cinnamon, conical with silvery-white umbo and ridged or beveled zonate; gills off-white, almost free; stem dark chocolate brown with silvery fibrils scattered over surface

CAP 0.5–2 cm wide, conical, with a rounded silvery-white central bump, dull pale cinnamon or pale pinkish-buff and with radiating glistening silvery-white silky fibrils when moist giving it a pale gray tint, translucent-lined at very margin, hygrophanous and becoming pale grayish-cinnamon, characteristically ridged or beveled zonate. **FLESH** colored as cap surface, thin. **GILLS** off-white, soon pink or flesh color, barely attached, close, narrow. **STEM** 4–6 cm long, 1–2 mm broad, equal, dark chocolate brown, with fine white fibrils scattered over surface, starkly white mycelioid over base, brittle, hollow. **SPORE PRINT** rose-salmon. **ODOR** faintly of chlorine when very fresh, then farinaceous when flesh is crushed. **TASTE** farinaceous.

Habit and habitat scattered or clustered often in soil or on the humus or mosses under mixed hardwoods, especially with sugar maple present, also found with hemlock present, and in boggy areas. July to October. **RANGE** widespread.

Spores: angular, pale straw color, 8–11 × 6–7.5 µm; 4- or 5-angled in side view, variable in one mount.

Comments This species is rather common in the Northeast and the Southeast as well. It is somewhat variable in cap color, ranging from pale cinnamon to pinkish-buff. The fleeting chlorinelike or nitrous odor, that becomes farinaceous when the flesh is crushed, is characteristic for fresh samples.

Inocephalus luteus (Peck)
T. J. Baroni

SYNONYM *Entoloma luteum* Peck, *Rhodophyllus luteus* (Peck) A. H. Smith
Cap dull brownish-yellow, conical-convex; gills white, often with greenish-yellow discolorations; stem similar color to cap, long and straight

CAP 1–3 cm wide, rounded conical, dull brownish-yellow, due to golden-yellow ground color overlain with brownish fibrils and scales, sometimes with green tints, especially over margin, obscurely translucent-lined to near center on some, smooth, moist, appearing glabrous (but use a lens). **FLESH** colored as cap surface or white, thin. **GILLS** white or greenish-yellow, turning yellow where bruised, soon pinkish, barely attached, close, broad, edges fringed. **STEM** 3–10 cm long, 2–5 mm broad, equal, similar colors as cap except white at the apex and base, finely fibrillose-striate, often twisted and grooved, white mycelioid over base, brittle, hollow. **SPORE PRINT** rose-salmon. **ODOR** not distinctive. **TASTE** not distinctive, or slightly bitter.

Habit and habitat single, scattered or in small clusters on the humus or mosses under mixed hardwoods or conifers. July to September. **RANGE** widespread.

Spores: angular, pale straw color, 9–13 × 8–12 μm; cube-shaped, 4-angled in side view.

Comments Not as frequently seen as the somewhat similar-statured, but bright yellow *Inocephalus murrayi* (syn. *Entoloma cuspidatum*) and the deep orange *I. quadratus* (syn. *E. salmoneum*). Another species with completely black or blackish-brown cap and stem, similar in shape to *I. luteus*, is the very rarely collected *E. peckianum* Burt. Unlike *I. luteus*, *I. murrayi*, and *I. quadratus*, the spores of *E. peckianum* have 5–6 angles in side view, not 4.

Inocephalus murrayi (Berkeley & M. A. Curtis) Rutter & Watling

YELLOW UNICORN ENTOLOMA

SYNONYMS *Entoloma murrayi* (Berkeley & M. A. Curtis) Saccardo (as *Nolanea murraii* in some publications), *Nolanea murrayi* (Berkeley & M. A. Curtis) Dennis, *Entoloma cuspidatum* (Peck) Saccardo

Cap, gills and stipe bright yellow, conical cap often with fingerlike projection (papilla), very fragile, spore print reddish-brown

CAP 1–4 cm wide, conical, becoming conical bell-shaped, often with a central fingerlike projection (papillate), bright yellow or mustard-yellow, slightly translucent-lined at very margin, hygrophanous and becoming paler and silky shiny, margin thin, often splitting. **FLESH** colored as cap surface, thin. **GILLS** bright yellow, soon with pink or flesh tints from spores, attached, close or subdistant, broad, with fringed edge. **STEM** 5–12 cm long, 2–5 mm broad, equal, bright yellow, often twisted striate, glabrous or with silky fibrils over surface, brittle, hollow. **SPORE PRINT** reddish-brown or vinaceous-cinnamon. **ODOR** and **TASTE** not distinctive.

Habit and habitat scattered or clustered often in damp soil or on the humus or mosses in swampy areas or bogs under mixed hardwoods and conifers. July to October. **RANGE** widespread.

Spores: angular, pale straw color, 9–12 × 8–12 µm; 4-angled mostly in all views, cheilocystidia cylindrical or clavate.

Comments This brightly colored delicate species is hard to miss. It is related to *Inocephalus quadratus* and *I. luteus*, as the microscopic features of these three are very similar and they mainly differ in color. The genus *Inocephalus* has many more species in the tropics than in the temperate zone. *Inocephalus murrayi* has been found in the cooler mountain climates of Jamaica and Costa Rica, and *I. quadratus* also in Costa Rica.

Inocephalus quadratus
(Berkeley & M. A. Curtis) T. J. Baroni

SALMON UNICORN ENTOLOMA
SYNONYMS *Entoloma quadratum* (Berkeley & M. A. Curtis) E. Horak, *Nolanea quadrata* (Berkeley & M. A. Curtis) Saccardo, *Entoloma salmoneum* (Peck) Saccardo, *Nolanea salmonea* (Peck) Pomerleau

Cap, gills and stem salmon-orange, conical cap occasionally with fingerlike projection (papilla) and mostly translucent-lined over margin when moist, spore print deep reddish-brown

CAP 2–4(–6) cm wide, conical or becoming bell-shaped, occasionally with a fingerlike projection (papillate) on the center, dark salmon-orange or apricot-orange, fading with age, occasionally developing olive discolorations, glabrous, shiny, translucent-lined, margin thin, often scalloped. **FLESH** colored as cap surface, thin. **GILLS** orange colored as cap and persistently so, soon with pink or flesh tints from spores, narrowly attached, distant or subdistant, broad, with fringed edges. **STEM** 4–10 cm long, 2–6 mm broad, equal, orange colored as cap or paler, with white or pale orange mycelioid base, occasionally with olive discolorations, glabrous or with silky fibrils over surface, twisted or not, brittle, hollow. **SPORE PRINT** reddish-brown or vinaceous-cinnamon. **ODOR** and **TASTE** not distinctive.

Habit and habitat scattered or clustered on soil and humus or mosses, well-rotted wood, in mixed hardwoods and coniferous forests, also in swampy areas or bogs. July to October. **RANGE** widespread.

Spores: angular, pale straw color, 8–12 × 8–12 µm; 4-angled mostly in all views and often bumpy angular.

Comments This species is fairly commonly encountered, especially in boggy areas. See comments under *Inocephalus murrayi* and *I. luteus*, as well.

Inocephalus lagenicystis
(Hesler) T. J. Baroni

SYNONYM *Entoloma lagenicystis* Hesler
Cap conical, dark brown with radiating dark fibrils and small scales over center; gills white becoming bright pink, very broad; stem white, discoloring pale brown when handled

CAP 3–6 cm wide, conical or becoming broadly conical or bell-shaped, some with low rounded bump on the center, dark brown or almost black at first, becoming lighter brown or brownish-olive with expansion except the center remaining almost black, flattened radially arranged fibrils or small scales around the center, some faintly lined over the margin. **FLESH** white, thin. **GILLS** white at first, soon bright pink or flesh colored, narrowly attached, close, very broad, with fringed edge. **STEM** 4–8 cm long, 3–6 mm broad, equal or enlarged somewhat below, whitish and fibrillose lined, sometimes twisted and furrowed, discoloring pale olive-brown when handled, brittle, narrowly hollow. **SPORE PRINT** reddish-brown or vinaceous-cinnamon. **ODOR** not distinctive. **TASTE** not distinctive, or somewhat astringent.

Habit and habitat single, but most often scattered or clustered on soil and humus, in mixed hardwoods and coniferous forests such as hemlock and spruce, also in swampy areas or bogs with sphagnum mosses. July to August. **RANGE** widespread.

Spores: angular, pale straw color, 9–12 × 6.5–8.5 µm; 5- to 6-angled in side view, pleuro- and cheilocystidia lageniform (with an enlarged base tapering to an elongated tip).

Comments A summer species typically found in rich moist woods in the Adirondack region, and especially in boggy areas when it has been dry for some time. The conical-shaped cap and general appearance reminds one of *Inocephalus luteus*, but *I. lagenicystis* is larger and lacks the yellow-green colors often encountered with *I. luteus*. See comments under *I. luteus*.

Trichopilus jubatus (Fries)
P. D. Orton

SYNONYMS *Entoloma jubatum* (P. D. Orton) P. Karsten, *Rhodophyllus jubatus* (Fries) Quélet, *Leptonia jubata* (Fries) Largent

Cap conical becoming plane, dark grayish-brown or reddish-brown, densely radiately fibrillose or velvety or minutely scaly; gills grayish-brown on faces and edges uniformly; stem grayish-brown or reddish-brown fibrillose, odor and taste not distinctive

CAP 1.5–6 cm, conical becoming conical-convex or bell-shaped, eventually plane with low bump on the center, dark grayish-brown or reddish-brown or some almost black, surface radially covered with fluffy pubescent velvety or matted dark brown fibrils, becoming fine scaly, especially on the center, opaque. **FLESH** colored as surface. **GILLS** clearly dark grayish-brown or chocolate-brown at first, becoming darker pinkish-brown, barely attached, close or crowded, broad, fringed on the edge and same color as the face of the gills. **STEM** 3–9 cm long, 2–8 mm broad, equal or enlarge toward base, grayish-brown or reddish-brown fibrillose-striate over a pale white ground color, except for white mycelioid base. **SPORE PRINT** reddish-vinaceous cinnamon. **ODOR** not distinctive. **TASTE** not distinctive, or slightly acrid.

Habit and habitat single, scattered, or clustered on humus under hardwood and mixed coniferous forests. July to August. **RANGE** widespread but not frequently collected.

Spores: angular, colorless, 7–10 × 5.5–7.5 μm, 6- to 9-angled in side view; cheilocystidia mostly bowling-pin-shaped, some also clavate or swollen at base with a long tapering neck (lageniform).

Comments Compare with *Trichopilus fuscomarginatus* which is even darker in color and has dark brown colored gill edges. Another similar species in shape, but with dark violet colors on the cap and stem, and sometimes with violet gills, has gone under the name *Entoloma violaceum* Murrill.

Trichopilus fuscomarginatus
(P. D. Orton) P. D. Orton

SYNONYMS *Entoloma fuscomarginatum* P.
D. Orton, *Rhodophyllus fuscomarginatus* (P.
D. Orton) Moser

Cap conical, very dark brown, radiately
fibrillose; gills pale gray-brown with darker
brown fringed edge, broad; stem dark
grayish-brown fibrillose, odor and taste
farinaceous, amongst *Sphagnum*

CAP 1.5–6 cm, conical-convex, becoming
conical bell-shaped, eventually plane with
low bump on the center, dark brown or
red-brown, surface radially covered with
matted dark brown fibrils, also sparkling
from micalike reflective fibrils covering
surface (micaceous), opaque. FLESH off-
white. GILLS pale gray at first, becoming
darker pinkish-brown, attached, close
or crowded, broad, brown fringed on
the edge. STEM 2.5–6 cm long, 2–7 mm
broad, equal or enlarge toward base,
grayish-brown fibrillose-striate except
for white or pale cream base, often
twisted. SPORE PRINT reddish-vinaceous
cinnamon. ODOR farinaceous. TASTE
farinaceous-rancid.

Habit and habitat single, scattered, or clustered on *Sphagnum* in boggy areas in hardwood and mixed coniferous forests. July
to August. RANGE Apparently restricted to
bogs or very wet riparian habitats in eastern North America where mats of *Sphagnum* occur.

Spores: angular, colorless, 9–12 × 7–9 µm,
6- to 8-angled in side view; cheilocystidia
clavate or swollen at base with a long
tapering neck (lageniform), bowling-pin-
shaped, filled with brown pigment.

Comments *Trichopilus elodes* (Fries) P. D.
Orton is very similar, occurs in similar
habitats, but lacks the brown-fringed
gill edges.

Entoloma violaceum Murrill

Cap conical-convex becoming plane, dark purple or purple-brown, densely fibrillose-scaly or minutely scaly; gills violet becoming white, then pink from spores; stem colored as cap, fibrillose-scaly

CAP 2.5–6 cm, conical-convex, eventually plane, variable in color but always with some violet-purple hues, at first dark purplish-brown, becoming grayish-brown with violet tints or, becoming lilac-brown or grayish-beige, surface densely matted woolly or minutely scaly, especially on the center, opaque. **FLESH** white or colored as surface. **GILLS** faintly lilac or deep violaceous at first, becoming white, then flesh-pink, barely attached, close, broad. **STEM** 3–10 cm long, 5–10 mm broad, equal or enlarge downward, similar color as cap or violet-gray, white mycelioid over base, fibrillose-scaly overall. **SPORE PRINT** reddish-vinaceous cinnamon. **ODOR** and **TASTE** not distinctive.

Habit and habitat single, scattered, or clustered on humus or soil in grassy areas or on well-decayed wood under hardwood and mixed coniferous forests. June to October. **RANGE** widespread but not frequently collected.

Spores: angular, colorless, 8–10 × 5.5–7.5 µm, 6-angled in side view; cheilocystidia mostly bowling-pin-shaped, some also clavate or fusoid-ventricose.

Comments The mushroom listed in field guides and known as *Entoloma violaceum* in eastern North America is a perfectly good species of *Trichopilus*. Unfortunately, the species concept needs careful reevaluation, since at least one entoloma expert in Europe, Machiel Noordeloos, considers the type of Murrill's species same as the European *Entoloma dichroum* (Persoon) P. Kummer [syn. *Leptonia dichroa* (Persoon) P. D. Orton], a taxon not recognized as existing in North America as yet. There is some variability in gill colors and cap colors for our violet entoloma that need to be documented more thoroughly. Some specimens have more red than violet hues in the cap, some specimens lack lilac or violet colors in the early gills. There may be more than one species involved.

Pouzarella nodospora
(G. F. Atkinson) Mazzer

SYNONYMS *Entoloma nodosporum*
(G. F. Atkinson) Noordeloos, *Nolanea
nodospora* G. F. Atkinson

Dark grayish to cinnamon-brown cap and
stipe that are both densely woolly-scaly;
gills gray-brown

CAP 1–3.5 cm, conical or conical-convex,
uniformly cinnamon-brown or dark
grayish-brown, surface densely thick
woolly or scaly fibrils, opaque. **FLESH**
grayish-brown, thin. **GILLS** grayish-brown
at first, becoming darker reddish-brown,
barely attached, subdistant or distant,
moderately broad. **STEM** 4–9 cm long,
2–7 mm broad, equal or with a bulbous
base, same color as cap, densely pubescent-
fibrillose overall, with dark brown erect,
bristling or matted down hairs covering
the base, hollow. **SPORE PRINT** reddish-
vinaceous cinnamon. **ODOR** and **TASTE**
not distinctive.

Habit and habitat single, scattered, or
clustered on dark rich wet humus or grass
covered areas, often in hardwood forest.
July to October. **RANGE** widespread but
not commonly seen, this would be an
exciting find.

Spores: angular, pale brown, 13–16 ×
7–9 µm, 6- to 8-angled in side view; scattered basidia filled with dark brown pigment; cheilocystidia clavate or clavate with
a bump or fingerlike projection at the top.

Comments The most robust of all the *Pouzarella* species and the type of the genus.
Pouzarella species often resemble *Inocybe*
species due to the fibrillose cap surface,
like the fiber heads, and the darkly colored gills. Also, because most are small
and delicate with conical caps, *Pouzarella*
species resemble small, delicate *Mycena*
species. However, *Pouzarella* differs from
Mycena by the dark grayish-brown gills,
reddish-brown spore deposits, the bristling
tufted hairs at the base of the stem, and the
habit of growing in wet mucky areas. They
can be difficult to find, are very fragile, but
to my eye beautiful and graceful, even if
they are mostly hairy little mushrooms.

Pouzarella flavoviridis (Peck)
Mazzer

SYNONYM *Entoloma flavovirde* Peck
Cap conical, translucent-lined, yellowish-green or olive-green, with tiny lemon-yellow fibrils; gills buffy brown; stem pale buffy brown under off-white striate fibers, whitish pubescent over base

CAP 1–3 cm, conical or conical-convex, becoming bell-shaped, yellowish-green or grayish-olive to olive-green, surface radially covered with lemon-yellow erect or matted silky fibrils, these easily rubbed off, then the color is yellowish-brown or grayish-brown, translucent-lined to near center when moist, becoming opaque. **FLESH** grayish-brown, thin. **GILLS** pale buffy-brown at first, becoming darker flesh-brown, barely attached, subdistant or distant, faces becoming wrinkled or intervenose with age, moderately broad. **STEM** 2.5–6 cm long, 1–3 mm broad, equal or with a subbulbous base, pale buffy brown, overlain with whitish striate fibers, often twisted and splitting, fine off-white pubescence covering the base, hollow.

SPORE PRINT reddish-vinaceous cinnamon. **ODOR** "soapy or fishy". **TASTE** bitter.

Habit and habitat single, scattered, or clustered on dark rich wet mucky leaves and mud in wet low areas, seepage areas and boggy areas in hardwood and mixed coniferous forests. July to October. **RANGE** widespread.

Spores: angular, pale brown, 10–14 × 7–9 µm, 5- to 7-angled in side view; scattered basidia filled with dark brown pigment; cheilocystidia clavate or swollen at base with a long tapering neck (lageniform).

Comments These small yellowish-green caps are almost iridescent when found on dark mucky substrates, very attractive and very fragile. Be prepared to get really messy if you wish to make an image of these mushrooms in their natural habitat. Notice the ones shown here were taken back in the laboratory on gray card, a standard scientific format used for publishing in peer reviewed journals in order to show clearly all details.

Entocybe nitida (Quélet)
T. J. Baroni, Largent & V. Hofstetter

> **SYNONYMS** *Entoloma nitidum* Quélet, *Rhodophyllus nitidus* (Quélet) Quélet
>
> Cap grayish-blue; gills white or pale cream, attached and broad; stem grayish-blue with white or yellowing tapered base

CAP 2–5 cm wide, broadly conical, then bell-shaped or broadly convex with a low central bump, some shallowly sunken, grayish-blue, some dark midnight blue over center, appearing glabrous or finely radiately fibrillose, moist or slightly slippery, opaque. **FLESH** white, fleshy. **GILLS** white or pale cream, becoming pinkish from spores, attached, subdistant, broad. **STEM** 3–10 cm long, 2–7 mm broad, equal with slight swollen then abruptly tapered base, colored as cap over upper half, white over base and may develop yellow tints, glabrous but longitudinally fibrillose-striate, sometimes twisted. **SPORE PRINT** pink. **ODOR** not distinctive. **TASTE** mild.

Habit and habitat single, scattered or in small clusters on humus or mosses under hardwoods mainly, but conifers may be present. July to October. **RANGE** widespread.

Spores subglobose, bumpy, weakly angled in profile view, 6–10 minute angles in polar view, colorless, 7–9 × 6–8 μm; clamp connections present on all hyphae.

Comments There are very few blue mushrooms in northeastern North America, this is one of the larger ones growing on the ground. It is visually striking, but rarely encountered. Compare with *Calliderma indigofera*. *Entocybe turbida* (Fries) T. J. Baroni, V. Hofstetter & Largent is very similar in size and stature with a pinched stipe base, but the cap is dark brown and the stem pale gray-brown. The spores are also similar in size, shape, with minute angles in polar view, and the hyphae have obvious clamp connections. *Entocybe turbida*, a European species, was originally discovered in South Carolina but should be expected in the southern areas of theNortheast.

Entocybe speciosa (Lennox ex T. J. Baroni) T. J. Baroni, Largent & V. Hofstetter

SYNONYM *Rhodocybe speciosa* Lennox ex T. J. Baroni

Cap medium-sized, pale brown to honey brown, opaque; gills white becoming pink, attached-notched and broad; stem yellow or pale ochre, with silvery flattened fibrils; on decaying logs

CAP 1.5–4 cm wide, convex becoming plane, pale brown or tan or honey brown, hygrophanous and becoming light tan, glabrous or silvery canescent over the margin, moist, opaque. **FLESH** colored as cap surface, 2–3 mm thick, tough. **GILLS** white, becoming pinkish, attached-notched, crowded, broad (to 7 mm). **STEM** 3–4.5 cm long, 3–8 mm broad, equal, light yellow or pale ochre, adorned with a coating of longitudinally flattened silver fibrils, powdery white-dotted over the apex, white mycelioid over the base. **SPORE PRINT** pink to salmon-brown. **ODOR** faint or distinct, farinaceous fragrant. **TASTE** strong farinaceous.

Habit and habitat scattered or clustered on decaying logs in conifer or mixed forests. July to October. **RANGE** widespread but rare.

Spores subglobose or short ellipsoid, scarcely bumpy, angled in profile view, 6–10 minute angles in polar view, colorless, 7–9 × 6–8 μm; clamp connections present on all hyphae, dextrinoid hyphae in the gill trama and cap flesh.

Comments Our northeastern version of this western species has darker brown caps, as shown here. It is unusual to find mushrooms with pink spores and attached gills on decaying wood. The attached gills rule out a species of *Pluteus*. Some *Leptonia*, with pink spores and attached gills, can be found on well-decayed wood, but the scales on cap and the subdecurrent gills are characteristic. *Entocybe priscua* (T. J. Baroni) T. J. Baroni, V. Hofstetter & Largent is somewhat similar to *E. speciosa*, but has a reddish-brown or cigar-brown cap and vinaceous-buff stem and grows on the ground under pines in the northernmost reaches of the Northeast.

Clitocella mundula (Lasch)
Kluting, T. J. Baroni & Bergemann

SYNONYMS *Agaricus mundulus* Lasch, *Clitopilus mundulus* (Lasch) P. Kummer, *Rhodocybe mundula* (Lasch) Singer, *Rhodopaxillus mundulus* (Lasch) Konrad & Maublanc

Large pale creamy-white or grayish with buff colored concentric lines or reticulations on the cap and with an inrolled margin; gills decurrent, crowded, narrow; stem colored as cap, tough; taste bitter

CAP 2–11 cm wide, convex, becoming plane, with a low central bump at first, eventually sunken, at times sharply so, margin inrolled, color variable, pale creamy-white with smoke-gray cast, some variants very dark gray, may stain black where bruised, usually with concentric or reticulate watery buff cracks, appearing glabrous but finely matted under a lens, dry. **FLESH** watery buff, fleshy. **GILLS** pale cream or buff, some blackening when injured, becoming pinkish from spores, decurrent, crowded, narrow. **STEM** 1–4 cm long, 6–14 mm broad, equal with swollen, tough, colored as cap, also pubescent overall, developing woolly patches over the base, white mycelium and chords at base binding substrate. **SPORE PRINT** clay-pink. **ODOR** none or slightly farinaceous. **TASTE** bitter and farinaceous.

Habit and habitat scattered or in bouquet-like clusters on humus under hardwoods, but also under conifers. July to October. **RANGE** widespread.

Spores subglobose, obscurely bumpy, obscurely angled in profile view, minutely angled in polar view, colorless, 5–6.5 × 4–5 μm; clamps absent; 3–5% aqueous KOH turning fresh or dried surfaces cherry or brownish-red.

Comments Molecular analysis by Kluting et al. (2014) indicates *Clitocella mundula* (previously placed in the genus *Rhodocybe*) and *Clitocella popinalis* are not the same species and that *C. mundula* occurs in Europe and North America, but *C. popinalis* only in Europe. A dark gray or purplish-gray form is often found under conifers. There is still work to be done on this group of species in North America.

Clitopilopsis hirneola (Fries) Kühner

SYNONYMS *Agaricus hirneolus* Fries, *Clitopilus hirneolus* (Fries) Kühner & Romagnesi, *Clitocybe hirneola* (Fries) P. Kummer, *Rhodocybe hirneola* (Fries) P. D. Orton, *Rhodophyllus hirneolus* (Fries) Singer

Small to medium grayish, with sunken and wrinkled cap with age; gills grayish becoming reddish from spores; stem colored as cap, pubescent at first, somewhat pliant

CAP 0.7–2 cm wide, convex, becoming plane, with a low bump over the center or becoming sharply sunken, smoke gray or mouse gray, surface finely fibrillose matted at first, becoming glabrous, dry. FLESH dark gray or grayish-brown, thin. GILLS pale mouse gray, developing reddish cast in age, subdecurrent, subdistant, edges white fuzzy under hand lens. STEM 1–5 cm long, 1–6 mm broad, equal, colored as cap, also with white matted pubescence overall, becoming glabrous, white mycelioid at base. SPORE PRINT pinkish-flesh. ODOR and TASTE not distinctive, or farinaceous.

Habit and habitat single or clustered on sandy soil, among grasses, mosses, on needle beds and humus, typically under or near conifers. July to October. RANGE widespread but rarely collected.

Spores globose or subglobose, with low, obscure pustulelike bumps, obscurely angled in profile view, but rounded angular in polar view, colorless or pale grayish from thickened wall, 5.5–8 × 5–6.5(–7) µm; cheilocystidia cylindrical inflated, with 1–4 cross walls, colorless, clamps absent.

Comments If you become fascinated with LBMs (little brown mushrooms), you will eventually need to have a microscope to achieve accurate identifications. The ecology and a good hand lens helps, since the cystidia on the gills edges of this species appear like fine white fuzz under the lens. However, the cheilocystidia with cross walls present is diagnostic for the species, and to view this you need a compound light microscope.

Rhodocybe caelata (Fries)
Maire

SYNONYMS *Agaricus caelatus* Fries, *Clitopilus caelatus* (Fries) Kühner & Romagnesi, *Tricholoma caelatum* (Fries) Gillet

Small to medium grayish, with sunken and wrinkled cap with age; gills grayish becoming reddish from spores; stem colored as cap, pubescent at first, somewhat pliant

CAP 0.5–3 cm wide, convex, becoming plane, sunken over the center, gray or grayish-brown or reddish-brown, with yellowish-brown discoloration with age, smooth or wrinkled with age, surface covered with silvery-white silky fibrils at first, becoming glabrous, dry. **FLESH** dark gray or grayish-brown, thin. **GILLS** gray-cream or grayish-brown, developing reddish cast in age, bluntly attached or subdecurrent, becoming short decurrent, close or subdistant. **STEM** 0.6–5 cm long, 1–7 mm broad, equal, colored as cap, also with white or silvery-gray pubescence overall at first, becoming glabrous, white myceloid at base, somewhat pliant. **SPORE PRINT** pinkish-flesh. **ODOR** not distinctive. **TASTE** not distinctive, or slightly bitter.

Habit and habitat single or clustered among grasses, mosses, lichens, sandy soil, disturbed soil, or humus under hardwoods or conifers. July to October. **RANGE** widespread but rarely collected.

Spores almond-shaped or ellipsoid, with irregular pustulelike bumps, minutely angled in polar view, colorless, 7–9 × 4–5 μm; cystidia on the gills variously shaped but with bright golden or ochre pigments in any colorless mounting medium, clamps absent.

Comments Compare with *Clitopilopsis hirneola* that is also rather small and grayish overall, but the cap on that species is not wrinkled and developing yellowish-brown discolorations with age. Also, the spores of *C. hirneola* are globose and the gills lack pigment-filled cystidia.

Rhodophana nitellina (Fries)
Papetti

SYNONYMS *Agaricus nitellinus* Fries, *Clitopilus nitellinus* (Fries) Noordeloos & Co-David, *Collybia nitellina* (Fries) Saccardo, *Rhodocybe nitellina* (Fries) Singer, *Rhodopaxillus nitellinus* (Fries) Singer

Small to medium, orange, smooth capped, fragile mushroom; gills buff but becoming pinkish-cinnamon from spores; stem colored as cap, white mycelioid at base; odor and taste farinaceous

CAP 1–4 cm wide, convex, becoming plane, sunken over the center, darker at first and orange-tawny or orange-cinnamon, hygrophanous and fading to ochre-buff or pinkish-cinnamon with age, smooth, glabrous, moist and translucent-lined at first, becoming dry and opaque. **FLESH** pinkish-buff, thin. **GILLS** warm buff or pinkish-cinnamon, bluntly attached or subdecurrent, close. **STEM** 2–6 cm long, 2–7 mm broad, equal, colored as cap, white mycelioid at base, white stuffed inside, soon hollow. **SPORE PRINT** pinkish-cinnamon. **ODOR** and **TASTE** farinaceous.

Habit and habitat single or clustered on soil or humus conifers or mixed with hardwoods. July to November. **RANGE** widespread but rarely collected.

Spores almond-shaped or ellipsoid, with irregular pustulelike bumps, minutely angled in polar view, colorless, 7–9 × 4–5 μm; cystidia absent, clamps present.

Comments The gills will darken with age from the pinkish-cinnamon spore production and make this species very different than any of the small to medium, orange-capped species like *Gymnopus* with which it might be confused, such as *G. dryophila*. *Rhodophana nitellina* is not a common species to find, but the pinkish-cinnamon spore deposit is a good diagnostic test if there is not a microscope available.

Alboleptonia sericella (Fries)
Largent & R. G. Benedict

SYNONYM *Entoloma sericellum* (Fries) P. Kummer

White overall with typically some yellow on the cap center, cap finely woolly or silky; gills white then soon pink; stem delicate translucent white

CAP 1–2.5 cm, convex or somewhat conical-convex, becoming bell-shaped and often with a narrow slightly sunken center with a small rounded bump, white or with pale cream tints on the center, becoming entirely yellow with age, surface finely woolly or covered with matted silky fibrils, opaque. **FLESH** white, thin. **GILLS** white at first, becoming delicate rosy-pink, attached, subdistant or distant, moderately broad, edge often fringed. **STEM** 2–6 cm long, 1–3 mm broad, equal or enlarged toward base, white or translucent colorless, can become tinted with yellow or ochre with age or handling, white-dotted over apex, finely white silky fibrillose elsewhere, becoming glabrous, smooth, thin white mycelioid over base, hollow. **SPORE PRINT** flesh pink. **ODOR** none or faintly herbaceous. **TASTE** not distinctive.

Habit and habitat single, scattered, or clustered on soil, humus or grass covered areas, often in hardwood forest. July to October. **RANGE** widespread.

Spores: angular, hyaline or pale straw color, 8–11.5 × 6–9 µm, 5- to 8-angled in side view; cheilocystidia scattered, cylindrical, fusoid, or swollen at base with a long tapering neck (lageniform).

Comments This species is very fragile, looking like a white species of *Mycena*, but the pink gills and silky-covered cap give it away.

Leptonia canescens (Hesler) Largent

SYNONYM *Entoloma canescens* Hesler

Cap and stem grayish-brown with dense covering of silvery-white hairs; gills white, short decurrent; odor distinctly fruity

CAP 1–5 cm wide, convex and with center shallowly depressed, grayish-brown with a silvery sheen, densely or moderately covered with silvery-white hairs or minute scales, center densely minute scaly, opaque. **FLESH** off-white or grayish, thin. **GILLS** white or very pale gray at first, becoming sordid pink or sordid flesh color from spores, attached with short decurrent tooth, subdistant or close, broad, edges fringed. **STEM** 3–7 cm long, 2–5 mm broad, equal or somewhat enlarged downward, white or pale ashy white or similar in color to the cap, often after handling, surface covered with silvery-white hairs, brittle, solid then hollow, white myceloid at base. **SPORE PRINT** flesh pink. **ODOR** distinctive, fruity fragrant. **TASTE** mild or slightly bitter.

Habit and habitat single, scattered, or clustered on soil and humus in deciduous woods. June to August. **RANGE** widespread but not commonly collected.

Spores: angular, pale straw color, 8–11 × 6–8 µm; 5- (to 6-) angled in side view, cheilocystidia clavate or inflated, some cylindrical.

Comments No other mushroom has this combination of features. It is found in the Northeast in northern New York in the Adirondack region, but is more commonly collected in the Smoky Mountains of the southeastern United States.

Leptonia odorifera (Hesler)
Largent

SYNONYM *Entoloma odoriferum* Hesler

Cap, gills and stem gray; cap broadly sunken, strongly translucent-lined and dotted with silvery fibrils and scales; odor strong of burnt rubber

CAP 2–4 cm, convex with broadly sunken center, dark gray or grayish-brown, darker over the center, with silvery sheen from colorless hairs and scales, concentrated over the center, with fine fibrils and scattered scales elsewhere, strongly translucent-striate-lined when moist. **FLESH** gray or grayish-brown, thin. **GILLS** pale grayish at first, becoming pink, attached and often short decurrent, subdistant or close, moderately narrow, edge smooth. **STEM** 4–9 cm long, 3–5 mm broad, equal, often flattened, gray, glabrous, smooth, white mycelioid over base, hollow. **SPORE PRINT** flesh pink. **ODOR** strong of burnt rubber. **TASTE** strong, unpleasant.

Habit and habitat scattered or clustered on soil or thin humus under hardwood or hardwood-coniferous forest, in areas of limestone bedrock. July to August. **RANGE** widespread but not commonly seen.

Spores: angular, hyaline or pale straw color, 9–12 × 7–9 µm, 5- (to 6-) angled in side view; cheilocystidia scattered, fusoid, clavate or subcapitate.

Comments This species is quite smelly, much like the white-spored and greenish-yellow-colored fruit bodies of *Tricholoma odorum*. Why do fungi have these strong, repulsive odors? Probably to keep insects and other animals from eating the flesh before the spores have been released.

Leptonia incana (Fries) Gillet

SYNONYMS *Entoloma incanum* (Fries) Hesler, *Leptonia euchlora* (Lasch) P. Kummer

Small lemon or greenish-yellow species with a shallow dark brown or black depression on the center of the cap; gills and stem bruise bluish-green quickly; odor of mouse urine

CAP 1–5 cm wide, convex and with center shallowly depressed, greenish-yellow or sordid yellowish with a black center, becoming darker grape green, eventually fading to pale yellow-brown, sparsely covered with black or dark brown radiating fibrils or fine scales on the center, faintly lined over the margin. **FLESH** yellow or greenish-yellow, thin, turning dark green or bluish-green when exposed. **GILLS** white at first or with some yellow tints, soon sordid pink or sordid flesh color from spores, bruising bluish-green when injured, attached, distant, very broad. **STEM** 2–6 cm long, 1–4 mm broad, equal, lemon-yellow, but rapidly discoloring bluish-green when handled, glabrous, brittle, hollow, white mycelioid at base, turning bluish-green when bruised. **SPORE PRINT** flesh pink. **ODOR** rather distinctive, of mouse urine or some say of buttered popcorn. **TASTE** not distinctive.

Habit and habitat single, but more often scattered or clustered on rich wet humus and mosses in mixed hardwoods and coniferous forests, also in swampy areas or bogs with sphagnum mosses. It seems to be confined or at least more common to areas where the soil is mostly limestone. June to August. **RANGE** widespread.

Spores: angular, pale straw color, 8.5–11 × 7–8 µm; 6-angled in side view.

Comments A very pretty and smelly small species that is rather photogenic. Not all that commonly collected. It seems to be found only on limestone soils, even in the Caribbean.

Leptonia flavobrunnea Peck

SYNONYM *Entoloma flavobrunneum* (Peck) Noordeloos

Cap dark brown, hairy or minute scaly over depressed center; gills pale lemon-yellow then flesh brown, short decurrent; stem pale tan with yellow apex, covered with tufts of silky hairs; taste farinaceous

CAP 0.5–4 cm wide, convex, becoming plane or arching up and margin wavy, all with center shallowly, but sharply depressed, dark brown over center, grayish-brown elsewhere, densely or moderately hairy overall, center densely minute scaly, opaque or faintly translucent-lined over margin with expansion. **FLESH** white or watery gray, thin. **GILLS** pale lemon-yellow or sordid cream color at first, especially near stipe or at cap margin, becoming sordid pink or sordid fleshy brown from spores but retaining some yellow color, attached with short decurrent tooth, close or crowded, broad (to 7 mm), edges fringed. **STEM** 3–6 cm long, 1.5–4 mm broad, equal or enlarged downward, pale tan or some dark grayish-brown, many distinctly lemon-yellow at apex and also in the base, white mycelioid over base, surface covered with tufts of silky hairs, fibrous under hairs, brittle, hollow, lemon-yellow or darker yellow within. **SPORE PRINT** flesh pink. **ODOR** mild and not distinctive. **TASTE** farinaceous and sometimes intensely bitter farinaceous.

Habit and habitat scattered or clustered on soil and humus in deciduous woods or on stream banks. June to August. **RANGE** widespread but not commonly collected.

Spores: angular, hyaline or pale straw color, 7–9 × 6–7 µm; 5- to 6- (to 7-) angled in side view, cheilocystidia broadly clavate or inflated.

Comments Originally described from Massachusetts, we find it occasionally in New York at moderate elevations (around 300 m above sea level), in older forests. Yellow gills in *Entoloma* species are rare; see also *E. subsinuatum* and *E. luridum*.

Leptonia foliomarginata
(Peck) T. J. Baroni

SYNONYMS *Agaricus foliomarginatus* Peck, *Leptoniella foliomarginata* (Peck) Murrill
Cap brown with blue margin and striate-lined; gills white with blue fringed edge; stem dark grayish-blue, smooth

CAP 2–5 cm, convex with a depressed center, grayish-brown with darker blue fibrils and scales scattered over the surface, center dark bluish-brown scaly or hairy, strongly striate-lined to disc, margin blue. **FLESH** white, thin. **GILLS** white, becoming flesh pink, with a dark blue edge, broad, close, edge fringed. **STEM** 3–6 cm long, 1–2 mm broad, equal, dark grayish-blue, white mycelioid over base, smooth, hollow. **SPORE PRINT** flesh pink. **ODOR** and **TASTE** not distinctive.

Habit and habitat single, scattered or in clusters on the ground, on decayed hardwood mulch or among mosses (especially *Sphagnum*) in rich moist forests. July to September. **RANGE** widespread, very common in the boreal forests.

Spores: angular, hyaline or pale straw colored, 8–12 × 6.5–7 µm, with 5–7 angles in side view, cheilocystidia cylindrical or inflated.

Comments Three other species of *Leptonia* with blue-black-edged gills (a feature called marginate) can be found in the Northeast: *L. serrulata*, *L. velutina*, and *L. subserrulata*. See comments under *L. serrulata*. *Leptonia foliomarginata* is very similar in appearance to the European *L. caesiocincta* (Kühner) P. D. Orton.

Leptonia formosa (Fries) Gillet

SYNONYMS *Entoloma formosum* (Fries) Noordeloos, *Leptonia fulva* P. D. Orton

Cap yellow with brown sunken disc, conical becoming plane, margin lined; gills white, attached, edges same color as faces (not marginate); stem yellow, smooth, white cottony mycelioid

CAP 1–5 cm, conical-convex becoming bell-shaped or eventually plane, always with a sunken center, yellow or yellowish-brown or copper color and darker date brown on the center, sparsely radiately fibrillose to margin, more densely brown erect scaly over center, margin striate-lined with age. **FLESH** white, thin. **GILLS** white at first or pale cream, but soon pink, attached or with short decurrent tooth, close or subdistant, broad, edge smooth. **STEM** 2.5–10 cm long, 1–4 mm broad, equal, yellow or yellowish-brown becoming darker bronze brown with age, glabrous, smooth, white cottony mycelioid over base, hollow.

SPORE PRINT flesh pink. **ODOR** and **TASTE** not distinctive.

Habit and habitat scattered or clustered on humus or on well-decayed soft crumbling logs in hardwood or hardwood-coniferous forest, in rich moist forests, often with beech and hemlock. July to August. **RANGE** widespread.

Spores: angular, hyaline or pale straw color, 9–14 × 6–9 µm, irregularly (5- to) 6- to 9-angled in side view; cheilocystidia cylindrical or club-shaped.

Comments Similar looking is the yellow *Leptonia xanthochroa* P. D. Orton, but it has more ochraceous pigments in the cap and stem, and the gills are yellowish-ochraceous with edges that are brown or dark yellow-brown fringed. *Leptonia xanthochroa* can be found in the Adirondack region of New York and most likely elsewhere in the Northeast, but it has not yet been reported in the technical literature.

Leptonia serrulata (Fries) P. Kummer

BLUE-TOOTHED ENTOLOMA

SYNONYM *Entoloma serrulatum* (Fries) Hesler

All parts deep blue or bluish-black; gill edges blue fringed, fragile

CAP 0.5–5 cm, convex with a depressed center, completely dark blue or blue-black at first, fading to violaceous-gray, finely hairy or minutely scaly overall, but distinctly blue-black scaly over center, margin becoming striate-lined to near center. **GILLS** white or often pale blue when young becoming flesh pink, attached or short decurrent, with dark blue fringed edge. **STEM** 1.5–6 cm long, 1–4 mm broad, dark blue or violaceous-blue with dark blue-dotted apex and blue fibrillose appressed elsewhere except for the white, mycelioid base, hollow. **SPORE PRINT** flesh pink. **ODOR** and **TASTE** not distinctive.

Habit and habitat single or clustered on the ground in deciduous forest, also on decayed logs among mosses in rich moist forests. July to September. **RANGE** widespread.

Spores: angular, hyaline or pale straw color, 9–12 × 6–8 µm, 5- to 7-angled in side view; cheilocystidia cylindrical or inflated with blue internal pigment.

Comments This is our most commonly collected bluish *Leptonia* species with blue-edged gills. Others with blued-edged gills are *L. foliomarginata*, *L. velutina*, and *L. subserrulata*.

Leptonia foliomarginata has a brown-striate cap with blue margin.

Leptonia subserrulata has a cap and stem that become creamy-white and the stem is sparsely dotted with blue scales and fibrils.

Leptonia velutina (Hesler) Largent is a striking species that might be confused with *L. serrulata*, but differs by its vase-shaped cap and heavily ornamented stem. In contrast, *L. serrulata* has a blue, convex (not vase-shaped) cap with nonvelvety surface ornamentation. Also the stem of *L. serrulata* is not as densely adorned and is dark blue.

Leptonia velutina

Leptonia subserrulata Peck

SYNONYMS *Entoloma subserrulatum* (Peck) Hesler, *Leptoniella subserrulata* (Peck) Murrill

Cap and stem pale cream or pinkish-buff, the cap with blue margin, the stipe with blue dots and fibrils scattered; gills white, then pinkish-buff and with dark blue-black edge

CAP 0.5–4 cm, convex becoming plane, always with a sunken center, dark bluish-gray-brown at first, but quickly fading to pale pinkish-buff or pinkish-gray with cream tints but the disc remaining brown and the margin faintly bluish, sparsely radiately brown fibrillose to margin, more densely brown scaly over center, margin striate-lined with age. **GILLS** white at first, but soon creamy pinkish-buff with dark bluish-black edges, attached or with short decurrent tooth, close, broad, edge fringed or granulate. **STEM** 3–8 cm long, 1–4 mm broad, pale blue at first, then pale creamy or pinkish-buff like the cap, but with scattered dark blue dots and scales over the apex and sparsely blue fibrillose appressed elsewhere except for the white, mycelioid base, hollow. **SPORE PRINT** flesh pink. **ODOR** not distinctive, or slightly fragrant. **TASTE** not distinctive.

Habit and habitat scattered or clustered on humus in hardwood or hardwood-coniferous forest, in rich moist forests, often with beech and hemlock. July to September. **RANGE** widespread but not commonly seen.

Spores: angular, hyaline or pale straw color, 8–12 × 6–8 μm, 5- to 6-angled in side view; cheilocystidia club-shaped individually but clumped together, with dark blue internal pigment.

Comments One of my favorite entoloma species, it is truly beautiful. The cap color is dark at first but soon turns pale cream color with a blue-rimmed margin as the cap grows and expands. Both young and mature stages are necessary to properly identify this species.

Leptonia lividocyanula
(Kühner) P. D. Orton

SYNONYMS *Entoloma lividocyanulum* Noordeloos, *Rhodophyllus lividocyanulus* Kühner

Cap reddish-brown turning yellowish-brown with age, strongly striate-lined; gills white, attached with decurrent tooth; stem bluish-gray, smooth, white cottony mycelioid over base

CAP 1–3.5 cm, convex with slight or distinctly sunken center, reddish-brown becoming paler yellowish-brown except for reddish-brown center with age, densely brown erect scaly over center, with fine fibrils and scattered scales elsewhere, but glabrous at margin, strongly translucent-striate-lined with age. **GILLS** white at first, but soon pink, attached or with short decurrent tooth, subdistant or close, broad, edge smooth. **STEM** 2.5–8 cm long, 1–4 mm broad, equal, bluish-gray at first, fading to gray then eventually brown, glabrous, smooth, white cottony mycelioid over base that often discolors yellow-orange, hollow. **SPORE PRINT** flesh pink. **ODOR** and **TASTE** not distinctive.

Habit and habitat scattered or clustered on humus in beds of sphagnum moss and grass in boggy areas under hardwood or hardwood-coniferous forest, in rich moist forests. July to August. **RANGE** Perhaps restricted to bogs with *Sphagnum* and grass in deep wet humus layers.

Spores angular, hyaline or pale straw color, 7–10 × 6.5–7.5 μm, irregularly 5- to 7-angled in side view; lacking cheilocystidia.

Comments In Europe this species is found in grasslands, but in eastern North America it seems to be restricted to fens and bogs with *Sphagnum* and grasses present. *Leptonia asprella* (Fries) P. Kummer and *L. gracilipes* Peck are similar due to the dark brown cap, white gills, and bluish stems, but they are found on humus under hardwoods. David Largent considers this group difficult to separate using morphological features.

Leptonia rosea Longyear

SYNONYM *Entoloma roseum* (Longyear) Hesler

Cap, dark rose-brown, densely scaly on center; gills white but with delicate rose tints; stem rosy pink, glabrous

CAP 1.5–4 cm, convex and remaining so, with a narrow barely sunken center, surface covered with deep rose-brown or wine-brown scales or becoming more fibrillose over the margin, not or only faintly translucent-striate-lined when moist. **FLESH** white, thin. **GILLS** white at first, becoming delicate rosy-pink, attached, subdistant or close, moderately narrow, edge fringed. **STEM** 6–9 cm long, 2–3 mm broad, equal or enlarged toward base, rosy-pink or with some brown hues, glabrous, smooth, white mycelioid over base, hollow. **SPORE PRINT** flesh pink. **ODOR** fragrant. **TASTE** mild.

Habit and habitat scattered or clustered on soil, humus or well-decayed wood under hardwood forest, in areas of limestone bedrock. July to August. **RANGE** widespread but not commonly seen.

Spores: angular, hyaline or pale straw color, 8–11 × 6.5–8 μm, 5- to 6-angled in side view; cheilocystidia fusoid, clavate or subcapitate.

Comments A few species of entolomatoid fungi seem to be mainly associated with limestone habitats in the eastern part of North America, this is yet another example of such a species along with *Leptonia odorifera* and *L. incana*.

Claudopus depluens (Batsch)
Gillet

SYNONYMS *Agaricus depluens* Batsch, *Entoloma depluens* (Batsch) Hesler, *Rhodophyllus depluens* (Batsch) Quélet

Small, white, shell-shaped, surface velvet silky; gills white, then pink, then fleshy-brown from spores; stipe stublike or absent, white cords radiating out from base; on decaying wood

CAP 0.6–2 cm wide, semicircular or kidney or shell-shaped, white or pale gray, densely erect velvety, dry. **FLESH** white. **GILLS** white, becoming fleshy-brown from spores, bluntly attached, close. **STEM** when present it is highly reduced, at the cap edge, equal, colored as cap, wispy hairy, with fine white spiderweb-like threads at base and spreading over the substrate. **SPORE PRINT** flesh or pinkish-brown. **ODOR** and **TASTE** not distinctive.

Habit and habitat single or in small clusters, on rotten wood of downed hardwood trees. July to September. **RANGE** widespread.

Spores angular, 5- to 7-angled in side view, 8.5–11 × 7–7.5 μm; cheilocystidia cylindrical.

Comments Compare with other shell-shaped (pleurotoid) species of *Clitopilus*, *Cheimonophyllum* (white spores), and *Crepidotus* (brown spores). If the spore print is pinkish-brown, a microscopic examination of the spores is necessary to be certain it is not a *Clitopilus* species.

Claudopus byssisedus (Persoon) Gillet differs from *C. depluens* by its grayish-brown, silky cap, lack of cheilocystidia, and its mealy odor and taste. Both species have pinkish-brown spores and are found in similar habitats. If the cap is white, you have *C. depluens*; if it's grayish brown, *C. byssisedus*.

Claudopus byssisedus

Claudopus parasiticus
(Quélet) Ricken

SYNONYMS *Leptonia parasitica* Quélet, *Entoloma parasiticum* (Quélet) Kreisel, *Rhodophyllus parasiticus* (Quélet) Quélet

Very small, white, shell-shaped; gills white, then fleshy-brown from spores, subdistant and broad; stipe stublike or absent; on decaying polypores or chanterelles or mosses

CAP 0.3–0.9 cm wide, circular or semicircular or kidney or shell-shaped, white, densely erect velvety, dry. **FLESH** white, very thin. **GILLS** white, becoming fleshy-brown from spores, bluntly attached, broad, subdistant or distant. **STEM** when present it is highly reduced, at the cap edge, equal, colored as cap or pale buff-brown, wispy hairy, with fine white spiderweb-like threads at base and spreading over the substrate. **SPORE PRINT** flesh or pinkish-brown. **ODOR** and **TASTE** not distinctive.

Habit and habitat in small clusters, on rotting fruit bodies of polypores, chanterelles and other fungi, also known to grow on mosses and occasionally on well-rotted bark of pines. July to September. **RANGE** widespread, not commonly found.

Spores angular, 9–12.5 × 8–10.5 μm, 5- to 6-angled in side view.

Comments This small pleurotoid species with pinkish-brown spores is the most common one known to grow on decaying polypores and chanterelles. However, an examination of the spores with a microscope is needed to confirm it is not *Entoloma pseudoparasiticum* Noordeloos, which has smaller, narrower spores (7.5–10 × 6–7.5 μm) and is gray becoming pale brown. *Entoloma pseudoparasiticum* has not been reported yet from eastern North America and is known to grow only on chanterelles.

Gilled Mushrooms with Brown Spore Prints

The color of the spores in this group can vary from dull clay-brown or yellow-brown to rusty-brown or orange-brown, depending upon the genus, but within a genus the color remains consistent for most species. For example, the species in the genus *Cortinarius* generally have rusty-brown spores, while the species in *Inocybe* have dull brown or dull yellow-brown spores. These mushrooms can range from small and delicate (*Conocybe*, *Crepidotus*, *Galerina*, *Bolbitius*, and *Flammulaster*) to quite large and fleshy (*Cortinarius*, *Gymnopilus*, *Pholiota*, *Tapinella*) with a host of species also medium-sized representing the classic LBM (little brown mushroom). In this group are several poisonous species, a few of which are covered, but not all. Generally, it is not a good idea to consume brown-spored mushrooms until you reach the level of expert in making accurate identifications of species.

The following key will help to identify the correct genus for mushrooms covered in this guide. This is not a key to all known genera, since some have not been included here for lack of images or space. It is out of the scope of any mushroom field guide to cover the hundreds of brown-spored mushrooms that occur in North America, but this guide presents the commonly collected ones that you will surely find, along with some not so common but very attractive species.

Growing on wood

STEM ABSENT OR OFF-CENTER (ECCENTRIC)
Crepidotus Mostly white, but some colored or with colored scales, fan-shaped, gills brown, not easily removed, small or medium, thin-fleshed
Pseudomerulius Cap smooth, gills orange-yellow and wavy-wrinkled, easily removed, odor spicy, like cinnamon, medium-sized, fleshy
Tapinella panuoides Cap velvety, gills yellow, frequently forked, easily removed from cap, medium to large, fleshy, often shelving, (spores red-brown dextrinoid in Melzer's reagent)

STEM PRESENT AND CENTRAL
Flammulaster Cap covered with rusty-brown granular-powdery layer, conical-convex, small
Galerina Cap glabrous, smooth, mostly conical or bell-shaped with translucent-lined margin, or convex and opaque, ring present of absent, mostly small and delicate, some medium-sized and fleshy, spore print orange-brown (spores bumpy-ornamented, with a plage)
Gymnopilus Cap and stem orange-brown or red-brown, small, medium or large, fleshy, bitter tasting, spore print bright orange-brown (spores bumpy-ornamented, lacking a plage)
Simocybe Fruit bodies with olive-brown colors, fleshy, cap velvety or smooth (spores smooth, bean-shaped)
Bolbitius Cap slimy, thin-fleshed, very fragile, gills ochre-brown and deliquescent, spore print rusty-brown (spores smooth, truncate with an apical pore)
Pholiota Cap fleshy, not fragile, sometimes slimy, often with brown scales but may be glabrous, with a membranous ring or a fibrillose ring zone (spores smooth and often kidney-shaped and with a minute apical pore or pore absent)

Growing on the ground on humus or grassy areas, including on wood chips, mosses, or dung

GILLS DECURRENT BUT EASILY REMOVED FROM THE CAP

Paxillus Cap brown, becoming olive or grayish-brown, with raised ribs on strongly inrolled margin, slimy or slippery when wet, matted hairy or smooth, gills yellow, turning brown when injured

Tapinella Cap orange-brown or yellow-brown, becoming black, densely velvety, gills not turning brown from injury

GILLS BLUNTLY ATTACHED OR NEARLY FREE AND NOT EASILY REMOVED FROM THE CAP

Agrocybe Small or medium, cap dry, fleshy, gills tobacco-brown, ring when present membranous

Bolbitius Cap slimy, thin-fleshed, very fragile, gills ochre-brown and deliquescent

Conocybe Cap dry, conical, thin-fleshed, fragile, gills cinnamon-brown

Cortinarius Small or large, variously colored, with cobweb ring when young, a large number of species, frequently encountered

Inocybe Gills grayish-brown, cap scaly or hairy

Hebeloma Gills grayish-brown, cap smooth, often slippery

Phaeocollybia Stem with long root (pseudorhiza), cap conical red-brown

Crepidotus mollis (Schaeffer) Staude

JELLY CREP

SYNONYMS *Agaricus mollis* Schaeffer, *Crepidotus fulvotomentosus* (Peck) Peck

Fan-shaped, lacking a stem, brown hairy or scaly on cap surface; flesh soft and gelatinous; on hardwood debris

CAP 1–5 cm wide, semicircular or fan-shaped, pale grayish or off-white but covered with orange-brown or olive-brown hairs and scales, surface densely hairy and fibrillose-scaly over all, occasionally almost glabrous, dry. **FLESH** white, soft and gelatinous. **GILLS** white, becoming grayish-cinnamon-brown, radiating from stublike lateral stem, close, soft, edge gelatinizing. **STEM** nonexistent or small stublike. **SPORE PRINT** yellowish-brown. **ODOR** and **TASTE** not distinctive.

Habit and habitat single, scattered, or clustered on decaying hardwood logs, stumps, branches. June to October. **RANGE** widespread.

Spores: ellipsoid, smooth, pale brown in KOH, 7–10 × 4.5–6 µm, lacking clamp connections.

Comments A highly variable species for the amount of hairs and scales found on the cap surface. If the gills are yellow or orange, and the cap also has yellow and orange tints, then you have *Crepidotus crocophyllus* (Berkeley) Saccardo, a rather commonly encountered species.

Crepidotus applanatus
(Persoon) P. Kummer

FLAT CREP

SYNONYMS *Agaricus applanatus* Persoon, *Crepidotus globiger* (Berkeley) Saccardo, *Crepidotus putrigenus* (Berkeley & M. A. Curtis) Saccardo

Small, white, stalkless and fan-shaped, on bark of hardwood logs, stumps, branches

CAP 1–4 cm wide, semicircular or shell-shaped or fan-shaped or spathulate, white or off-white, becoming pale brownish or pale cinnamon, surface glabrous or minutely hairy near attachment, moist and translucent-lined over margin at first. **FLESH** white, thin. **GILLS** white, becoming brown, attached and subdecurrent, radiating from short lateral stem, close or crowded, narrow. **STEM** very small stublike, equal. **SPORE PRINT** brown or cinnamon-brown. **ODOR** and **TASTE** not distinctive.

Habit and habitat shelving and overlapping clusters on decaying hardwoods or more rarely conifers. July to October. **RANGE** widespread.

Spores: globose, minutely spiny, pale brown in KOH, 4–5.5 µm, clamp connections present.

Comments A white, shelving, stalkless mushroom commonly encountered on decaying hardwood debris. White forms of *Pleurotus ostreatus* and *P. pulmonarius*, that also grow on similar bark-covered hardwood substrates, have decurrent gills that do not turn brown, are generally larger, and produce grayish-lilac spore drops. If the gills (and spore print) are reddish or pinkish-brown, see *Claudopus depluens*, a much smaller species that generally fruits only in small numbers and not typically overlapping on well-decayed barkless hardwood debris. Also see *Cheimonophyllum candidissimum*.

Crepidotus nyssicola (Murrill) Singer

SYNONYM *Pleuropus nyssicola* Murrill
Medium, white, vase-shaped, glabrous cap; gills white decurrent, close; stem often off-center (eccentric), white, fleshy-tough, in small bouquetlike clusters; on well-decayed wood

CAP 3–6 cm wide, deeply sunken or vase-shaped with arching and downturned margin, white or pale creamy-white, may turn dark red-brown where injured, also may develop dark brown areas irregularly over the surface from spores, surface minutely hairy at first, soon glabrous, moist and faintly translucent-lined over margin. **FLESH** white, thin. **GILLS** white, becoming brown from spores, decurrent, close. **STEM** 1–3 cm long, 4–8 mm wide, central or off-center (eccentric), equal or tapered downward, glabrous or slight scaly, fleshy-tough. **SPORE PRINT** brown or cinnamon-brown. **ODOR** not distinctive, or like the giant puffball. **TASTE** not distinctive.

Habit and habitat scattered or clustered and bouquetlike on decaying hardwoods and conifers, originally described from a decaying southern swamp tupelo, a hardwood. July to October. **RANGE** widespread but rare.

Spores: subglobose, minutely spiny, pale brown in KOH, 5–7 × 5–6 µm; cheilocystidia cylindrical with swollen head (capitate) or club-shaped; clamp connections present.

Comments One might mistake this species for a pleurotus, but the spore print will correct that mistake quickly.

Pseudomerulius curtisii
(Berkeley) Redhead & Ginns

SYNONYMS *Paxillus curtisii* Berkeley, *Meiorganum curtisii* (Berkeley) Singer, J. García & L. D. Gómez, *Paxillus corrugatus* G. F. Atkinson, *Tapinella corrugata* (G. F. Atkinson) E.-J. Gilbert

Shelving on conifer wood, bright orange-yellow wavy-wrinkled gills; cap fan-shaped, yellow then olive-yellow, suede-like; stem absent; odor spicy pungent, disagreeable to most

CAP 2–5 cm wide, laterally attached, convex with margin strongly inrolled, becoming plane, from the top fan-shaped or semi-circular, maize-yellow or canary-yellow at first, turning brownish-yellow or olivaceous-yellow especially around the margin, some developing reddish-brown stains, dry, smooth, suedelike or slightly woolly. **FLESH** pale yellow, spongy. **GILLS** bright golden-yellow or orange-yellow, developing olive-brownish hues with age, no obvious stipe present, close or crowded, easily removed from cap, regularly and repeatedly forked, wavy-wrinkled. **STEM** absent. **SPORE PRINT** yellowish-brown. **ODOR** characteristic, disagreeable spicy pungent, like slightly off cinnamon. **TASTE** not distinctive, or slightly bitter.

Habit and habitat usually clustered and shelving on barkless downed conifer logs or stumps, especially pine or hemlock. July to October. **RANGE** widespread.

Spores ellipsoid, smooth, pale yellow or colorless in 3% KOH, 3–4 × 1.5–2 µm.

Comments A brightly colored species if you see the gills. I personally like the pungent odor, but the insects seem not to, thus allowing longer time periods for reproduction for this fungus. Many insects like eating fungal tissues, cutting down the time they get to release spores and reducing their chances of reproductive success.

Tapinella panuoides (Fries)
E.-J. Gilbert

STALKLESS PAXILLUS
SYNONYMS *Agaricus panuoides* Fries, *Crepidotus panuoides* (Fries) Pilát, *Paxillus panuoides* (Fries) Fries

Stemless, often shelving, yellow-brown, velvety, then smooth caps; gills yellow or with orange hues, frequently forked, easily removed from the cap

CAP 2–10 cm wide, fan- or petal- or shell-shaped with margin strongly inrolled and becoming scalloped, convex then plane, olive-brown, yellow-brown, or tan, dry, fine velvety at first, then glabrous. **FLESH** dirty white or sordid yellow, thin. **GILLS** yellowish or with dull orange hues, radiating from attachment to substrate, close, frequently forked, some cross-veined, easily removed from cap. **STEM** absent or very reduced. **SPORE PRINT** yellowish-brown. **ODOR** and **TASTE** not distinctive.

Habit and habitat single, scattered, or in overlapping clusters on decaying wood of conifers, also on milled lumber. May to November. **RANGE** widespread.

Spores ellipsoid, smooth, almost colorless in 3% KOH, many orange-brown (dextrinoid) in Melzer's reagent, 3.5–6 × 2.5–4 µm.

Comments The easily removed gills, like tubes from a bolete cap, make a good field character since very few mushrooms show this feature. There are many pleurotoid (stemless or stems off-center on cap) mushrooms, but compare this species with *Pseudomerulius curtisii* and *Phyllotopsis nidulans*. All three have distinctive features that will allow you to distinguish them in the field.

Flammulaster erinaceellus
(Peck) Watling

SYNONYMS *Agaricus erinaceellus* Peck, *Flocculina erinaceella* (Peck) P. D. Orton, *Phaeomarasmius erinaceellus* (Peck) Singer, *Pholiota erinaceella* (Peck) Peck, *Agaricus detersibilis* Peck

Small rusty-brown caps with orange-yellow ground color, covered with rusty-brown granular scales, margin with patches of white fibrillose ring; gills dull yellow, bluntly attached; stem similar to cap; on rotting hardwood logs

CAP 1–3 cm wide, conical-convex, becoming plane, dark rusty-brown with yellow or ochraceous ground color, dry, granular-scaly overall, margin briefly with white fibrillose ring fragments. **FLESH** pale yellow, thin. **GILLS** dull yellow or pale orange-brown, bluntly attached, close or crowded, edges crenulate. **STEM** 3–4(–6) cm long, 2–4 mm wide, similar in color and covering with cap, becoming paler and glabrous with age, with evanescent fibrillose ring. **SPORE PRINT** dingy cinnamon-brown. **ODOR** not distinctive. **TASTE** not distinctive, or slightly bitter and metallic.

Habit and habitat single or clustered on rotting hardwood logs. July to October. **RANGE** widespread.

Spores ellipsoid or somewhat bean-shaped in side views, smooth, dingy cinnamon in 3% KOH, 6–9 × 4–4.5 μm; cheilocystidia abundant, cylindrical with a rounded, inflated head (capitulate), clamp connections present.

Comments A delicate but attractive LBM found throughout the Northeast. There are many species of small, wood-inhabiting, brown-spored (and white-spored) mushrooms, but this particular one is fairly common and easily recognized.

Galerina tibiicystis
(G. F. Atkinson) Kühner

SPHAGNUM-BOG GALERINA
SYNONYMS *Galerula tibiicystis*
G. F. Atkinson, *Galera tibiicystis*
(G. F. Atkinson) A. Pearson, *Galerula lasiosperma* G. F. Atkinson

Growing in *Sphagnum*, light brown or orange-brown, translucent-lined cap; stem long and fragile, lacking any veil material, but dotted with white pruina over most of stem

CAP 1–3.5 cm wide, conical-convex, becoming convex with low broad bump over the center, light brown or orange-brown and translucent-lined to near center, hygrophanous and fading to orange-buff from center outward, becoming opaque, glabrous, moist. **FLESH** color of surface, thin. **GILLS** light brown as cap, becoming rusty-brown from spores, bluntly attached, broad, subdistant. **STEM** 5–10(–20) cm long, 1.5–3 mm broad, equal, long and thin, colored as cap, or becoming pale buff-brown or darker brown on some, upper stem covered with fine white dots, becoming glabrous, hollow and very fragile, without a ring. **SPORE PRINT** rusty-brown. **ODOR** and **TASTE** not distinctive.

Habit and habitat scattered or in small clusters, on *Sphagnum* in bogs. May to November. **RANGE** widespread.

Spores elliptical, rusty-brown in water mounts, bumpy overall except near apiculus on the face of the spore where it is smooth (called a plage), 8–11(–14) × 5–6(–7) µm; cheilocystidia and caulocystidia similar, having an inflated base and narrow elongate neck with small round head (tibiiform).

Comments One of the larger moss-loving species of *Galerina*. *Galerina sphagnicola* (G. F. Atkinson) A. H. Smith & Singer occurs in the same habitat on *Sphagnum* and is similar in appearance but differs by fine white ring remnants on the cap margin and stipe surface. Also the spores of *G. sphagnicola* are calyptrate, with a loose skirtlike outer wall, and those of *G. tibiicystis* are not calyptrate. See also *G. paludosa* that is found on *Sphagnum*.

Galerina paludosa (Fries)
Kühner

SYNONYMS *Galerula paludosa* (Fries) A. H. Smith, *Tubaria paludosa* (Fries) P. Karsten, *Hydrocybe paludosa* (Fries) M. M. Moser

In *Sphagnum* bogs or fens, long thin orange-brown fragile mushroom with white torn rings of veil tissue distributed over most of stem

CAP 1–3 cm wide, conical-convex, becoming bell-shaped or eventually plane, orange-brown, hygrophanous and fading to ochre-buff, covered at first with thin layer of white veil fibrils that disappear after cap expansion, margin translucent-lined, often with sparse white veil tissue hanging from cap margin. **FLESH** colored as surface, thin. **GILLS** pale yellow or pale brownish, attached with decurrent tooth, close, broad, edges white-fringed. **STEM** 7–16(–20) cm long, 1–4 mm broad, equal and fragile, watery hazel, covered with white submembranous torn rings of veil from near apex and over most of stem, becoming glabrous with age. **RING** white, submembranous. **SPORE PRINT** rusty-brown. **ODOR** and **TASTE** not distinctive.

Habit and habitat single or more frequently in scattered clusters on *Sphagnum* in bogs or fens. **RANGE** widespread.

Spores elliptical or ovate, pale rusty-brown in aqueous mounts, very slightly dotted or wrinkled except near apiculus on the face of the spore where it is smooth (called a plage), 9–10.5 × 6–7 μm; cheilocystidia fusiform with rounded apex mostly.

Comments This species of *Galerina* is the most common found in sphagnum bogs in the Northeast. Is it toxic? No one is sure, but do not eat this small fragile mushroom. It has relatives that are deadly poisonous, see *Galerina marginata*.

Galerina marginata (Batsch) Kühner

DEADLY GALERINA

SYNONYMS *Galerina autumnalis* (Peck) A. H. Smith & Singer, *Galerina unicolor* (Vahl) Singer

Small or medium-sized, tacky or sticky cap with faint translucent lines; gills yellow then rusty; stem with ring or bandlike ring zone covered with rusty-brown spores; on well-decayed wood

CAP 2.5–6.5 cm wide, convex, becoming plane with low bump over the center, dark brown or orange-brown, hygrophanous and fading to orange-buff from center outward, glabrous, sticky, faintly translucent-lined. **FLESH** watery brown, thin. **GILLS** yellow, becoming rusty-brown from spores, bluntly attached, broad, close. **STEM** 2–10 cm long, 3–6 mm broad, equal, off-white from fibrils covering surface, becoming orange-brown where fibrils rubbed off, or pale buff-brown, off-white membranous ring covering gills, mostly collapsing on upper stem and covered with rusty-brown spores, leaving a flattened bandlike ring, this sometimes disappearing. **SPORE PRINT** bright rusty-brown. **ODOR** not distinctive, or faintly farinaceous. **TASTE** not recommended.

Habit and habitat scattered or in small clusters, on downed tree logs, often well-rotted and very wet hardwood or conifer wood. June to September, but more common in the fall. **RANGE** widespread, very common.

Spores elliptical, pale rusty-brown, bumpy overall except near apiculus on the face of the spore where it is smooth (called a plage), 8–11 × 5–6.5 μm.

Comments Deadly poisonous. Be very careful if you collect honey mushrooms (*Armillaria*) for consumption, since these two can occur in similar habitats and have a vaguely similar brownish coloration. *Armillaria* species are larger and produce white spore deposits, and they really look very different with the black stiff hairs on the cap.

Galerina stylifera
(G. F. Atkinson) A. H. Smith & Singer

SYNONYM *Galerula stylifera* G. F. Atkinson

Small or medium, slippery conical-convex cap with translucent lines; gills orange-tan, bluntly attached; stem dark brown from base upward, covered with silvery flattened fibrils, leaving evanescent ring zone at apex; on woody debris along unimproved roads

CAP 1.5–5 cm wide, conical-convex, becoming plane with low bump over the center, orange-brown or cinnamon-brown, strongly translucent-lined when fresh, hygrophanous and fading to pale orange-buff from center outward, becoming opaque, glabrous, slippery, becoming dry. **FLESH** watery brown, thin. **GILLS** orange-tan, becoming rusty-brown from spores, bluntly attached, subdistant, broad. **STEM** 4–6 cm long, 3–6 mm broad, equal, dark brown or reddish-brown below, orange-tan above, mostly covered with silvery flattened fibrils up to an evanescent ring zone in upper region, brown and hollow inside. **RING** filmy-fibrillose, leaving thin and easily lost ring zone of fibrils, this ring zone usually disappearing. **SPORE PRINT** bright rusty-brown. **ODOR** and **TASTE** not distinctive.

Habit and habitat scattered or clustered on buried sticks or around woody debris on road edges, or on logs, mostly conifer. July to September. **RANGE** widespread, very common.

Spores elliptical, pale orange-brown, faintly bumpy or nearly smooth, plage difficult to see, 6.5–8.7 × 4–5 µm; cheilocystidia bowling-pin-shaped, some with swollen round cap on top.

Comments Much larger than the typical moss-inhabiting galerina, but the cap looks just the same. I find this species along unimproved roadsides just as originally described by Atkinson.

Gymnopilus luteus (Peck)
Hesler

SYNONYM *Pholiota lutea* Peck

Large pale yellow, scaly cap; gills pale yellow; stem with ring, flattened scaly below ring; taste bitter; on decaying wood of hardwoods

CAP 5–10 cm wide, convex, pale cream-yellow or warm buff, covered with small flattened fibrillose scales, sometimes these a little darker than the ground color, dry, opaque. **FLESH** pale yellow, fleshy-firm. **GILLS** pale yellow, becoming rusty-brown from spores, attached, close. **STEM** 4–7.5 cm long, 6–16 mm broad, equal but enlarged at base, same color as cap, but turning rusty-brown in areas when handled, mostly covered with flattened fibrillose scales up to the ring near the top of the stem, solid inside. **RING** membranous or fibrillose-membranous, as a distinct skirt with rusty-brown spores on the upper surface. **SPORE PRINT** bright rusty-brown. **ODOR** pleasant. **TASTE** bitter.

Habit and habitat scattered in clusters or bouquetlike bunches on decaying wood, logs, trunks of hardwoods. July to September. **RANGE** widespread, but not common.

Spores elliptical, bumpy (verruculose) overall, reddish-brown in KOH, darker reddish-brown in Melzer's reagent (dextrinoid), 6–9 × 4–5 µm; cheilocystidia flask-shaped and also with swollen head (capitate).

Comments *Gymnopilus junonius* (Fries) P. D. Orton, formerly known as *G. spectabilis* in most field guides, can be similar in size and has a ring but differs by the brighter rusty-orange colors and the cap that is not uniformly scaly. It appears *G. junonius* also more easily stains rusty on the cap, gills, and stem. Of the two, *G. junonius*, the big laughing gym, is more frequently encountered.

Simocybe centunculus (Fries)
P. Karsten

SYNONYMS *Agaricus centunculus* Fries, *Naucoria centunculus* (Fries) P. Kummer, *Ramicola centunculus* (Fries) Watling

Small olive-brown, densely woolly caps with translucent-lined margins; gills brown with olive tint, attached; stem colored as cap or with more olive hues, densely pubescent at first; on well-decayed hardwood trees

CAP 0.5–2 cm wide, convex, becoming plane, some with low broad central bump, dark grayish-brown with olive tints, but soon paler bronze or yellowish-brown, hygrophanous from center outward and even paler yellowish-tan, translucent-lined over margin at first, minutely woolly (tomentulose) overall (use a lens). **FLESH** brown, thin. **GILLS** pale yellowish-brown with olive tint, attached, close, broad, edges white, frayed. **STEM** 0.5–3 cm long, 1–2 mm broad, equal, same color as cap or with darker olive tints, white mycelioid at base, minutely white fibrillose overall at first, becoming glabrous over the base. **SPORE PRINT** brown. **ODOR** not distinctive, or pungent (geranium odor for me). **TASTE** bitterish but with floral hints.

Habit and habitat clustered on well-decayed, crumbling wood of hardwood stumps or downed logs. August to November. **RANGE** widespread.

Spores broadly ellipsoid or slightly bean-shaped, smooth, pale tan in KOH, 6–8 × 4–5 μm.

Comments One of the LBMs that can be identified using solely field characters with some degree of confidence. The habitat and colors of the fruit bodies are distinctive. Compare with *Flammulaster erinaceellus* that is found in similar habitats.

Bolbitius titubans (Bulliard)
Fries

SYNONYMS *Agaricus titubans* Bulliard, *Bolbitius variicolor* G. F. Atkinson, *Bolbitius vitellinus* (Persoon) Fries

Very fragile, with yellow, slimy, strongly ridged-lined cap; gills orange-brown, flabby, nearly free; stem lemon-yellow, covered with tufted fibrils, very fragile; on rich humus or fertilized soil

CAP 2–4 cm wide, conical at first or bell-shaped, becoming convex then plane with a rounded central bump (umbo), gray-olivaceous with lemon-yellow buff colors mixed in, slimy at first, wrinkled on the center, ridged-lined over the margin. **FLESH** thin, very soft-fragile. **GILLS** pale lemon-yellow at first, then orange-brown, nearly free, close, narrow. **STEM** 4–10 cm long, 3–8 mm broad, equal, dry, mostly lemon-yellow becoming straw color, very fragile, minutely dotted with tufts of fibrils over all. **SPORE PRINT** orange or bright rusty-brown. **ODOR** and **TASTE** not distinctive.

Habit and habitat single, scattered, or clustered on freshly fertilized soil in grassy fields or margins of forested areas. July to September. **RANGE** widespread.

Spores elliptical, smooth, truncate and with large germ pore, 10–15 × 6–7 μm.

Comments Formerly split into two species commonly found in field guides, *Bolbitius variicolor* and *B. vitellinus*. Current taxonomic work considers the colors variable and that they all belong to a single overlooked species, *B. titubans*.

Pholiota aurivella (Batsch) P. Kummer

GOLDEN PHOLIOTA

SYNONYMS *Agaricus aurivellus* Batsch, *Lepiota squarrosa* var. *aurivella* (Batsch) Gray

Bouquetlike clusters, ochre-yellow, medium to large caps with slimy surface and flattened brown scattered spotlike scales; stem yellowish with curled scales and an evanescent ring

CAP 4–18 cm wide, convex or broadly convex, often with broad central bump, ochraceous-orange or ochre-yellow, becoming dark brownish-orange with age, with flattened large brown scales that become gelatinous, rest of cap slimy, radially brown embedded fibrillose, shiny when dry. **FLESH** yellow, thick. **GILLS** yellowish becoming rusty-brown age, bluntly attached or notched, close, broad. **STEM** 5–10 cm long, 5–15 mm wide, yellow to pale yellowish-brown, cottony-pubescent above ring, large curled scales below ring. **RING** creamy-white, fibrillose mostly, leaving a thin zone on upper stem, soon lost. **SPORE PRINT** brown. **ODOR** and **TASTE** not distinctive.

Habit and habitat usually clustered in bouquetlike bunches on various living or dead hardwoods, such as maple, elm, beech, but also on conifers. September to November. **RANGE** widespread.

Spores oblong or kidney-bean-shaped, with apical pore, smooth, pale brown in 3% KOH, 7–11 × 4.5–6 μm.

Comments *Pholiota squarrosoides* (Peck) Saccardo has pointed, erect, dry scales on a white to pale-cinnamon, slimy cap. These two slimy species are listed as edible in some field guides. Truthfully though, these scaly, large, clustered pholiotas need to be studied with a microscope to make accurate identifications, since the microscopic characters are less variable than the colors and forms of the fruit bodies. *Pholiota squarrosa*, yet another scaly capped pholiota with dry scales and dry cap, is known to cause gastrointestinal distress in some people.

Pholiota malicola var. *macropoda* A. H. Smith & Hesler

SYNONYM *Flammula malicola* Kauffman

Cap yellow, slippery, smooth, margin with fibrillose ring remnants at first; gills yellowish; stem becoming rusty-brown upward, otherwise pale yellow; odor of green corn; in bouquetlike clusters on wood

CAP 4–12 cm wide, convex, becoming plane, pale yellow or yellowish-green with dingy ochre-tan or orange-buff center at first, becoming paler with age, glabrous, slippery, often with sparse whitish fibrillose ring remnants around margin. **FLESH** yellow, thick. **GILLS** yellowish, becoming rusty-brown with age, attached, close, broad. **STEM** 6–10 cm long, 4–10 mm wide, equal, off-white or pale yellow and somewhat silky covered above the ring, fibrillose-striate and staining dark rust-brown after handling from base upward below the ring zone. **RING** off-white or buff, fibrillose mostly, leaving a thin zone on upper stem, soon lost. **SPORE PRINT** brown. **ODOR** fragrant, like husked green corn. **TASTE** mild.

Habit and habitat mostly in bouquetlike clusters on standing, decaying conifer trees, stumps or buried wood. August to November. **RANGE** widespread.

Spores ellipsoid or ovate, with apical pore, smooth, pale brownish-orange in 3% KOH, 7.5–11 × 4.5–5.5 µm.

Comments This is a fairly common species in the fall. It is distinctive by the clustered growth on conifer wood and its fragrant odor. If a similar-looking species is found on hardwoods and the taste is bitter then you probably have *Pholiota alnicola* (Fries) Singer, if mild tasting then *P. flavida* (Fries) Singer or *P. malicola* var. *malicola*. Unfortunately, a microscope is needed to accurately distinguish these latter two species.

Pholiota multifolia (Peck)
A. H. Smith & Hesler

SYNONYMS *Flammula multifolia* Peck, *Gymnopilus multifolius* (Peck) Murrill, *Pleuroflammula multifolia* (Peck) E. Horak

Cap bright yellow with obscure darker tawny flattened scales, ring patches hanging from margin; gills yellow, staining tawny, crowded; stem colored as cap, with cottony scales below ring, all turning tawny; on woody materials

CAP 5–8 cm wide, convex, with or without a low central bump, bright yellow when young, becoming tawny orange-brown and mostly darker centrally, with obscure darker tawny, flattened, fibrillose scales and patches scattered over the surface, with tawny ring tissue as flaps hanging from the margin, dry. **FLESH** yellow, turning rusty-brown, thick. **GILLS** pale yellow, turning rusty-brown when injured, barely attached, close or crowded, broad. **STEM** 3–7(–10) cm long, 4–10 mm wide, equal, colored as cap, glabrous above the ring, sheathed from base to ring zone with yellow cottony scales that turn tawny with injury, solid, tough. **RING** yellow, turning tawny, fibrillose mostly, leaving torn pieces on the cap margin and stem apex. **SPORE PRINT** brown. **ODOR** not distinctive. **TASTE** bitter.

Habit and habitat scattered or in dense clusters on sawdust, wood chips, logs of hardwoods such as ash, poplar, maple, and others. June to October. **RANGE** widespread.

Spores ellipsoid or bean-shaped, with minute apical pore, smooth, tawny in 3% KOH, 6.5–9 × 4.5–5 µm; cheilocystidia cylindrical with swollen, capitate head.

Comments Not as eye-catching as the scaly *Pholiota* species, but beautiful just the same. Compare with *P. flammans* which has a bright yellow cap that is slippery under the shaggy yellow scales, the gills are not crowded and do not stain rusty-brown, and the taste is mild.

Pholiota lucifera (Lasch) Quélet

SYNONYM *Agaricus luciferus* Lasch

Cap golden-yellow with tawny center and flattened scales, slippery, ring patches on margin; gills dull cream-yellow, spotting cinnamon-brown; stem colored as cap, with cottony scales below ring, turning tawny; on the ground on woody materials

CAP 1.5–6 cm wide, convex, becoming plane, golden-yellow with tawny orange-brown center and tawny flattened, fibrillose scales and patches scattered over the surface, with tawny ring tissue as flaps hanging from the margin, dry, opaque. **FLESH** creamy-white, thick. **GILLS** dull cream-yellow, developing cinnamon-brown spots, barely attached, close, broad, edges fringed. **STEM** 1–5 cm long, 3–8 mm wide, equal but with a slight bulb at base, colored as cap from ring down, powdery white above the ring, sheathed from base to ring zone (peronate) with golden-yellow cottony scales that turn tawny with injury, solid then narrowly hollow, yellow inside turning tawny. **RING** yellow, turning tawny, fibrillose mostly, leaving torn pieces on the cap margin and stem apex. **SPORE PRINT** brown. **ODOR** not distinctive. **TASTE** bitter.

Habit and habitat scattered or in clusters that can be bouquetlike on the ground from wood fragments and woody debris, wood chips in gardens or on lawns from buried wood. September to October.
RANGE Known from central New York State and Montreal area.

Spores ellipsoid or bean-shaped, with very obscure minute apical pore, smooth, tawny in 3% KOH, 6.5–9(–10.5) × 4–5.5 µm; cheilocystidia variously shaped, but often cylindrical with swollen, capitate head.

Comments *Pholiota multifolia* is somewhat similar because of the yellow colors, rusty-tawny staining, and bitter taste. However, it differs by its larger dry cap, its preference for decaying logs, and its more brightly colored and more intensely staining crowded gills.

Pholiota squarrosoides (Peck) Saccardo

SHARP-SCALY PHOLIOTA
SYNONYM *Agaricus squarrosoides* Peck
Fruit bodies in bouquetlike clusters, caps white, slippery, with dry tawny erect scales; stem densely covered with down turned scales and mostly fibrillose ring

CAP 3–11 cm wide, convex, becoming plane some with central broad bump, white or off-white, some with ochre tints and slippery at first, covered with dry tawny erect scales, margin with scattered fibrillose remnants of ring tissue, opaque. **FLESH** white, thick. **GILLS** off-white, becoming rusty-brown from spores, bluntly attached, close or crowded, broad. **STEM** 5–10(–14) cm long, 5–10(–15) mm wide, equal, white or buff, becoming brown in base, covered with large, down turned ochre-tawny scales below ring, silky white above ring, fleshy-pliant. **RING** off-white, fibrillose mostly, leaving torn pieces on the cap margin and stem apex. **SPORE PRINT** brown. **ODOR** and **TASTE** not distinctive.

Habit and habitat single, scattered or most often in bouquetlike clusters on the decaying wood or hardwoods. September to October. **RANGE** widespread.

Spores ellipsoid, apical pore lacking, smooth, reddish-brown in 3% KOH, 4–5.5 × 2.5–3.5 µm; pleuro- and cheilocystidia swollen with an apical fingerlike projection (mucronate), colorless.

Comments *Pholiota squarrosa* (Vahl) P. Kummer is somewhat similar because of the scaly cap and bouquetlike habit, but the cap is dry, the gills turn greenish with age, and some collections have a strong odor of garlic. Microscopically they also differ, with *P. squarrosa* having larger spores (6–8 × 4–4.5 µm) and cystidia with refractive contents.

Pholiota lenta (Persoon) Singer

SYNONYMS *Agaricus lentus* Persoon, *Hebeloma glutinosum* Saccardo, *Flammula betulina* Peck, *Flammula lenta* (Persoon) P. Kummer, *Gymnopilus lentus* (Persoon) Murrill

In the fall, a very slimy pinkish-buff cap with fibrillose ring remnants around the margin; gills creamy-white becoming gray-brown; stem white above brownish below, with faint fibrillose ring zone; on leaf duff under hardwoods

CAP 3–7 cm wide, convex, becoming plane, pinkish-buff with off-white margin or with smoky grayish hues mixed in, center darkening to pale cinnamon-brown, glutinous-slimy, glabrous, with fibrillose ring remnants around margin. **FLESH** white, thick. **GILLS** white or pale creamy-white, becoming gray-tawny or pale gray-brown, attached with decurrent tooth, close, broad. **STEM** 3–8 cm long, 4–12 mm wide, equal or some with subbulbous base, white above, brownish over the lower half, fibrillose, except white-dotted above the ring zone. **RING** white, silky-fibrillose weblike, leaving a faint zone or not. **SPORE PRINT** brown. **ODOR** not distinctive. **TASTE** mild.

Habit and habitat single or scattered on humus, leaf litter in hardwood forests. September to November. **RANGE** widespread.

Spores ellipsoid or oblong and slightly kidney-bean-shaped, with a minute apical pore, smooth, ochraceous in 3% KOH, 5.5–7 × 3.5–4.5 µm.

Comments Look for this slimy-capped species in the fall. It looks somewhat like a species of *Hebeloma*, but the very glutinous-slimy cap with pale creamy-white gills should help distinguish it from the gray-brown-gilled and radish-smelling *Hebeloma* species found at the same time.

Tapinella atrotomentosa
(Batsch) Šutara

VELVET-FOOTED PAX

SYNONYMS *Agaricus atrotomentosus* Batsch, *Paxillus atrotomentosus* (Batsch) Fries

Large velvety, cap and stem; cap yellow or orange-brown and stem dark brown; gills decurrent, forked; on conifer wood

CAP 4–14 cm wide, convex becoming plane or shallowly vase-shaped, with margin strongly inrolled at first, orange-brown or yellowish-brown or reddish-brown at first, becoming darker brown to almost black with age, dry, fine or densely velvety at first. **FLESH** dirty white or sordid yellow, thick, firm. **GILLS** white becoming yellowish, decurrent, close or crowded, frequently forked, some cross-veined, easily removed from cap. **STEM** 4–10 cm long, 10–30 mm wide, equal and very thick, central or off-center attachment to cap, covered with brown or blackish-brown velvety or matted hairs. **SPORE PRINT** yellowish-brown. **ODOR** and **TASTE** not distinctive.

Habit and habitat single, scattered, or clustered on conifer stumps, downed logs. July to November. **RANGE** widespread.

Spores ellipsoid, smooth, almost color less in 3% KOH, many orange-brown (dextrinoid) in Melzer's reagent, 4–6 × 3–4 μm.

Comments Considered poisonous. A darkly colored, fleshy mushroom in conifer woods that is frequently encountered. See comments under *Paxillus involutus*.

Paxillus involutus (Batsch) Fries
POISON PAXILLUS

SYNONYMS *Agaricus involutus* Batsch, *Omphalia involuta* (Batsch) Gray

Medium to large, orange-brown or grayish-brown, hairy capped mushroom, with strongly inrolled and often ribbed margin; gills yellowish turning slowly brown when injured, decurrent and easily peeling off cap

CAP 4–15 cm wide, convex with margin strongly inrolled, cottony and often ribbed or lined, becoming plane and broadly sunken, ochraceous-brown or yellow-brown at first, becoming olive-brown or grayish-brown with age, slimy or slippery when wet, becoming dry, densely matted hairy or smooth. **FLESH** yellowish turning brown, thick. **GILLS** yellowish-olive at first and staining brown when bruised, becoming dark cinnamon brown with age, decurrent, close or crowded, easily removed from cap. **STEM** 2–8 cm long, 4–20 mm wide, mostly central or sometimes eccentric, equal, colored as cap or paler, staining when injured, glabrous, tough. **SPORE PRINT** yellowish-brown. **ODOR** not distinctive. **TASTE** not distinctive, or acidic and sour.

Habit and habitat single, scattered, or clustered on the ground or on well-decayed wood under hardwoods and conifers, often near seepage areas and bogs. July to November. **RANGE** widespread.

Spores ellipsoid, smooth, pale brown in 3% KOH, 7–10 × 5–6 μm.

Comments The easily removed gills, like tubes from a bolete cap, make a good field character. This species is poisonous and clearly documented in the medical literature as the cause of mild to extremely severe gastrointestinal and hepatorenal problems. Two of the 17 patients treated for poisoning from *Paxillus involutus* or *Tapinella atrotomentosa* died in a 2002 study. Learn to identify these mushrooms.

Phaeocollybia jennyae
(P. Karsten) Romagnesi

SYNONYM *Naucoria jennyae* P. Karsten

Cap dark reddish-brown becoming orange-brown, with sharply conical bump, smooth; gills yellowish becoming reddish-brown, barely attached; stem colored as cap below and long rooting; taste like radish

CAP 1.5–6 cm wide, conical with sharply conic central bump, becoming plane with conical bump prominent, dark reddish-brown at first, hygrophanous and becoming ochre-brown from the center toward the margin, smooth, shiny, slightly slippery, glabrous. FLESH white, thin. GILLS yellowish at first, developing reddish-orange hues with age, and reddish-spotted or colored like cap with age, barely attached, close or crowded, narrow. STEM 5–9 cm long above the ground, but that length and more deeply rooting into the substrate (a pseudorhiza), 1.5–8 mm wide, pale ochre-buff over apex, dark reddish-brown like cap below, glabrous. SPORE PRINT ochre or rusty-brown. ODOR radish-like. TASTE radishlike, becoming bitter.

Habit and habitat usually clustered on the ground in mosses and reindeer lichens under conifers, especially spruce. July to September. RANGE widespread but usually at higher elevations.

Spores broadly ellipsoid or apple-seed-shaped, finely punctate ornamented or almost smooth, pale orange-yellow in 3% KOH, 4–6.5 × 3–4 µm.

Comments Typically found under higher elevation spruce stands in the Adirondack Mountains. Also found in the Northeast in similar habitats, but often at lower elevations, is *Phaeocollybia christinae* (Fries) R. Heim, named by Fries for his wife. It is similar in form, but the cap is brighter orange-brown or copper-red, the odor is of sweet marzipan, and the spores are much larger (8–14 × 4.5–6 µm). *Phaeocollybia rufipes* H. E. Bigelow also is found in similar habitats, has a slightly smaller, orange-cinnamon cap that fades to pale ochre-buff, and the stipe base is brick-red. The odor and taste is farinaceous and the spores are 7–8.5 × 4–5 µm. Contrary to some field guides, *P. rufipes* is not a synonym of *P. christinae*.

Inocybe geophylla (Bulliard)
P. Kummer

WHITE FIBER HEAD

SYNONYMS *Agaricus geophyllus* Bulliard, *Agaricus candidus* Fries, *Inocybe lilacina* (Peck) Kauffman

Cap white, silky, with central bump; gills white then grayish-brown; stem white, equal with a bulbous base, white fibrous covered or dotted; immature caps with white cobweb veil; odor spermatic

CAP 1–3 cm wide, conical-convex, becoming bell-shaped then plane with a central bump, whitish but with center often pale buff, with age discoloring buff or brownish or ochraceous, dry, matted silky overall. **FLESH** whitish. **GILLS** white at first but soon grayish-brown, attached, close, broad, edges white fringed. **STEM** 1.5–5 cm long, 1.5–5 mm broad, equal or slightly club-shaped or with enlarged base, white, surface densely white fibrous covered or dotted, especially over the apex, otherwise nearly smooth, white inside. **RING** white, cobweblike and quickly disappearing upon expansion. **SPORE PRINT** brown. **ODOR** spermatic. **TASTE** similar to smell.

Habit and habitat single, scattered, or clustered, on humus or on the ground under conifers or in mixed woods with hardwoods. July to November. **RANGE** widespread.

Spores ellipsoid, brown, smooth, 7.5–11 × 5–7 μm; cheilocystidia and pleurocystidia fusiform, thick-walled, with crystals over apex.

Comments Considered poisonous. Several variations are described for this species, especially from Europe. C. H. Peck described a lilac-colored variety from North America in 1874, *Inocybe geophylla* var. *lilacina* Peck. It is sometimes considered an independent species, *Inocybe lilacina* (Peck) Kauffman, but really only differs by the lilac colors.

Inocybe calamistrata (Fries) Gillet

GREEN-FOOT FIBER HEAD

SYNONYMS *Agaricus calamistatus* Fries, *Inocybe hirsuta* Quélet

Cap brown with recurved scales; gills rusty-brown; stem brown with greenish or olivaceous-blue base; odor fishy

CAP 1–4 cm wide, broadly conical-convex, becoming bell-shaped, dark brown or reddish-brown, dry, densely recurved scaly over the center, more fibrillose and sparser scaly over margin. **FLESH** whitish, but slowly becoming pinkish or vinaceous. **GILLS** brown, cinnamon-brown or rusty-brown, attached, close, broad, edges white fringed. **STEM** 4–8 cm long, 2–5 mm broad, equal, dark brown or reddish-brown similar to cap but paler lower half turning distinctly dull greenish or olivaceous-blue, becoming black with age, surface covered with fibrous recurved scales, white over apex, bluish-green in base. **SPORE PRINT** brown. **ODOR** odd smell, often more like rancid fish odor, but some claim it is spermatic. **TASTE** not distinctive.

Habit and habitat single, scattered, or clustered, on humus, sometimes sandy or gravelly soils under conifers or in mixed woods with hardwoods. August to November. **RANGE** widespread but infrequently encountered.

Spores ellipsoid or slightly bean-shaped, brown, smooth, 9–12 × 4.5–6.5 µm; cheilocystidia inflated and pleurocystidia absent.

Comments Considered poisonous. Do not confuse with *Psilocybe* species that have dark purple-brown spore prints, are more delicate in stature, and typically are not scaly over the cap and stem. Yes, some *Psilocybe* species do turn blue at the base of the stem when handled, but the similarities end there.

Inocybe tahquamenonensis
D. E. Stuntz

Cap dark purplish-brown with small, erect pointed scales; gills vinaceous; stem brown and scaly like cap; flesh of cap and stem vinaceous turning slowly brown when exposed

CAP 1–3 cm wide, broadly conical-convex, becoming plane with or without a low central bump, very dark chocolate-brown or reddish-brown or purplish-brown, dry, densely covered with small erect, pointed scales. **FLESH** vinaceous turning brown. **GILLS** bright pinkish-vinaceous or purplish-red, turning brown with age, bluntly attached, subdistant, broad, edges purplish-brown fringed. **STEM** 3–8 cm long, 3–7 mm broad, equal or enlarged downward, same color as cap, surface densely covered with erect or matted down, pointed brown scales, bright vinaceous inside, slowly brown. **RING** cobweblike and quickly disappearing upon expansion. **SPORE PRINT** brown. **ODOR** not distinctive, or perhaps radishlike. **TASTE** not distinctive.

Habit and habitat single, scattered, or clustered, on soil, humus or on the ground under hardwoods or in mixed woods, often under older beech trees. August to October. **RANGE** widespread, not uncommon in some areas of the Adirondacks.

Spores angular-nodulose, brown, bumpy, 6–8.5 × 5–6 µm.

Comments *Pouzarella nodospora* is somewhat similar to this species of *Inocybe*, but *P. nodospora* is separated by the pink or fleshy-brown spore print and the cap and stem flesh that is not wine-red, turning brown when exposed.

Inocybe albodisca Peck

WHITE-DISC FIBER HEAD

Cap pinkish or grayish or pale lilac brown with pale cream rounded bump on the center; gills white then grayish-brown; stem paler than cap, equal with abrupt bulb at base; odor spermatic

CAP 2–5 cm wide, broadly conical-convex, becoming bell-shaped with a broad rounded bump over the center, pale lilac-flesh color except for the white center at first, soon grayish-brown or yellowish-brown or pinkish-brown with a white or pale cream center constant, moist or dry, minutely fibrillose becoming radially cracked and splitting to near center. **FLESH** white, moderately thin. **GILLS** white, becoming pale grayish-brown, barely attached, close, narrow, edges white fringed. **STEM** 3–5 cm long, 3–5 mm broad, equal with slightly and abruptly bulbous base, colored as cap but paler, surface glabrous or finely white-dotted over apex, solid white inside. **SPORE PRINT** brown. **ODOR** spermatic, nauseous. **TASTE** unknown.

Habit and habitat scattered or clustered, on humus or in grassy areas on roadsides under conifers, especially hemlock or in mixed woods with beech, birch, and aspen. August to November. **RANGE** widespread but infrequently encountered.

Spores angular and nodulose, brown, bumpy, 6–8 × 4.5–6 µm; cheilocystidia and pleurocystidia abundant, ventricose, apex crystalline encrusted.

Comments Considered poisonous. One of the easier to identify fiber head mushrooms because of the pale cap center.

Inocybe rimosa (Bulliard)
P. Kummer

STRAW-COLORED FIBER HEAD
SYNONYMS *Agaricus rimosus* Bulliard,
Inocybe fastigiata (Schaeffer) Quélet,
Inocybe schista (Cooke & W. G. Smith)
Saccardo, *Inocybe umbrinella* Bresadola

Cap pale yellow or brownish, conical
bell-shaped and strongly splitting with
age; gills with olive-yellow tint at first;
stem paler than cap, white mealy over
apex, silky or glabrous elsewhere; odor
spermatic

CAP 2–8 cm wide, conical-convex, becoming
bell-shaped with a conical bump over
the center, clear pale yellow or ochre-
brown, radially silky fibrillose, but soon
radially cracked and splitting through
entirely through the fleshy from the margin to near center. **FLESH** off-white, thin.
GILLS with olive-yellow tint at first, becoming pale grayish-brown, barely attached,
close, edges white fringed. **STEM** 3–10 cm
long, 3–10 mm broad, equal, white or pale
cream, surface glabrous or finely silky,
also white mealy-dotted over apex. **SPORE
PRINT** brown. **ODOR** spermatic, nauseating.
TASTE disagreeable.

Habit and habitat scattered or clustered,
on humus under hardwoods or conifers.
August to November. **RANGE** widespread.

Spores elliptical, smooth, pale brown in
KOH, 9–15 × 6–8 μm; cheilocystidia cylindrical or club-shaped.

Comments Poisonous. *Inocybe rimosa* has
the typical aspect of a fiber head with the
radially fibrillose and splitting cap. In
some older guides it is known as *I. fastigiata*. If your mushroom is very similar
but has an odor of husked green corn,
then you have *I. sororia* Kauffman. Unfortunately, there are many other *Inocybe*
species with a similar look, and you really
have to rely on microscopic features to
distinguish them accurately. If your mushroom has a dark brown splitting cap with
smaller spores, it most likely is *I. fastigiella*
G. F. Atkinson.

Hebeloma crustuliniforme
(Bulliard) Quélet

POISON PIE

SYNONYMS *Agaricus crustuliniformis*
Bulliard, *Hebeloma diffractum* (Fries) Gillet,
Hebeloma longicaudatum (Persoon) P.
Kummer

Large bulky mushrooms with white or pale
cream slippery cap; pale grayish-tan gills;
stem off-white with scales or flakes over
upper half; odor of radish when cap flesh
crushed

CAP 3–11 cm wide, convex, becoming
broadly convex or plane and often with low
broad bump over center, white or very pale
cream with buff-tan mixed in, tacky or
slimy when wet, glabrous, opaque. **FLESH**
white, thick. **GILLS** off-white becoming
grayish-tan or pale brown, attached, close,
broad, edges white fringed and with clear
liquid droplets when young. **STEM** 4–7 cm
long, 5–14 mm broad, equal or enlarged
downward, off-white or colored as cap, over
apex to middle covered with distinctive
white scales or flakes. **SPORE PRINT** dull
brown. **ODOR** of radish when crushed.
TASTE bitter and radishlike.

Habit and habitat mostly scattered or clustered, often in fairy rings, on the ground
under oaks or other hardwoods. September to November. **RANGE** widespread.

Spores almond-shaped or elliptical, pale
yellow-brown in aqueous mounts, very
slightly dotted or roughened, 9–13 ×
5–7.5 μm; cheilocystidia club-shaped.

Comments A toxic mushroom causing
severe gastrointestinal distress if eaten.
Hebeloma species are readily identified by
their grayish-tan or pale brown lamellae
and slimy caps. None of the species should
be consumed.

Hebeloma mesophaeum
(Persoon) Quélet

SYNONYMS *Agaricus fastibilis* var. *mesophaeus* Persoon, *Inocybe mesophaea* (Persoon) P. Karsten, *Hebeloma flammuloides* Romagnesi, *Hebeloma strophosum* (Fries) Saccardo

Large mushrooms with brown and white slippery cap; pale grayish-tan gills; stem whitish with fibrillose ring zone; odor of radish when cap flesh crushed, under conifers

CAP 2–7 cm wide, convex, becoming plane, brown or reddish-brown or pinkish-brown, often becoming two-toned and white around the outer margin, tacky or slimy when wet, glabrous, opaque, margin sometimes with white veil fibrils. **FLESH** white, thick. **GILLS** pale creamy-white or with pinkish-gray, becoming pale brown, attached-notched, close or crowded, broad, edges white fringed, not beaded with liquid when young. **STEM** 3–9 cm long, 3–10 mm broad, equal, some with slight bulb as base, off-white but discoloring brown from base upward, minute scaly or mealy over apex, fibrillose-striate below, often with fibrillose ring zone present in upper area. **RING** fine white fibrillose. **SPORE PRINT** dull brown. **ODOR** of radish when crushed. **TASTE** radishlike or bitterish, or both.

Habit and habitat solitary, scattered or clustered, on the ground under conifers, especially spruce and pine. September to November. **RANGE** widespread.

Spores almond-shaped or elliptical, pale yellow-brown in aqueous mounts, very slightly dotted or almost smooth, 8–11 × 5–7 µm; cheilocystidia cylindrical or fusiform.

Comments Probably better referred to as the *Hebeloma mesophaeum* group as there is quite a lot of variability for this morphological type. A detailed study of these fungi in the Northeast is sorely needed.

Agrocybe firma (Peck) Singer

SYNONYM *Naucoria firma* Peck

Cap dark brown or almost black at first, turning yellow-brown, smooth; gills gray-brown; stem pale olive-gray and densely covered with pale granular scales; on well-decayed wood in old-growth hardwood forests

CAP 2–8 cm, convex then plane, dark brown or nearly black, becoming grayish-brown and finally a yellow-brown with loss of moisture, with broad, low umbo, glabrous. **GILLS** earth brown or gray-brown, becoming yellow-brown, attached, close. **STEM** 3–8 cm long, 5–10 mm broad, equal, firm, pale olive-gray or colored like lamellae, densely covered with pale granular scales (dots), also with distinct longitudinal lines when fresh, darker brown (as cap color) where scales rubbed off, white cords (rhizoids) at base. **SPORE PRINT** cinnamon brown. **ODOR** not distinct. **TASTE** mild or faintly mealy.

Habit and habitat Scattered or many clustered together (gregarious) on well-decayed hardwood debris in dense forests usually. July to October. **RANGE** widespread.

Spores elliptical, smooth, not truncate, 6–9(–10) × 4–5 μm.

Comments Most species of *Agrocybe* have an annulus on the stem; only a few do not. The dark cap colors, densely scurfy and lined stipe, lack of a ring, and growth on well-decayed wood make *A. firma* easy to recognize in the field.

Agrocybe sororia (Peck) Singer has a tawny fading to pale yellow-buff cap, sometimes wrinkled or cracked or pitted. The stem is the same color as the cap and lacks a ring. This species is found in grassy areas and on wood mulch. It has a mealy taste like *A. firma*.

Agrocybe amara (Murrill) Singer might be confused with *A. sororia* since it lacks a veil and has an orange to pale yellowish-brown cap. However, *A. amara* grows on dung or rich, fertilized soil, and more importantly has a bitter taste. Of the two, *A. amara* seems to be rarely collected, at least from reports in the literature.

Agrocybe sororia

Agrocybe praecox (Persoon)
Fayod

SPRING AGROCYBE

SYNONYMS *Agaricus praecox* Fries, *Agrocybe sphaleromorpha* (Bulliard) Fayod, *Pholiota praecox* (Persoon) P. Kummer, *Pholiota togularis* (Bulliard) P. Kummer, *Conocybe togularis* (Bulliard) Kühner

On lawns and wood chips, cap clay-color (yellow-brown), glabrous, often cracking, often with torn ring fragments hanging from margin; stem colored like cap, with white cords at base

CAP 3–11 cm, convex then plane and with low broad central bump, cream or becoming more clay-color, glabrous, soft, center cracking like a mudflat with age, dry, margin with white ring patches hanging down (appendiculate). **GILLS** pale at first, finally clay-color, bluntly attached, broad. **STEM** 3–10 cm long, 3–12 mm broad, equal, same color as cap, glabrous, white cords at base, with white, shredding ring over apex if not on margin of cap. **RING** white below, dark brown above from spores. **SPORE PRINT** dark brown. **ODOR** and **TASTE** mild.

Habit and habitat scattered or clustered on humus or wood chips in flower beds, lawns, field, open woods, rather common. April to June. **RANGE** widespread.

Spores elliptical, smooth, apex truncate, with a pore, 8–11 × 5–6.5 μm.

Comments There is some mating and ecological evidence to support the use of the name *Agrocybe molesta* (Lasch) Singer for the species growing on grass and imaged here. A different strain of *A. praecox* grows on woody litter. They both look very similar and are not easy to distinguish. *Agrocybe pediades* (Fries) Fayod is similar in colors and also frequents lawns, but it is small and slender. If the cap is white or creamy-white and viscid, the stem has a ring, and it is on lawns, you have *Agrocybe dura* (Bolton) Singer.

Conocybe apala (Fries) Arnolds
WHITE DUNCE CAP
SYNONYMS *Conocybe albipes* (G. H. Otth) Hausknecht, *Conocybe lactea* (J. E. Lange) Métrod

On lawns in the mornings, showing pointed milk-white caps among blades of grass; gills white but soon cinnamon-brown; stem milk-white, very fragile

CAP 0.5–2.5 cm wide, 1–4 cm high, conical, narrowly or obtusely so, pale yellow-brown at first, but rapidly fading to cream or milky-white with only the center remaining brown, surface glabrous, smooth or with radiating lines or wrinkled radially from center to margin, dry. **FLESH** white, very thin, less than 1 mm. **GILLS** white at first, soon cinnamon-brown, barely attached, close, very narrow. **STEM** 4–11 cm long, 1.5–3 mm broad at apex, equal and with swollen bulb at the base, mostly straight and very fragile and splitting, milky-white, covered with white tiny fuzzy dots overall, glabrous with age, hollow. **SPORE PRINT** rusty-brown. **ODOR** and **TASTE** not distinctive.

Habit and habitat scattered or clustered in lawns and playing fields mainly, ephemeral and lasting only a day or less. May to June. **RANGE** widespread.

Spores ellipsoid, smooth, with apical germ pore, orange-brown in KOH, 10–16 × 7–10 μm.

Comments Usually only found in the morning as the heat of the day causes the thin fragile stems to collapse and the cap to shrivel up.

Conocybe macrospora
(G. F. Atkinson) Hausknecht

SYNONYMS *Galerula macrospora*
G. F. Atkinson, *Conocybe tenera* f. *bispora*
J. E. Sass

Parabolic, reddish-brown cap rapidly fading to tan; gills crowded, broad with minutely hairy edges; stem long, straight with bulbous base, surface striate-lined; on nitrogen-rich substrates, such as lawns

CAP 1–3 cm wide, broadly conical-parabolic or bell-shaped, reddish-brown at first, hygrophanous and rapidly fading to tan or yellowish-brown, surface glabrous, smooth or with very fine translucent radiating lines over margin, dry. **FLESH** reddish-brown, very thin, less than 1 mm. **GILLS** cinnamon-brown, barely attached, close or crowded, broad (3–4 mm), edge minutely hairy. **STEM** 3.5–8.5 cm long, 2–3 mm broad at apex, equal with abruptly swollen bulb at the base (to 6 mm), mostly straight, very fragile, surface same color as cap or paler, covered with white fine dots over apex, also fine striate-lined apex to base, hollow. **SPORE PRINT** rusty-brown. **ODOR** and **TASTE** not distinctive.

Habit and habitat scattered or clustered in lawns, pastures, on dung, on wood chips in flower gardens, compost, or in rich soil on leaves and conifer needles. May to July. **RANGE** widespread but not commonly collected.

Spores elongate-ellipsoid, smooth, with thick walls and apical germ pore, orange-brown in KOH, 14–21 × 7.5–11 µm, basidia 2-sterigmate; cheilocystidia bowling-pin-shaped.

Comments Probably mistaken for *Conocybe tenera* (Schaeffer) Fayod in the northeastern United States since the macroscopic characters are very similar. The large spores and 2-sterigmate basidia are distinctive for *C. macrospora*, as *C. tenera* produces spores that only reach 12 µm long, with basidia 4-sterigmate. There are no studies published to suggest which is more frequently found, but perhaps because I live near where Professor Atkinson worked, Cornell University, I tend to find this species most often.

Cortinarius acutus (Persoon)
Fries

SYNONYM *Agaricus acutus* Persoon

Small, yellow-brown mushroom, with a conical bump on conical cap and strongly translucent-striate cap margin; gills ochre tan with white fringed margin; stem clothed with white flattened fibrils and scales

CAP 0.6–1.5 cm wide, 0.7–1.2 cm high, conical or conical bell-shaped, with conical bump over center, this bump occasionally bent, yellowish-brown or pale reddish-brown and strongly translucent-striate to margin, hygrophanous and turning clay-color, surface glabrous or becoming silvery-white hairy or minutely scaly, margin scalloped, moist. **FLESH** pale ochre, thin, 0.5 mm. **GILLS** ochre tan or pale brown becoming darker rusty-brown, attached, close or subdistant, broad (to 4 mm), edges white fringed. **STEM** 2–6 cm long, 1–3 mm broad at apex, equal, yellowish-brown, but covered with white silky fibrils and flattened scales of universal veil tissue, hollow. **RING** white, cobweb-like when attached to cap margin, leaving thin line on upper stem and silky fibrils on cap margin. **SPORE PRINT** brown. **ODOR** of iodine. **TASTE** not distinctive.

Habit and habitat clustered on humus in needle beds and humus in coniferous forests, often with spruce or hemlock present. August to October. **RANGE** widespread.

Spores almond-shaped or ovoid, minutely warty, pale brown in KOH, 8–9 × 4.5–5.5 μm.

Comments A very small cortinarius that resembles a galerina.

Cortinarius armillatus (Fries)
Fries

BRACELET CORT

SYNONYMS *Agaricus armillatus* Fries, *Hydrocybe armillata* (Fries) M. M. Moser

Large, reddish-brown overall, with distinctive cinnabar-red bands or bracelets on stem, usually near birch

CAP 4–12 cm wide, convex and often with low broad bump over center, becoming plane, reddish-brown or reddish-orange, becoming yellowish-orange over margin, surface glabrous but with small reddish flattened scales or fibrils over center, also with darker radiating streaks from center to margin, moist. **FLESH** yellowish-brown. **GILLS** pale brown becoming darker rusty-brown, attached, subdistant, broad. **STEM** 7–20 cm long, 8–15 mm broad at apex, equal and enlarged downward to a bulb or club-shaped, white silky-fibrous at first and overlain with one or more bands of cinnabar reddish universal veil tissue over lower half, ground color eventually brown, hollow. **RING** white, cobweblike when attached to cap margin, leaving thin line on upper stem. **SPORE PRINT** brown. **ODOR** faintly of radish or not detectable. **TASTE** not distinctive, or bitterish.

Habit and habitat single, scattered, or clustered on humus in hardwood forest or mixed hardwood and coniferous forest, often with birch present. August to October. **RANGE** widespread.

Spores almond-shaped or ellipsoid, minutely warty, pale brown in KOH, 7–12 × 6–7 µm.

Comments One of the more common and easily identified *Cortinarius* species in the late summer and early fall.

Cortinarius camphoratus
(Fries) Fries

SYNONYM *Agaricus camphoratus* Fries
Large, fleshy, pale lilac colored overall, but with pileus becoming yellowish to golden-brown with age; odor strong of raw or rotting potatoes, unpleasant

CAP 1–10 cm wide, convex, becoming plane, silvery-white mixed with pale lilac at first, but soon yellowish-brown or golden-brown over center with margin pale lilac, surface matted hairy, becoming flattened scaly over the center. **FLESH** lilac or purplish, thick. **GILLS** lilac or purple at first, soon pale brown becoming cinnamon-brown, attached, subdistant or close, broad. **STEM** 2–8 cm long, 1–3 mm broad at apex, equal enlarging to bulb at base or club-shaped, lilac as cap margin, covered with white silky fibrils of universal veil tissue, hollow. **RING** white, cobweblike when attached to cap margin, leaving thin line on upper stem and silky fibrils on cap margin. **SPORE PRINT** rusty-brown. **ODOR** strong of raw or rotting potatoes, unpleasant. **TASTE** unpleasant.

Habit and habitat clustered in needle beds and humus in coniferous forests, often with spruce, fir or hemlock present. August to October. **RANGE** widespread.

Spores broadly almond-shaped or ellipsoid, minutely warty, pale brown in KOH, 9.5–11 × 5.5–6.5 µm.

Comments *Cortinarius lilacinus* Peck is similar looking, but lacks a strong, unpleasant odor.

Cortinarius caperatus
(Persoon) Fries

GYPSY

SYNONYMS *Agaricus caperatus* Persoon, *Pholiota caperata* (Persoon) Gillet, *Rozites caperatus* (Persoon) P. Karsten

Large mushroom with pale orange-tan, wrinkled cap, coated with white hoary fibrils; gills white then cinnamon-brown, attached and broad; stem white with thick membranous ring attached at midstipe

CAP 5–15 cm wide, oval becoming convex, then broadly bell-shaped with a low central bump, pale orange-tan but with silvery-white hoary coating over the center, smooth or radiately wrinkled, opaque. **FLESH** white, fleshy. **GILLS** off-white becoming cinnamon-brown, attached, close, broad. **STEM** 5–12 cm long, 10–25 mm broad, equal or the base slightly swollen, off-white or pale tan, scaly over the apex above the ring, glabrous and smooth below, or sometimes with hoary coating over base. **RING** white, thick, membranous, tightly attached at midstipe. **SPORE PRINT** rusty-brown. **ODOR** and **TASTE** not distinctive.

Habit and habitat single, scattered, or clustered on humus under hardwoods or conifers. September to November. **RANGE** widespread.

Spores ellipsoid or almond-shaped, bumpy, orange-brown in KOH, 10–15 × 7–10 µm.

Comments This mushroom is considered a choice edible, but it typically is not found in great numbers. The name gypsy seems to fit its nomenclatural history of being moved around from *Pholiota* to *Rozites* and now into *Cortinarius*. The last move was based on molecular evidence. Most *Cortinarius* species have a cobweblike ring, a cortina, but the molecules indicate the gypsy, with its membranous ring, belongs with other species of *Cortinarius*.

Cortinarius corrugatus Peck

Cap is rusty-orange, slippery and radially furrowed; stem long, straight with abrupt bulb at the base, golden-yellow but staining rusty-brown when handled

CAP 3–10 cm wide, rounded conic, becoming plane, rusty-orange or yellowish-orange with a reddish-orange center, surface slippery and strongly radially furrowed and wrinkled. **FLESH** white. **GILLS** pale grayish-violet at first, soon rusty-cinnamon, attached, close, broad. **STEM** 5–12 cm long, 6–20 mm broad at apex, equal with an abrupt bulb at base, pale yellow, becoming golden-yellow, but quickly turning rusty-brown when handled, typically yellow scaly or dotted overall, narrowly hollow. **RING** tawny, cobweblike when attached to cap margin, very thin and barely leaving fibrils on upper stem. **SPORE PRINT** rusty-brown. **ODOR** and **TASTE** not distinctive.

Habit and habitat scattered or clustered on humus under hardwoods like beech, oak and chestnut. July to October. **RANGE** widespread.

Spores ellipsoid, minutely warty, pale brown in KOH, 12–15 × 7–9 μm.

Comments This is a fairly common species where beech, oaks, and chestnuts are found in mixed forests in warmer wet weather.

Cortinarius evernius (Fries)
Fries

SYNONYMS *Agaricus evernius* Fries, *Hydrocybe evernia* (Fries) M. M. Moser

In wet or boggy areas, dark violet-brown slippery cap with conical bump; gills dark violet at first; stem silvery-violet, long, covered with white fibrils and patches

CAP 3–11 cm wide, rounded conical or bell-shaped, becoming plane with small conical bump over the center, dark violet at first, hygrophanous and changing to reddish-brown and eventually fading to ochraceous-brown, surface silky and radially streaked with fibrils, slippery when wet, also faintly translucent-lined over margin. **FLESH** violaceous-brown, thin. **GILLS** dark violet with white edge at first, soon cinnamon-brown, attached, subdistant or distant, broad. **STEM** 7–20 cm long, 5–20 mm broad at apex, equal, violet but covered with white fibrils and patches of universal veil, solid and dark violet in upper areas. **RING** white, cobweblike when attached to cap margin, very thin and leaving a ring of fibrils on upper stem. **SPORE PRINT** rusty-brown. **ODOR** faintly of radish or not distinctive. **TASTE** not distinctive.

Habit and habitat scattered or clustered on humus or in *Sphagnum* moss or other moss beds in wet areas or bogs under conifers, especially spruce. July to October. **RANGE** widespread.

Spores ellipsoid or slightly almond-shaped, minutely warty, pale brown in KOH, 8–11 × 5–6 µm.

Comments A tall colorful mushroom. Compare with *Cortinarius violaceus* that can occur in similar habitats, but it is a much bulkier species with a scaly cap and with evenly distributed and longer-lasting violet colors on all parts. The odors are quite different as well.

Cortinarius flexipes (Fries) Fries

SYNONYMS *Cortinarius paleifer* Svrček, *Hydrocybe flexipes* (Persoon) M. M. Moser

Cap conical-convex, dark chocolate brown with white scales and hairs; stem chocolate with violaceous tints, densely white cottony ringed; odor of geranium

CAP 1–3 cm wide, convex with low conical bump on the center, becoming plane with conical bump over the center, dark chocolate brown, surface covered with silvery-white scales and hairs. **FLESH** brown. **GILLS** dark brown at first, soon rusty-brown from spores, attached, close or subdistant, edges white fringed. **STEM** 3–10 cm long, 3–8 mm broad at apex, equal with a tapered base, dark brown or violet brown, especially over base, surface sheathed with white cottony veil that breaks up into partial rings or patches on most of the stem below the ring. **RING** white, cobweblike when attached to cap margin, leaving a thickened ring of white cottony and hairy fibrils on upper stem from the combination of the universal and partial veils, often coated with rusty-brown spores. **SPORE PRINT** rusty-brown. **ODOR** of geranium mainly, but sometimes other fruity odors. **TASTE** not distinctive, mild.

Habit and habitat single, scattered, or clustered on humus and mosses under conifers and mixed hardwoods. August to October. **RANGE** only reported, so far, from high elevations in New York.

Spores elliptical, faintly warty, pale brown in KOH, 7.5–10 × 5–6 µm.

Comments A fairly commonly collected species in late summer and early fall in coniferous woods. *Cortinarius hemitrichus* (Persoon) Fries is similar but larger with cap sizes reaching 5 cm and the fruit bodies lacking any odor.

Cortinarius distans Peck

SYNONYM *Phaeomarasmius distans* (Peck) Singer

Cap conical-convex with rounded conical central bump, cinnamon-brown but fading to tan, densely scurfy overall; gills cinnamon-brown, distant; stem with white ring zone on young specimens; under hardwoods

CAP 3–8(–10) cm wide, convex with rounded conical bump on the center, becoming broadly bell-shaped, orange-brown or cinnamon-brown, fading to tan, often with thin white margin, surface covered with fine scales (scurfy), may become glabrous with age. **FLESH** cinnamon-brown, thin. **GILLS** pale brown becoming cinnamon-brown, bluntly attached or notched, distant, thick and ridged. **STEM** 4–9 cm long, 5–15 mm broad at apex, equal or narrowly club-shaped, colored as cap or paler, surface glabrous or somewhat silky, with apical ring or ring zone which falls apart. **RING** white, cobweb-fibrillose (a cortina), leaving a white fibrillose zone, basal mycelium white. **SPORE PRINT** rusty-brown. **ODOR** and **TASTE** not distinctive.

Habit and habitat single, scattered, or clustered on humus under mixed hardwoods, often with oak present. May to September. **RANGE** widespread.

Spores elliptical, warty, brown in KOH, 7–9 × 5–6 μm.

Comments Peck's species has been considered the same as *Cortinarius hinnuleus* Fries in some field guides. However, the cap of *C. hinnuleus* is not scurfy, the gills are not as distant, and from images posted on the Internet, the central bump on the cap is much smaller and pointed conical.

Cortinarius flavifolius Peck

Bulky, moderately large creamy-white fruit bodies with felted-silky covered cap and stem; gills yellow then ochre-yellow becoming rusty-brown from spores; stem typically club-shaped

CAP 4–10(–15) cm wide, convex, becoming plane, cream-buff, becoming buff or pale tawny-yellow with age, surface smooth, dry, covered with felted-silky layer. FLESH white, thick. GILLS pale yellow becoming ochre-yellow then cinnamon or rusty-brown from spores, bluntly attached, subdistant, broad. STEM 4–18 cm long, 6–12 mm broad at apex, to 40 mm in base, club-shaped, often bulbous with a short root, white, covered with a white silky, thin, sheath from the base to the ring zone (peronate), surface with fine flattened hairs. RING white, dense, cobweblike when attached to cap margin, leaving a fine fibrillose zone on the upper stem that is often coated with rusty-brown spores. SPORE PRINT rusty-brown. ODOR and TASTE not distinctive.

Habit and habitat single or a few scattered on humus under hardwoods, often with beech present. July to October. RANGE widespread.

Spores subglobose or broad elliptical, warty, brown in KOH, $7–9 \times 5–6$ µm.

Comments A handsome mushroom that is not so commonly encountered and therefore not featured in field guides. If you do find a collection, it would be a good candidate to post on MushroomObserver.org.

Cortinarius gentilis (Fries) Fries
DEADLY CORT
SYNONYM *Agaricus gentilis* Fries
Small, reddish-brown or orange-brown overall, with ephemeral yellow fibrillose ring zone erratically adorning the stem; under conifers

CAP 2–6 cm wide, broadly conical, becoming convex then plane, often with low broad bump over center, rich reddish-brown or orange-brown or yellow-brown, glabrous, moist but hygrophanous and fading to yellow or amber-yellow. **FLESH** yellow, thin. **GILLS** yellow-brown to reddish-brown, bluntly attached, subdistant, thick, broad. **STEM** 3–11 cm long, 3–8 mm broad, equal, similar color to cap, glabrous, but with a ring zone of yellowish fibrils scattered over the surface. **RING** yellow, cobweblike when attached to cap margin, leaving an ephemeral, sparse fibrillose zone on the upper stem, often coated with rusty-brown spores. **SPORE PRINT** rusty-brown. **ODOR** and **TASTE** not distinctive.

Habit and habitat scattered or clustered on humus under conifers. August to October. **RANGE** widespread.

Spores subglobose or short ellipsoid, warty, brown in KOH, 7–9 × 6–7 μm.

Comments Deadly poisonous. In this group of orange-brown *Cortinarius* species are some of the deadliest toxic mushrooms, like *C. orellanus* and *C. rubellus*. All should be assiduously avoided. These latter two species are generally darker reddish-orange colored, with woolly or finely scaly caps and larger spores.

Cortinarius limonius (Fries) Fries is moderately large, orange or orange-brown overall, with golden-yellow or yellow veil material on the stem. Do not eat this or any other orange-brown *Cortinarius* species.

Cortinarius limonius

Cortinarius odorifer

Britzelmayr

Cap coppery yellow-brown with greenish-yellow margin, slimy; gills and stem yellowish-green, stem with flared bulb; strong odor of anise when cut open

CAP 4–11 cm wide, convex, becoming plane, coppery yellow-brown with greenish-yellow margin, becoming reddish-brown over center, surface smooth, slimy. **FLESH** bright greenish-yellow, thick. **GILLS** yellowish-green becoming cinnamon-brown, attached, close, edges white fringed. **STEM** 5–10 cm long, 10–30 mm broad at apex, equal but with flared out bulb at base, silvery-yellow-green or olivaceous green, surface with fine flattened hairs. **RING** greenish-yellow, cobweblike when attached to cap margin, leaving a fine fibrillose zone on the upper stem often coated with rusty-brown spores. **SPORE PRINT** rusty-brown. **ODOR** strong, distinctly of anise. **TASTE** not distinctive, mild.

Habit and habitat single or a few scattered on humus under conifers, such as hemlock or spruce. August to October. **RANGE** Not widespread, rarely collected, associated with limestone soils.

Spores almond-shaped or lemon-shaped, warty, pale brown in KOH, 10.5–12 × 6.5–7.5 µm.

Comments This mushroom is a rare find in northeastern North America. The odor of anise makes it immediately distinctive. It is another species not generally presented in North American field guides.

Cortinarius scaurus (Fries)
Fries

SYNONYM *Cortinarius herpeticus* Fries
Cap olive or greenish-brown, darker mottled over margin with watery spots, slimy or slippery; gills yellowish-green at first; stem with yellowish-green and also pale violet fibrils

CAP 2–8 cm wide, convex, becoming plane with low broad bump over the center, greenish-black becoming olive-brown or olivaceous-green and mottled with darker watery brown spots around margin, surface glabrous, slimy or merely slippery. **FLESH** pale olive-green, becoming watery buff tan. **GILLS** yellowish-green or olivaceous-green, soon rusty-brown from spores, attached, close. **STEM** 4–13 cm long, 8–12 mm broad at apex, equal but with a bulbous base, pallid ground color from silvery flattened fibrils, but overlain with sulfur yellowish-green or greenish-lemon fibrils at apex and also pale violaceous or lavender fibrils especially over apex, but also at the base. **RING** greenish-yellow, cobweblike when attached to cap margin and also covered with thin slimy layer, leaving a thin ring of yellowish fibrils on upper stem, often coated with rusty-brown spores. **SPORE PRINT** rusty-brown. **ODOR** and **TASTE** not distinctive.

Habit and habitat scattered or clustered on humus under conifers like balsam fir. August to October. **RANGE** Only known and reported, so far, from mountains in Adirondack Park in New York.

Spores elliptical, densely warty, pale brown in KOH, 9–11.3 × 6–8 μm.

Comments *Cortinarius herpeticus* is kept as a separate species in some European guides, but they are difficult to separate on morphology. Also, even though not listed as such in Index Fungorum, *C. olivaceous* Peck appears to be the same species as well.

Cortinarius semisanguineus
(Fries) Gillet

SYNONYMS *Dermocybe semisanguineus* (Fries) M. M. Moser, *Cortinarius cinnamomeus* f. *semsanguineus* (Fries) Saccardo

Cap and stem yellowish and silky; gills dark blood-red

CAP 2–8 cm wide, convex, soon plane with broad conical bump over the center, yellowish-brown or ochre-yellow often with olive tints, surface radially silky fibrillose, dry. **FLESH** white or pale yellow. **GILLS** dark blood red or cinnabar red, attached, crowded. **STEM** 2.5–10 cm long, 5–10 mm broad at apex, equal, pale yellowish or darker ochre-yellow or reddish over the base, silky covered as cap. **RING** yellow, cobweblike when attached to cap margin, leaving a thin ring of yellow fibrils on upper stem coated with rusty-brown spores. **SPORE PRINT** rusty-brown. **ODOR** not distinctive, or like radish. **TASTE** not distinctive, mild.

Habit and habitat scattered or clustered on humus under conifers or hardwoods. July to November. **RANGE** widespread.

Spores elliptical, minutely warty, pale brown in KOH, 6–8 × 3.5–5 μm.

Comments *Cortinarius sanguineus* (Wulfen) Fries is similar looking but has a blood-red cap, stem, and gills and is more often found only in coniferous woods.

Cortinarius sphaerosporus
Peck

Cap bright yellow, glutinous; gills violet; stem yellow and slimy, slender club-shaped

CAP 1.5–7 cm wide, convex, becoming plane, straw-yellow or ochre-yellow, surface glabrous, thick glutinous when wet. **FLESH** violet at first, soon whitish. **GILLS** violet in earliest stages, soon whitish then cinnamon-brown, attached, close, broad. **STEM** 3.5–10 cm long, 5–10 mm broad at apex, slender club-shaped or equal and slightly enlarged downward, pale violet over apex, elsewhere white beneath the yellowish thick, slimy universal veil, veil as drying leaves thin yellowish patches over base of the stem. **RING** white, cobweb-like when attached to cap margin, covered with yellowish, thick slimy universal veil at first, very thin and leaving a ring of white fibrils on upper stem coated with rusty-brown spores. **SPORE PRINT** rusty-brown. **ODOR** and **TASTE** not distinctive, mild.

Habit and habitat scattered or clustered on humus under conifers such as fir, white pine, and spruce. July to September. **RANGE** widespread.

Spores subglobose, minutely warty, pale brown in KOH, 6–7.5 × 5–6.5 μm.

Comments *Cortinarius delibutus* Fries, a European species, is similar looking but has larger spores, 7–9 × 6–8 μm. If you are finding this species in the Northeast, it is most likely *C. sphaerosporus* and not the European look-alike, although reports of *C. delibutus* can be found on foray checklists in the Northeast. A thorough comparative study using morphological and molecular techniques is most likely needed to determine if these are genetically different species.

Cortinarius squamulosus
Peck

Cap purple-brown, scaly; stem purple-brown with large bulb, and often a flat, bandlike ring on the upper stem above the bulb

CAP 2–12 cm wide, convex, becoming plane with low broad bump over the center or slightly sunken, dark reddish-brown or purplish-brown, becoming chocolate-brown, surface covered with fibrils over the margin and scales over the center. **FLESH** purplish-brown, thick. **GILLS** purple or purple-brown, becoming dark chocolate-brown, soon rusty-brown from spores, attached, close, edges white fringed. **STEM** 8–15 cm long, 10–20 mm broad at apex, bulb 4–6 cm broad, upper apex equal but most of stem club-shaped or markedly bulbous, dull purple-brown, becoming dark chocolate-brown, surface with fine flattened hairs. **RING** white, cobweblike when attached to cap margin, leaving a distinct flattened white band on the upper stem from the combination of the universal and partial veils, often coated with rusty-brown spores. **SPORE PRINT** rusty-brown. **ODOR** spicy or rather pungent and disagreeable. **TASTE** not distinctive, mild.

Habit and habitat single, scattered, or clustered on humus under hardwoods, such as oak. August to October. **RANGE** widespread.

Spores elliptical or subglobose, warty, pale brown in KOH, 6–8 × 5–7 μm.

Comments Not commonly collected. The species name refers to the rough, erect scales on the cap surface, but the markedly bulbous stem base is also diagnostic. Compare with *Cortinarius violaceus* which is a much darker, richer violet color.

Cortinarius violaceus
(Linnaeus) Gray

VIOLET CORT
SYNONYM *Agaricus violaceus* Linnaeus
Dark violet overall, cap scaly, odor fragrant, under conifers

CAP 4–12 cm wide, convex, becoming plane and often with low broad bump over center, dark violet, surface densely hairy and fibrillose-scaly over all, dry. **FLESH** dark violet, thick. **GILLS** dark violet as pileus, becoming rusty-tinged from spores, attached, subdistant. **STEM** 6–14 cm long, 10–20 mm broad at apex, club-shaped or base more bulbous, dark violet, surface matted fibrous, with incomplete dark grayish-violet rings and patches from universal veil. **RING** dark violet, cobweb-like when attached to cap margin, leaving a fine fibrillose zone on the upper stem often coated with rusty-brown spores. **SPORE PRINT** rusty-brown. **ODOR** fragrant, of cedar or freshly sharpened pencils, or highly scented flowers. **TASTE** not distinctive.

Habit and habitat single, scattered, or clustered on humus under conifers. August to October. **RANGE** widespread.

Spores almond-shaped or ellipsoid, finely warty, pale brown in KOH, 11.5–13.5 × 7.5–8.5 µm (European specimens) or 11.5–18 × 7–10 (reported for North American specimens).

Comments A truly beautiful species with a fragrant odor. *Cortinarius alboviolaceus* (Persoon) Fries is very pale lilac, and the other violet or purplish species of *Cortinarius* lack the highly ornamented scaly cap and cedarlike odor. If the fruit bodies have a violet slippery to slimy cap with yellow spots and are small to medium-sized, you have *C. iodes* Berkeley & M. A. Curtis.

Gilled Mushrooms with Dark-Colored Spore Prints

Spore deposits in this group can range from dark chocolate-brown to sepia or from violet-black to black. The mushrooms can be large (*Agaricus*, *Hypholoma*, *Stropharia*) or small and delicate (*Coprinopsis*, *Deconica*, *Melanophyllum*, *Panaeolina*, *Psilocybe*) though most species are moderate in size. In this group are some fine edible mushrooms, some hallucinogenic mushrooms, and some that cause gastrointestinal problems if eaten. Although not as abundant and diverse as the white- and brown-spored mushrooms, this group has some very interesting colorful species.

Gills free (or barely attached)

GILLS NOT DELIQUESCING (TURNING TO INK)
Agaricus Cap glabrous or at most with flattened fibers, flesh thick, white, sometimes changing color to red or yellow, medium to large fruit bodies, gills remote from stem, leaving an obvious channel (spores smooth)
Melanophyllum Cap granular-powdery, dark gray brown, flesh thin, white turning red, small fruit bodies, gills free or barely attached, (spores minutely bumpy)
Parasola Cap glabrous but strongly corrugate-grooved, flesh very thin, small and fragile, gills free, edges only, but not completely, deliquescing on some species

GILLS DELIQUESCING (TURNING TO BLACK INK)
Coprinus Membranous ring present, gills and flesh turning pink just before black spore production, white cord present in hollow stem
Coprinellus Granular white sugarlike coating on cap, dense cottony mat at base of stem
Coprinopsis Flattened scales on cap, thin ring zone at stem base

Gills decurrent

Chroogomphus Gills thick and waxy-looking, cap slimy when wet, flesh orange (hyphae of context amyloid)
Gomphidius Gills thick and waxy-looking, cap slimy, flesh white (hyphae of context not amyloid)

Gills bluntly attached

STEM WITH A RING (SOMETIMES FIBRILLOSE AND NOT SO OBVIOUS)
Stropharia Ring membranous, obvious, fruit bodies large and fleshy (with spiny acanthocyte cells on basal mycelia)
Hypholoma Ring fibrillose, sometimes not so obvious, medium or small, fleshy (lacking acanthocytes on basal mycelium)

STEM LACKING A RING
Deconica On mosses, at higher elevations, small, delicate, spores purple-brown or black, not turning blue when injured

Panaeolus Gills similar color as cap, with darker spots on the faces, cap colors grayish or dark browns, on lawns or dung, small, fragile fruit bodies, spores black (cap skin composed of inflated cells, spores smooth)

Panaeolina Similar to *Panaeolus* but spores warted, a "weedy" species on lawns

Psilocybe Gills evenly black, fruit bodies small, delicate, turning blue when injured, spores purple-brown or black (cap skin composed of cylindrical cells)

Psathyrella Gills evenly brown, stem easily snapping cleanly, cap strongly changing color with loss of moisture, spores brown or purple-brown, often on wood or woody debris, rarely on dung (cap skin composed of inflated cells), hundreds of species, very common

Chroogomphus vinicolor
(Peck) O. K. Miller Jr.

WINE-CAP CHROOGOMPHUS
SYNONYM *Gomphidius vinicolor* Peck

Ochre-brown mushroom with conical-convex cap eventually burgundy-red colored; flesh orange; gills ochre, decurrent, thick, becoming smoky gray from the spores, stipe ochre-buff turning wine-red over the base; under pines

CAP 2–8 cm wide, conical-convex, becoming plane and with conical bump, quite variable and ochre or yellow-brown at first, becoming burgundy-red, surface smooth, slimy when wet, shiny when dry. **FLESH** orange or salmon. **GILLS** buff or pale orange, turning ochre, then smoky-ochraceous, decurrent, distant, thick. **STEM** 5–10 cm long, 5–20 mm broad, equal but tapering to base, surface ochre-buff becoming wine-red over base, surface nearly smooth. **RING** delicate, hairy, ochre, best seen when attached to cap margin, leaving thin layer of fibrils on upper stem after collapsing. **SPORE PRINT** smoky-gray or nearly black. **ODOR** and **TASTE** not distinctive.

Habit and habitat single, scattered, or clustered on the ground under pines. August to November. **RANGE** widespread.

Spores canoe-shaped (subfusiform), smooth, smoky-brown, 17–23 × 4.5–7.5 μm, hymenial cystidia with thickened walls.

Comments Look for this species in the fall under pines. If the cystidia have uniformly thin walls, you have *Chroogomphus ochraceus* (Kauffman) O. K. Miller, Jr. that generally is a larger mushroom with a yellow-brown and convex, not conical-convex pileus that does not stain wine-red on the base of the stem.

Agaricus bitorquis (Quélet)
Saccardo

SPRING AGARICUS
SYNONYMS *Psalliota bitorquis* Quélet, *Agaricus rodmanii* Peck

Moderately large and firm, cap white with fine velvety nap on surface that traps sand and soil, thus often dirty looking; stem typically shorter than cap diameter, with distinctive U-shaped, bandlike ring; in hard packed soil, on lawns as well

CAP 5–20 cm broad, convex then plane, white at first with fine velvety nap, this often dirty from soil and sand embedded in the fibrils, becoming gray-brown with age. **FLESH** white, very firm. **GILLS** pale grayish-pink at first, then chocolate-brown, free, close or crowded, broad. **STEM** 2–5 cm long, 1–3 cm broad, equal, or tapered at base, short, squat, thick, white becoming pale grayish-brown, smooth, with bandlike ring, **RING** white, membranous, persistent on mid or lower stem and U-shaped with upper edge flaring out more than the lower edge, lower edge with brown granular patches. **SPORE PRINT** chocolate or purple-brown. **ODOR** and **TASTE** pleasant, mushroomy.

Habit and habitat scattered or clustered in hard packed soil on lawns and other urban areas with plants, usually several to many fruiting at one time. It has broken through paved driveways on occasion. May to June, also September. **RANGE** widespread.

Spores nearly round or oval, smooth, no pore, dark brown in KOH, 5–6 × 4–5 μm.

Comments A choice edible. The solid-fleshed cap is larger than the squat stipe and the U-shaped, bandlike ring on the stem is distinctive. The flesh is very firm and hard. The cap surface is often dirty from sandy particles. Wash this mushroom carefully and enjoy a very fine flavor. *Agaricus arvensis* Schaeffer, another choice edible, also occurs on lawns but in the late summer and fall and is a large white mushroom with a long stem and smooth cap that slowly stains yellow. The ring on the stem has thick, wedge-shaped or coglike tissue on its lower surface, and the odor is fragrant or faintly of anise. Compare also *A. campestris* and *A. placomyces*.

Agaricus campestris Linnaeus
MEADOW MUSHROOM

Short squat mushroom in grassy areas with pink then chocolate colored free gills; cap surface white but may have grayish or brownish colors, and flattened scales

CAP 4–7(–11) cm broad, convex becoming plane, white overall becoming buff colored or grayish in maturation, surface smooth or finely woolly, occasionally with flattened brown scales, margin thin, brittle. **FLESH** white, unchanging or turning brown. **GILLS** pink at first, soon chocolate-brown, free, crowded. **STEM** 2–5 cm long, 1–2 cm broad, generally shorter than width of cap, equal but with tapered base, white, smooth above ring, matted fibrillose below ring. **RING** delicate, thin, membranous, soon collapsing or falling off. **SPORE PRINT** chocolate-brown. **ODOR** pleasant, typical mushroom. **TASTE** mild or nutty.

Habit and habitat on lawns and grassy areas, often in pastures of horses or cows, often on playing fields of schools, usually many fruit bodies making fairy rings, small to medium-sized fruit bodies can be overlooked in moderately high grass. September to November. **RANGE** widespread.

Spores elliptical, smooth, no pore, brown in KOH, average-sized for *Agaricus* species, 6–9 × 4–6 µm.

Comments There appear to be several different "campestroid" *Agaricus* species, small squat mushrooms with pink then chocolate-colored free gills, that are found in grass habitats. These can be difficult to separate from the actual *Agaricus campestris*. Interestingly, *A. campestris* has not been confirmed from the Northeast using molecular techniques. If you have a microscope, you can identify *A. bisporus* (J. E. Lange) Imbach by its 2-sterigmate basidia (basidia producing two spores, not the typical four). *Agaricus bisporus* is not uncommon in our area. If the cap has brown flattened scales you might have *A. argenteus* Braendle. If the cap is white, turning cream-colored as it dries, and the surface hyphae below the stem ring are inflated, it is most likely *A. andrewii* A. E. Freeman. In any case, they are all edible, just make certain your collection has free gills and is producing chocolate spores, not flesh-brown or salmon-brown spores—see *Entoloma subsinuatum*.

Agaricus placomyces Peck

SYNONYM *Psalliota placomyces* (Peck) Lloyd

Tall, large, fleshy mushroom with white cap covered with brown or black flattened scales and fibrils; flesh white turning yellow; stem white bruising bright yellow; odor antiseptic, phenol-like

CAP 5–15 cm broad, convex then plane, white under the brown or blackish-brown flattened scales and fibrils. **FLESH** white, turning yellow, thick. **GILLS** pale grayish-pink at first, then chocolate-brown, free, crowded. **STEM** 6–12 cm long, 1–2 cm broad, equal, or enlarged at base, white, smooth, bruising bright yellow over base on surface and inside flesh, with apical ring. **RING** membranous, persistent, large, flaring, often with yellow or brown watery droplets on lower surface before it breaks to reveal the lamellae. **SPORE PRINT** chocolate or purple-brown. **ODOR** and **TASTE** unpleasant, like phenol, antiseptic.

Habit and habitat Scattered or gregarious under hardwoods or mixed coniferous and hardwood forests, also frequently on lawns with scattered trees present. July to September. **RANGE** widespread.

Spores elliptical, smooth, no pore, brown in KOH, 5–7 × 3.5–5 µm.

Comments The odor should warn you of the possible unpleasant side effects of eating this mushroom. *Agaricus meleagris* Withering, known from western North America, and *A. pocillator* Murrill, from the southeastern United States, are similar and in fact may be just variations of *A. placomyces*. The scales and fibrils on the cap can be a pale brown as well, the odor can be quite faint, and the brown watery droplets may be lacking in dry weather, but the bruising reaction is consistent.

Agaricus sylvaticus Schaeffer

SCALY WOOD MUSHROOM, BLUSHING WOOD MUSHROOM

SYNONYMS *Agaricus haemorrhoidarius* Schulzer, *Psalliota sylvatica* (Schaeffer) P. Kummer

Cap with contrasting, flattened, reddish-brown scales over white ground color; stem white becoming grayish, with persistent membranous ring; flesh turning bright red when exposed; in wooded areas

CAP 2–10 cm broad, convex then plane, dark reddish-brown flattened scaly or fibrillose over white ground color, dry. **FLESH** white, turning red when exposed then slowly brown. **GILLS** pale at first, then pinkish-red, eventually purplish or chocolate-brown, free, close, broad. **STEM** 5–12 cm long, 1–2 cm broad, equal, or slightly broader at base, white becoming pale grayish-brown, bruising red, smooth, with apical ring, white but quickly turning red inside. **RING** membranous, persistent on upper stem. **SPORE PRINT** chocolate or purple-brown. **ODOR** and **TASTE** not distinctive.

Habit and habitat single or in small groups on the ground under conifers. July to October. **RANGE** widespread.

Spores elliptical, smooth, no pore, brown in KOH, 4.5–6 × 3–3.5 μm.

Comments There are several species of *Agaricus* that stain red when the flesh is exposed. The image shown here is most likely not the European *A. sylvaticus*, but our northeastern species displays the typical features—reddish-brown flattened scales on the cap, reddening flesh, and persistent ring on the stem. *Agaricus haemorrhoidarius*, also described from Europe, is considered a synonym. The intensely contrasting scales on the cap and the slender stem are morphological differences from the European species that indicate our northeastern version is different.

Although *A. sylvaticus* is listed as edible in some books, there are also reports of individuals developing gastric distress from eating *Agaricus* species that look similar to this mushroom. Always try small portions at first if you just have to try using this group for food. Also, avoid all collections of *Agaricus* that smell of iodine; they will not taste good and your chances of having a case of gastric distress go up.

Melanophyllum haematospermum (Bulliard) Kreisel

RED-GILLED AGARICUS

SYNONYMS *Agaricus haematospermus* Bulliard, *Lepiota haematosperma* (Bulliard) Boudier, *Psalliota haematosperma* (Bulliard) S. Lundell & Nannfeldt, *Melanophyllum echinatum* (Roth) Singer

Cap gray-brown with granular powdery coating, margin hung with ring fragments; gills bright wine red at first, free or nearly so; stem covered with same material as cap; odor pungent

CAP 1–4 cm broad, broadly conical, becoming plane, gray-brown, covered with a granular-powdery coating, margin with hanging gray-brown ring fragments with age. **FLESH** white, turning pink, thin. **GILLS** bright vinaceous-pink at first, becoming pinkish-brown or dark brown, free or barely attached, close. **STEM** 2–7 cm long, 1–3 cm broad, equal with basal bulb from mycelium binding substrate, often curved, flesh-pink under the gray-brown powdery covering that is easily removed, with apical ring zone. **RING** gray-brown, granular-powdery over thin, delicate membrane that soon collapses or leaves fragments on cap margin. **SPORE PRINT** greenish or olive-gray when very fresh, then reddish, drying dark brown. **ODOR** pungent. **TASTE** not pleasant.

Habit and habitat single, scattered, or clustered on humus or leaf litter in mixed forests, in bogs, compost heaps, greenhouses. July to September. **RANGE** widespread but rare.

Spores elliptical-elongate, minutely bumpy, no pore, colorless, 5–6.5 × 2.3–4 µm; cap with globose cells making up the granules.

Comments The free gills and their bright pink color at first remind one of an *Agaricus* species, but the very small size, the greenish-colored spores and the granular cap surface clearly set this species apart from the genus *Agaricus*. If the caps are granular and the spores are white, see *Cystoderma* with clearly attached gills or *Lepiota* with clearly free gills.

Coprinus comatus (O. F. Müller) Persoon

SHAGGY MANE

SYNONYMS *Agaricus comatus* O. F. Müller, *Agaricus cylindricus* Sowerby, *Agaricus ovatus* Scopoli, *Agaricus fimetarius* Bolton

Large, fleshy, cylindrically capped mushroom, with tan, shaggy-fibrous scales; cap easily removed from stem; stem hollow with stringlike strand; in clusters on lawns and wood chips

CAP 3–5 cm wide, 4–15 cm high, cylindrical or elongate-parabolic at first, becoming bell-shaped as margin curls up, white with center and shaggy scales light-brown, covered by cottony scales and fibrils, splitting easily over the margin, dry. **FLESH** white but becoming reddish when gills edges turn red. **GILLS** white, then red, soon dark gray then black, becoming inky (deliquescing) progressively from cap margin to top of stem, melting, free, very crowded, broad. **STEM** 5–20 cm long, 10–20 mm broad at apex, equal, often with bulb at base and a ring on the lower stem, white, glabrous, narrowly hollow with white stringlike strand inside the chamber. **RING** white, membranous, movable and easily lost. **SPORE PRINT** black. **ODOR** and **TASTE** not distinctive.

Habit and habitat single, scattered but mostly clustered, often in fairy rings on lawns, wood chips, hard-packed soil. May through October. **RANGE** widespread.

Spores ellipsoid, smooth, with apical germ pore, nearly black in KOH, 13–18 × 7–9 µm.

Comments A fine edible species that you can do many things with in the kitchen. I prefer to sauté the mushrooms in olive oil and then make a meatless red pasta sauce, very nice. If you like wine with your meal, be sure to compare this species with *Coprinopsis atramentaria* that is vaguely similar in shape but lacks the shaggy scales and other features. Why is one species of these inky caps in *Coprinus* while the other is in *Coprinopsis*? Molecular phylogenetics indicate the two are not closely related.

Coprinopsis atramentaria
(Bulliard) Redhead, Vilgalys & Moncalvo

ALCOHOL INKY, TIPPLER'S BANE
SYNONYM *Coprinus atramentarius* (Bulliard) Fries

Large, fleshy, gray or gray brown, smooth, egg-shaped caps in cluster in grassy areas from buried wood or roots or decaying trees; with thin ring at base of white stem

CAP 3–7.5 cm wide, 3–9 cm high, conical or egg-shaped at first, becoming bell-shaped or convex, gray or grayish-brown, surface glabrous but occasionally with flattened scales over center, smooth or with radiating lines over margin, dry. **FLESH** white. **GILLS** white at first, soon dark gray then black, becoming inky, melting, free, crowded, broad. **STEM** 5–17 cm long, 10–15 mm broad at apex, equal, white, silky-fibrous, hollow. **RING** white, fibrous, leaving thin line near stem base. **SPORE PRINT** black. **ODOR** and **TASTE** not distinctive.

Habit and habitat single, scattered or mostly clustered in lawns near dead trees or from buried wood or dead roots. May to September. **RANGE** widespread.

Spores ellipsoid, smooth, with apical germ pore, nearly black in KOH, 7–11 × 4–6 µm.

Comments It can be edible, but if you drink any alcohol during consumption or even 1–2 days after eating this mushroom, you can become ill with headache, nausea, tingling of extremities, and flushing of skin around the face from the coprine toxin in the mushrooms. The symptoms go away on their own within a few hours, and if you do not imbibe alcohol, there is nothing to worry about. Compare with *Coprinus comatus*.

Coprinopsis variegata (Peck)
Redhead, Vilgalys & Moncalvo

SCALY INKY CAP

SYNONYMS *Coprinus atramentarius* var. *variegatus* (Peck) Rick, *Coprinus variegatus* Peck, *Coprinus quadrifidus* Peck

Large, fleshy, pale gray mushroom, with large, flat, tan, fibrous scales on an egg-shaped caps; thin ring zone at base of white woolly-scaly stem; in clusters on decaying hardwood logs

CAP 2–7.5 cm wide, 5–8 cm high, elongate-conical or egg-shaped at first, becoming bell-shaped, pale gray or gray-brown, covered by buff or pale tan flattened cottony scales and fibrils, lacking lines over margin, dry. **FLESH** white. **GILLS** white, soon dark gray then black, becoming inky (deliquescing), melting, free, crowded, broad. **STEM** 4–12 cm long, 5–10 mm broad at apex, equal, white, woolly-scaly below the ring zone, hollow. **RING** white or pale tan, felted-woolly, leaving thin line on the lower stem. **SPORE PRINT** black. **ODOR** and **TASTE** not distinctive, or unpleasant.

Habit and habitat scattered but mostly clustered on decaying hardwoods. June to August. **RANGE** widespread.

Spores ellipsoid, smooth, with apical germ pore, nearly black in KOH, 7.5–10 × 4–5 μm.

Comments One of the larger species of inky caps that can fruit rather prolifically. Although listed as edible in some field guides, the variations with off odors are known to cause mild poisoning cases. Compare with *Coprinus comatus*.

Coprinellus micaceus (Bulliard) Vilgalys, Hopple & Jacq. Johnson

MICA CAP

SYNONYMS *Agaricus micaceus* Bulliard, *Coprinus micaceus* (Bulliard) Fries

Small to medium-sized tawny-colored, strongly lined caps, densely covered with shiny micalike granules at first; gills white then inky black; stem white; fruit bodies fragile; on decaying wood, often on lawns from buried wood

CAP 2–5 cm wide, broadly conical at first, becoming bell-shaped, tawny or pale tan, becoming grayish as spores mature, covered at first by minute sugarlike shiny granules (mica), these granules quickly lost, margin strongly lined to near center, dry. **FLESH** white, fragile. **GILLS** white, soon dark gray then black, becoming inky (deliquescing) from the edges at first, melting but not completely, free, close or crowded. **STEM** 2–8(–12) cm long, 3–6 mm broad at apex, equal, white, with fine granules over apex, silky smooth elsewhere, hollow, fragile. **SPORE PRINT** black. **ODOR** and **TASTE** not distinctive.

Habit and habitat densely clustered, around stumps or woody debris. April to October. **RANGE** widespread.

Spores ellipsoid and slightly flattened in one view, mitriform, or shaped like a bishop's hat, in another view, smooth, with apical germ pore, nearly black in KOH, 8–11 × 5–6 × 5.5–6.5 µm.

Comments *Coprinellus* and *Parasola* are two more genera separated from *Coprinus* based on molecular phylogenetic evidence. From outward appearances, a similar-looking species to *Coprinellus micaceus* is *C. truncorum* (Scopoli) Redhead, Vilgalys & Moncalvo. It can be identified only after examination of the spores that are elliptical and not mitriform. Both are edible and fairly common. Two common *Parasola* species are *P. auricoma* (Patouillard) Redhead, Vilgalys & Hopple and *P. plicatilis* (Curtis) Redhead, Vilgalys & Hopple with reddish-brown, strongly corrugate caps. The latter species has paper-thin delicate caps and occurs on lawns, while *P. auricoma* produces thicker fleshy caps and prefers wood chips.

Stropharia hardii G. F. Atkinson

HARD'S STROPHARIA

Medium to large orange-yellow-capped mushroom with gray then purplish-brown gills; stem equal with a bandlike ring that is smooth on the undersurface; growing on decaying conifer wood

CAP 3–10 cm wide, convex, becoming plane, bright orange-yellow or brownish-yellow, with darker brownish spots on some, also margin with thin white patches of ring tissue hanging down (appendiculate), slippery at first, then dry, glabrous, opaque. **FLESH** white. **GILLS** off-white becoming increasingly grayish-brown then purplish-brown, attached, close, broad. **STEM** 5–12(–14) cm long, 5–15 mm broad at apex, equal but often with abrupt bulb at base, white off-white or pale yellow, white chords at base. **RING** white above and below, bandlike with upper part flaring, membranous, smooth on both sides.

SPORE PRINT purplish-brown or black. **ODOR** strong mushroomy. **TASTE** mild.

Habit and habitat scattered or clustered on woody debris in hardwood forests or on wood-chip beds. September to October. **RANGE** widespread.

SPORES ellipsoid, with apical pore, smooth, dark reddish-brown in KOH, 6–7 × 3.5–4 µm; pleurocystidia with pale yellowish content in 3% KOH (chrysocystidia).

Comments See *Stropharia hornemannii* with its slimy purplish-brown or smoky reddish-brown cap and scaly stem growing on decaying conifer wood. *Stropharia rugosoannulata* f. *lutea* is somewhat similar in color, but is easily separated by the much larger size, with a club-shaped stem and the cogwheel-like teeth on the underside of the ring. The spores are much larger and thick-walled as well.

Stropharia hornemannii
(Fries) S. Lundell & Nannfeldt

LACERATED STROPHARIA

SYNONYMS *Agaricus hornemannii* Fries, *Naematoloma hornemannii* (Fries) Singer, *Psilocybe hornemanii* (Fries) Noordeloos, *Stropharia depilata* (Persoon) Saccardo

Medium to large smoky reddish-brown or purplish-brown, slimy capped mushroom with gray then purplish-brown gills; stem white to cream, equal, with conspicuous curved scales below bandlike ring; growing on decaying conifer wood

CAP 4–15 cm wide, convex or broadly bell-shaped, becoming plane, purplish-brown or grayish-brown or smoky reddish-brown at first, but mostly fading to tan or yellowish-brown with age, sometimes with thin white patches of ring tissue hanging down (appendiculate) from margin, slimy at first, then dry, glabrous, opaque. **FLESH** white. **GILLS** pale gray becoming increasingly grayish-brown then purplish-brown, attached, close, broad. **STEM** 5–15 cm long, 5–25 mm broad at apex, equal, off-white or pale yellow, large recurved scales below ring in young stages, finely fibrillose above it, white mycelioid chords at base. **RING** white, bandlike, collapsing. **SPORE PRINT** purplish-brown or black. **ODOR** not pleasant. **TASTE** disagreeable.

Habit and habitat scattered or clustered on well-decayed woody debris in coniferous forests. August to November. **RANGE** widespread in northern coniferous forests.

Spores ellipsoid, with apical pore, smooth, dark reddish-brown in KOH, 10–14 × 5–7 µm; pleurocystidia with pale yellowish content in 3% KOH (chrysocystidia).

Comments Compare with *Stropharia rugosoannulata* that has a red-wine-colored cap and might look somewhat similar. The underside of the ring, the habitat, and the odor should help you make the right decision about which fungus you have collected.

Stropharia rugosoannulata
Farlow ex Murrill

WINE-CAP STROPHARIA
SYNONYMS *Stropharia ferrii* Bresadola, *Naematoloma rugosoannulatum* (Farlow ex Murrill) S. Ito, *Psilocybe rugosoannulata* (Farlow ex Murrill) Noordeloos

Large wine-red-capped mushroom with gray then purplish-gray gills; ring on upper stem thick fleshy, with ragged triangular toothlike scales on the undersurface; stem white; on wood chips

CAP 5–15 cm wide, convex or broadly bell-shaped, becoming plane, wine-red or reddish-brown, fading to tan with age, margin sometimes with fibrillose patches of creamy-white ring tissues (appendiculate), slippery at first, then dry, glabrous but occasionally becoming scaly from disrupted surface, opaque. **FLESH** white, thick (6–12 mm). **GILLS** white becoming increasingly grayish, then dark purplish-gray or black, attached, close or crowded, broad. **STEM** 5–18 cm long, 10–30 mm broad at apex, equal or at times with slightly enlarged base, white discoloring pale yellowish or creamy-tan, fine fibrillose-scaly above ring, glabrous or fibrous below ring, white mycelioid chords at base. **RING** white, thick, membranous, lined above, ragged toothlike or wedge-shaped thick scaly below. **SPORE PRINT** purplish-brown or black. **ODOR** pleasant. **TASTE** mild.

Habit and habitat scattered or clustered and in large numbers sometimes on wood chips or woody mulch in gardens. May to October. **RANGE** widespread and fairly common.

Spores ellipsoid, with apical pore, thick-walled, smooth, dark reddish-brown in KOH, 10–14 × 6–9 µm; pleurocystidia and cheilocystidia with pale yellowish content in 3% KOH (chrysocystidia); spiny crystalline balls (acanthocytes) encrusting the white mycelial chords.

Comments This mushroom, also called the king stropharia, because of the crownlike ring, is considered a choice edible and can be cultivated. See various online sources of information, but one of the better ones is on the mushroom blog at Cornell University.

Stropharia rugosoannulata f. *lutea* Hongo

SYNONYMS *Stropharia ferrii* var. *lutea* Hongo, *Naematoloma rugosoannulatum* f. *luteum* S. Ito, *Stropharia bulbosa* f. *lutea* (Hongo) Hongo

Large golden-capped mushroom with gray gills, a clublike golden stem and a thick ring with ragged triangular scales on the undersurface; odor intensely mushroomy

CAP 4.5–15 cm wide, convex, becoming plane then margin arching up, golden-yellow with slight olive cast as first, becoming paler yellow with expansion, margin with patches of yellow ring (appendiculate), slippery at first, then dry, glabrous, opaque. **FLESH** white, thick (6–12 mm). **GILLS** white becoming increasingly grayish, attached, close or crowded, broad. **STEM** 5.5–11 cm long, 12–30 mm broad at apex, club-shaped, white over apex and base otherwise yellowish or creamy-tan, glabrous, white mycelioid chords at base. **RING** golden-yellow, thick, membranous, lined above, ragged toothlike scaly below. **SPORE PRINT** purplish-brown or black. **ODOR** strong mushroomy. **TASTE** mild.

Habit and habitat scattered or clustered on wood chips or woody mulch in gardens. **RANGE** Known, so far, from Japan, New York, and Europe, apparently not common.

Spores ellipsoid, with apical pore, thick-walled (1.6 µm) smooth, dark reddish-brown in KOH, 12–14.6 × 7.5–9 µm; pleurocystidia and cheilocystidia, ventricose-rostrate mainly and with pale yellowish content in 3% KOH (chrysocystidia); spiny crystalline balls (acanthocytes) encrusting the white mycelial chords.

Comments Essentially a golden-colored *Stropharia rugosoannulata*. I think the odor is much more enticing for this golden-yellow form.

Psathyrella candolleana
(Fries) Maire

COMMON PSATHYRELLA
SYNONYMS *Agaricus candolleanus* Fries,
Psathyrella appendiculata (Bulliard) Maire
& Werner

Cap pale buff, becoming white, with white
membranous ring patches hanging from
margin; gills white, then grayish then dark
brown, crowded; stem white, glabrous,
easily snapping; most often on lawns, but
in forests as well

CAP 2–10 cm wide, broadly conical
becoming convex, then plane, with a low
broad central bump, with white, fragile,
ring patches hanging from the margin,
honey-yellow at first, hygrophanous and
quickly changing to pale pinkish-buff or
grayish-buff, then off-white from center
outward, glabrous, with shallow, fine,
radial lines at margin, moist becoming
dry. **FLESH** white or becoming brown, thin
and fragile. **GILLS** white becoming grayish
then dark brown, barely attached, close
or crowded, narrow. **STEM** 4–10 cm long,
3–8 mm broad, equal, white, glabrous, dry,
hollow, easily snapping. **RING** white, rarely
on stem, patches found on cap margin, but
easily lost. **SPORE PRINT** purplish-brown.
ODOR and **TASTE** not distinctive.

Habit and habitat single, scattered or
densely clustered on lawns, pastures, but
may also be found in forests around decaying hardwoods. May to September. **RANGE**
widespread, but rare.

Spores ellipsoid, truncate with broad apical pore, smooth, dark brown in 3% KOH,
7–10 × 4–5 μm.

Comments If you have buried decaying
wood in your yard, such as old roots from
recently or long-removed trees, you will
see this moderately sized, white, fragile
mushroom, often growing in fairy rings.
It is so fragile it falls apart if not collected
carefully. The hanging ring fragments on
the cap margin are diagnostic.

Psathyrella delineata (Peck)
A. H. Smith

SYNONYMS *Hypholoma delineata* Peck, *Drosophila delineata* (Peck) Murrill

Large mushroom with cap dark red-brown, fading to tan, radially wrinkled, with white ring patches on margin; gills same color as cap with age; on decaying hardwood logs and debris

CAP 3–10 cm wide, convex, then plane, with white membranes ring patches hanging from the margin, dark red-brown at first, fading to tan, with thin coating of white fibrils in early stages, glabrous, mostly radially wrinkled from center outward, moist becoming dry. **FLESH** white or becoming brown, thin and fragile. **GILLS** pale brown becoming dark brown, bluntly attached, close, broad. **STEM** 6–10 cm long, 5–15 mm broad, equal, white from coating of silky fibrils, brown under fibrils, smooth, dry, hollow, easily snapping. **RING** white, rarely on stem, patches found on cap margin, but easily lost. **SPORE PRINT** purplish-brown. **ODOR** and **TASTE** not distinctive.

Habit and habitat single, scattered or densely clustered on wood of decaying hardwoods. July to September. **RANGE** widespread.

Spores ellipsoid or somewhat bean-shaped in side view, inconspicuous apical pore, smooth, dark brown in 3% KOH, 6.5–9 × 4.5–5.5 μm.

Comments Of the hundreds of species of *Psathyrella* already documented in North America, this is one of the easier ones to identify with solely macroscopic features.

Psathyrella piluliformis
(Bulliard) P. D. Orton

CLUSTERED PSATHYRELLA
SYNONYMS *Agaricus piluliformis* Bulliard, *Agaricus spadiceus* var. *hydrophilus* (Bulliard) Fries, *Psathyrella hydrophila* (Bulliard) Maire

Small to medium-sized rusty-brown caps that quickly fade to tan, with white ring fibrillose patches over margin; gills pale then dark brown, crowded; stem white, easily snapping; densely clustered on decaying hardwood

CAP 2–5 cm wide, broadly conic then convex, becoming plane, with thin white ring patches often hanging from the margin but these easily lost, dark rusty-brown at first, fading to tan or honey-brown, glabrous, moist becoming dry. **FLESH** thin, fragile. **GILLS** pale buff, becoming dark brown, bluntly attached, crowded, narrow. **STEM** 3–7(–15) cm long, 2–6(–10) mm broad, equal, white, smooth, dry, hollow, easily snapping. **RING** white, thin fibrillose, rarely on stem, patches found on cap margin, but easily lost. **SPORE PRINT** dark brown. **ODOR** and **TASTE** not distinctive.

Habit and habitat densely clustered and in small bouquetlike bunches on wood of decaying hardwoods. July to September. **RANGE** widespread.

Spores ellipsoid, with inconspicuous apical pore, smooth, dark brown in 3% KOH, 4.5–6 × 3–3.5 μm.

Comments These small, dark brown, wood loving *Psathyrella* species are numerous and for the average mushroomer without a microscope, impossible to identify accurately. Even with a good microscope, it still is not so easy because of the large numbers of species with a great deal of variability in form and size of structures. *P. piluliformis*, formerly known as *P. hydrophila* in field guides, is one of the more common species, so you should run into it at some point.

Psathyrella minima Peck

Tiny fruit bodies, cap cinnamon-brown but quickly fading to pinkish-buff, densely frosted on cap and stem; on deer dung

CAP 1–4 mm wide, conic then bell-shaped, cinnamon-brown at first, fading quickly to tan or pale pinkish-buff, densely covered with minute hairs as though frosted, moist and translucent-lined at first, becoming dry and opaque. **FLESH** very thin, very fragile. **GILLS** pale honey colored, becoming brown, bluntly attached, distant, broad. **STEM** 1–2 cm long, 0.5–1 mm broad, equal, pale honey colored or with grayish densely covered with white tufts of minute hairs, dry, hollow, easily snapping. **SPORE PRINT** probably dark brown. **ODOR** not distinctive. **TASTE** unknown.

Habit and habitat scattered or densely clustered on deer dung. July to September. **RANGE** widespread.

Spores ellipsoid, with apical pore, smooth, dark brown in 3% KOH, 6–8 × 3–4 μm; cap surface a mixture of inflated cells with fusoid-ventricose pointed pileocystidia mixed in.

Comments Deer dung usually forms pellets, but sometimes the excrement is soft and forms fair-sized pats, much like cow droppings. The image shown here is one of those deer pats with many fruit bodies covering the surface. The species is most likely more common than the literature indicates, one just has to be very observant, and also have some luck. Herbivore dung can yield a number of different species of dark-spored agarics. If you find small tawny-capped mushrooms on porcupine dung, you could well have *Psathyrella galericolor* A. H. Smith.

Panaeolina foenisecii
(Persoon) Maire

HAY MAKER'S MUSHROOM
SYNONYMS *Agaricus foenisecii* Persoon, *Panaeolus foenisecii* (Persoon) J. Schröter, *Psathyrella foenisecii* (Persoon) A. H. Smith, *Psilocybe foenisecii* (Persoon) Quélet

A small brown mushroom on lawns with dramatically changing cap color; cap dark brown turning buff; gills brown, attached, white fringed; stem buffy-brown, very fragile

CAP 1–3 cm wide, convex or conical or bell-shaped at first, becoming broadly convex, dark date-brown at first, hygrophanous and becoming paler cinnamon-brown or tan or vinaceous buff in zones or bands or evenly colored, smooth, glabrous, faintly translucent-lined when moist, becoming dry and opaque, cracking when very dry. **FLESH** colored as cap, thin, very fragile. **GILLS** colored as cap or reddish-brown, becoming mottled over the faces with darker dots, attached, close, broad, edges white fringed. **STEM** 4–8 cm long, 2–5 mm wide, equal, strict, brittle, pale buffy-brown, but becoming brown from the base upward, smooth except for white powdered apex, hollow inside. **SPORE PRINT** dark date-brown. **ODOR** and **TASTE** not distinctive.

Habit and habitat scattered or clustered in small or large numbers on lawns or in grassy areas. May to September. **RANGE** widespread.

Spores ellipsoid, coarsely roughened, thick-walled, dark brown in 3% KOH, apex protruding and truncate, with apical pore, 12–17 × 7–9 μm.

Comments A common lawn mushroom, it comes up early at or just before the first cutting of hay in late May and fruits off and on over several months if the weather is warm and humid.

Panaeolus papilionaceus
(Bulliard) Quélet

SYNONYMS *Agaricus papilionaceus* Bulliard, *Copelandia papilionacea* (Bulliard) Bresadola, *Panaeolus campanulatus* (Linnaeus) Quélet, *Panaeolus sphinctrinus* (Fries) Quélet, *Panaeolus retirugus* (Fries) Gillet

Medium-sized with conic to bell-shaped grayish cap with white toothlike patches around the margin; gills mottled black; stem long and straight, brittle; grassy areas, gardens, or on herbivore dung

CAP 1–5 cm wide, parabolic or broadly conical at first, becoming bell-shaped, with white toothlike ring fragments around the margin, reddish-brown but rapidly becoming grayish-brown, brown or even pale ashy white with center retaining some brown pigment, smooth or finely wrinkled, glabrous, opaque, slippery when wet, soon dry. **FLESH** colored as cap, thin. **GILLS** grayish, becoming mottled over the faces with black dots with age, eventually black, attached, close or crowded, edges white fringed. **STEM** 4–16 cm long, 2–5 mm wide, equal, straight, brittle, colored as cap, with dense grayish fuzzy coating, hollow inside. **SPORE PRINT** black. **ODOR** and **TASTE** not distinctive.

Habit and habitat single or in small numbers on aged horse or cow dung, in vegetable or flower gardens, on lawns or in grassy areas. June to September. **RANGE** widespread.

Spores ellipsoid, smooth, thick-walled, dark brown in 3% KOH, apex protruding and truncate, with apical pore, 11–18.5 × 7.5–12 µm.

Comments Notice among the synonyms that four distinct species used to be recognized, based on cap color, slipperiness of the cap, or whether it wrinkled or not. These variations are now considered normal for this high-nitrate, dung-loving species.

Psilocybe caerulipes (Peck) Saccardo

BLUE-FOOT PSILOCYBE

SYNONYM *Agaricus caerulipes* Peck

Small inconspicuous on decaying hardwoods with white flecked, cinnamon-brown, slippery cap; gills attached, cinnamon-brown; stem cinnamon-brown, but white-dotted above, and white fibrillose downward; stem and cap slowly turning greenish-blue

CAP 1–3.5 cm wide, broadly conical becoming convex, then plane, occasionally with broad central bump, cinnamon or pale reddish-brown, bruising greenish-blue, with white fibrillose scales scattered about, becoming glabrous, margin faintly lined, slippery or sticky. **FLESH** watery brown, thin and fragile. **GILLS** tan becoming cinnamon, attached, close, broad, edges white fringed. **STEM** 3–6 cm long, 1.5–3 mm broad, equal, cinnamon brown but covered with white fibrils and scales, densely white-dotted over apex, turning greenish-blue slowly over base when handled, dry,

hollow. **RING** white, fine white fibrillose, rarely on stem but may leave a zone, mostly left on cap margin, easily lost. **SPORE PRINT** purplish-brown to black. **ODOR** and **TASTE** not distinctive.

Habit and habitat single or in small clusters on decaying hardwood logs and debris. August to October. **RANGE** widespread but not common.

Spores ellipsoid, with apical pore, smooth, dark brown in 3% KOH, 7–10 × 4–5.5 μm.

Comments One of the many LBMs. Some are very toxic and it pays to learn those species thoroughly (see *Galerina autumnalis*) because they can be deadly. The bluing stem base helps with identification of *Psilocybe caerulipes*, which means literally blue foot. *Psilocybe quebecensis* Ola'h & R. Heim, described from Quebec, is another species found on decaying hardwood debris, that has a stem turning slowly greenish-blue when injured. It differs by its cap that is straw-yellow turning silver-gray, and the fruit bodies are generally thinner overall.

Deconica montana (Persoon) P. D. Orton

SYNONYMS *Agaricus montanus* Persoon, *Psilocybe montana* (Persoon) P. Kummer, *Psilocybe ochreata* (Saccardo) E. Horak, *Psilocybe atrorufa* (Schaeffer) Quélet

Small, translucent-lined, orange-brown, slippery cap; gills bluntly attached, subdistant, reddish-brown or black; stem colored as cap or paler, dry, glabrous; on moss beds in mountains

CAP 0.5–1.5 cm wide, rounded-convex, becoming plane, sometimes with small low broad central bump, dark orange-brown or reddish-brown and translucent-lined from center to margin, hygrophanous and fading to yellowish-brown or eventually to pale buff, slippery or sticky, glabrous. **FLESH** watery brown, fragile. **GILLS** tan becoming dark reddish-brown or nearly black, bluntly attached or with decurrent tooth, subdistant, broad. **STEM** 1–5 cm long, 1–2.5 mm broad, equal, same color as cap or paler yellowish-tan, glabrous, dry, hollow, fragile. **RING** white, wispy-fibrillose, quickly lost. **SPORE PRINT** purplish-brown. **ODOR** and **TASTE** not distinctive.

Habit and habitat single or in small clusters on mosses in mountainous areas. August to October. **RANGE** widespread but not common.

Spores ovoid or ellipsoid, thick-walled and with apical pore, smooth, brown in 3% KOH, 7–9 × 5–6 μm.

Comments There are many species of Psilocybe, most of which are small and difficult to identify. The ones that stain blue or at least have psilocin or psilocybin compounds in the fruit bodies are now considered the true members of the genus. The nonbluing ones are placed in Deconica. If you like to hike above timberline, you most likely will run into this species. You can find it below timberline as well, on moss beds.

Hypholoma capnoides (Fries)
P. Kummer

SMOKY-GILLED NAEMATOLOMA
SYNONYMS *Agaricus capnoides* Fries, *Naematoloma capnoides* (Fries) P. Karsten, *Psilocybe capnoides* (Fries) Noordeloos

Thick-fleshed, medium-sized, found on conifer wood in the fall, often in bouquet-like clusters, cap brick-red, with paler margin; gills gray then purple-brown; stem with ring zone

CAP 2.5–7.5 cm wide, convex or broadly convex, then plane, cinnamon or rusty-cinnamon overall, soon paler yellowish or orange-brown around the margin with center darker, glabrous, dry, often with delicate white ring fragments around the margin. **FLESH** white or pale cream, thick. **GILLS** white or creamy-white at first, soon gray, then becoming purple-brown when spores mature, bluntly attached, close or crowded. **STEM** 2–8(–10) cm long, 3–10 mm broad, equal, pale yellowish but slowly turning rusty-brown from the base upward with age, glabrous or finely fibrillose, also with a ring zone. **RING** off-white, cobweblike, covering gills at first, then as a faint fibrillose zone on upper stem or lost, also pieces on cap margin. **SPORE PRINT** purplish-brown. **ODOR** and **TASTE** not distinctive.

Habit and habitat usually clustered into bouquetlike bunches on decaying conifer logs and stumps. September to December. **RANGE** widespread.

Spores ellipsoid, smooth, yellowish-brown in KOH, 6–7 × 4–4.5 μm; pleurocystidia inflated and with yellow amorphous pigment body in KOH (a chrysocystidium).

Comments If the gills are greenish and the cap also has some yellow-green hues, then you have *Hypholoma fasciculare* (Hudson) P. Kummer. It also grows in bouquetlike clusters on decaying conifer wood, but *H. fasciculare* tastes bitter and makes people ill if they consume it.

Hypholoma lateritium
(Schaeffer) P. Kummer

BRICK TOPS

SYNONYMS *Agaricus lateritius* Schaeffer, *Hypholoma sublateritium* (Fries) Quélet, *Naematoloma sublateritium* (Fries) P. Karsten, *Psilocybe lateritia* (Schaeffer) Noordeloos

Thick-fleshed, medium-sized, found on conifer wood in the fall, often in bouquet-like clusters, cap brick-red, with paler margin; gills gray then purple-brown; stem with ring zone

CAP 3–10 cm wide, convex becoming broadly convex, then plane, brick red overall or with paler yellowish or pinkish-buff margin, glabrous, dry, often with delicate yellowish veil fragment around the margin. **FLESH** white or pale cream, thick. **GILLS** white at first, soon pale gray becoming dark purple-brown when spores mature, bluntly attached, close or crowded. **STEM** 5–10 cm long, 5–15 mm broad, equal, pale yellowish or creamy-white, often rosy-brown over lower portions, sometimes staining yellow where bruised, glabrous or finely fibrillose above ring zone. **RING** yellowish, thin, fibrillose, soft, covering gills at first, soon only as a faint fibrillose zone on upper stem or lost, also pieces on cap margin occasionally. **SPORE PRINT** purplish-brown. **ODOR** not distinctive. **TASTE** not distinctive, or slightly bitter.

Habit and habitat usually clustered into bouquetlike bunches on decaying hard-wood logs and stumps. August to November. **RANGE** widespread.

Spores ellipsoid, smooth, with apical pore, pale yellow-brown in KOH, 6–7 × 4–4.5 µm; pleurocystidia inflated and with golden-yellow pigment bodies in KOH (chrysocystidia).

Comments Listed as a good edible if the young caps are used. Compare with *Hypholoma capnoides*. The species has been listed as *Hypholoma* or *Naematoloma sublateritium* for many years, and is listed in more American field guides as such. It appears the name *lateritium* (*-us*) takes precedence in use.

Hypholoma elongatum
(Persoon) Ricken

SYNONYMS *Agaricus elongatus* Persoon, *Naematoloma elongatum* (Persoon) Konrad, *Psilocybe elongata* (Persoon) J. E. Lange

On *Sphagnum*, with yellowish-green, moist, smooth caps; gills pale yellow, becoming grayish-pink then brown; stems very long, yellowish or olive-yellow with scattered white fibrils, thin and fragile

CAP 0.5–2 cm wide, convex or broadly bell-shaped at first, becoming broadly convex, then plane, pale yellow or yellowish-green, with honey-yellow or sordid olive-yellow centrally, hygrophanous and becoming paler, glabrous, moist, translucent-lined or obscurely ridged over the margin. **FLESH** pale yellow or white, thin, soft. **GILLS** white or pale yellowish at first, soon pale grayish-pink then grayish-brown, bluntly attached, subdistant, broad. **STEM** 4–15 cm long, 1–3 mm broad, equal, pale yellowish or dingy olive-yellow, paler over apex, often becoming brown over lower portions, with widely scattered silky white fibrils or patches, these more concentrated over apex. **SPORE PRINT** dull cinnamon-brown. **ODOR** and **TASTE** not distinctive.

Habit and habitat scattered or clustered in *Sphagnum* in bogs or fens. July to October. **RANGE** widespread.

Spores ellipsoid, smooth, pale cinnamon-brown, with small apical pore, 8–12 × 5–7 µm; some pleurocystidia with internal golden-yellow pigment body in KOH mounts (chrysocystidia).

Comments Of the several different species of thin spindle-stemmed LBMs, this one with the yellow-green cap, growing on sphagnum moss is fairly common if you frequent boggy areas. To identify most of the other species, you need a good microscope to measure spores and look for different types of cystidia.

Boletes: The Fleshy Pore Fungi

This group of fungi, like all others, is going through a reevaluation of the evolution of the diverse forms, using DNA molecules to determine phylogeny or relatedness among the species. In the boletes in particular, there has been a significant series of changes put forth based on the DNA data, with many new genera being described to accommodate the patterns that are emerging from a phylogenetic perspective. This work is not completed and when it is over, it looks like the porcini—*Boletus edulis* and its close relatives—will be the only species remaining in the genus *Boletus*.

Most form mycorrhizae with woody plants, but not all (see *Boletinellus*). Some can be very specific about the tree partner with which they form symbioses; for example, most species of *Suillus* partner with *Pinus*, but a few choose *Larix*. For many, the exact biology is not fully known as yet, whether they form mycorrhizae, are parasites, or have some other unknown relationship with plants or animals. These unknowns will surely make for some very interesting future research.

Besides the porcini, a number of boletes are considered fine edible gourmet foods, for example some species of *Baorangia*, *Gyroporus*, *Hemileccinum*, and *Leccinum*. On the other hand, some boletes can cause severe gastric distress, again some species of *Leccinum*, a number of red-pored boletes, and especially those that turn bright blue when the flesh is exposed. Take care to learn species thoroughly and carefully if you plan to use them for food.

The following key is only a guideline to help one determine the genus. Please read the entire description and comments until you have developed some expertise at field identifications. The more common species can be identified with macro features, but a microscope will be essential when trying to make completely accurate identifications of many species of boletes.

Pores and tubes bright yellow

Aureoboletus Stem long, lacking any veils, stem smooth or coarsely netted, often slimy at base

Pulveroboletus Stem normal length, powdery bright yellow covering (veils) present, pores staining greenish-blue in one species

Boletinellus Stem off-center, pores radially elongate, slowly blue when injured under ash

Baorangia Pores and tubes turning blue rapidly when bruised, tubes very short, cap and stem bright to dull red

Hemileccinum Cap cinnamon-brown, stem pale yellow with yellow or reddish dots and scales

Retiboletus Stem coarsely bright yellow netted, taste bitter

Boletus Some species large, with thick stem, stem netted or not, tubes and pores turning blue when bruised

(Note: see also *Xerocomus*, *Xerocomellus*, *Boletellus*, part of the genus *Boletus*, *Pseudoboletus*, *Caloboletus*, and *Suillus*, all with yellowish, but not bright yellow pores and tubes)

Pores and tubes white, becoming pale yellow

Gyroporus Stem stuffed at first, soon hollow, brittle, breaking cleanly, spore print yellow

Xanthoconium Stem solid, not brittle but fleshy-fibrous, spore print rusty-yellow-brown

Boletus separans Large, porcini-like with lilac-violet colors on the cap and stem that turn bright blue in household ammonia, stem bearing fine white reticulum, spore print orange-brown

Pores and tubes white or pale yellow, becoming yellow-green

Boletus Large species with pores and tubes white then yellow-green, not turning blue when bruised, large species with netted or not netted stem

Boletellus Tubes pale yellow then yellow-green or olive-green, most turn blue when bruised, spores are longitudinally striate or winged

Caloboletus Pores and tubes pale yellow, then yellow green, cap white or pale gray, flesh yellow then instantly blue when exposed, taste bitter

Pseudoboletus Growing on *Scleroderma citrinum*, an earthball

Suillus Cap slimy or slippery, stem often dotted with sticky bumps, under pines or larch

Xerocomus, *Xerocomellus* Small boletes with radially angular pores

Pores and tubes white, becoming pale grayish-brown

Retiboletus griseus Stem coarsely pale white or yellow netted, cap gray-brown, taste mild

Pores and tubes white, becoming flesh-pinkish or pink

Austroboletus Cap and stem maroon brown, stem typically long, cap small in diameter compared to stem length

Harrya Bright pink cap and dots on stem, stem base bright yellow

Tylopilus Cap and stem brown, black, orange, purple, violet, or purple-violet, often bitter tasting, but some mild, spore deposits flesh-brown or pinkish

Pores and tubes off-white, becoming brown or grayish; or pores dark brown or black from the first

Bothia Pores and tubes yellow-brown, staining rusty-brown, radially elongate, compound, decurrent

Leccinum, *Leccinellum* Pores off-white then grayish-brown or olive-tan, stem with obvious black or dark brown dots (scabers), sometimes scabers pale

Sutorius Pores dark chocolate-brown, stem bluish-gray with purple-brown scales

Strobilomyces Pores pale grayish but soon black, staining red when injured, cap and stem scaly and black, flesh turning orange-red when exposed

Pores dark red, pinkish, reddish-brown, or orange-brown

Boletus Mostly large species, the red-pored boletes, tubes yellow, pores, tubes, and flesh turn blue when injured, usually under oak

Chalciporus Small, tubes and pores similarly rose, pinkish- or reddish-brown, or orange-brown, elongate-angular, one species with yellow-brown pores and tubes that instantly turn blue when injured, taste peppery or mild

Aureoboletus auriporus
(Peck) Pouzar

SYNONYMS *Boletus auriporus* Peck, *Pulveroboletus auriporus* (Peck) Singer

Cap pinkish-brown, slippery; pores bright golden-yellow, staining pinkish or reddish; stem long with yellow fibrils embedded in slime over base

CAP 2–9 cm wide, convex, pinkish-cinnamon or pinkish-brown or reddish-brown, fading with age, surface smooth, slippery when fresh, coated with flattened fibrils, becoming dry. **FLESH** white or pale yellow. **PORES** bright golden-yellow, turning pinkish or reddish when bruised, mostly angular, 1–3 per mm. **TUBES** same color, to 15 mm deep. **STEM** 4–12 cm long, 6–17 mm broad, equal but enlarged gradually over base, pale yellow over apex, streaked with pinkish-brown downward, white mycelium at base, surface glabrous, slimy over lower half and bright golden fibrils embedded in slime, flesh white or pale yellow. **SPORE PRINT** olive-brown. **ODOR** not distinctive. **TASTE** flesh mild, slime on cap and stem acidic.

Habit and habitat solitary or scattered on the ground under hardwoods with oak present. July to October. **RANGE** widespread.

Spores subfusiform, smooth, 11–16 × 4–6 µm.

Comments The cap turns deep red with household ammonia, not fleetingly green. Compare with *Aureoboletus innixus* that is more frequently found in the Northeast than *A. auriporus*. Singer considered the slime layer at the base of the stem, with the golden fibrils embedded in the slime, as a type of universal veil for this species and thus placed it in *Pulveroboletus*.

Aureoboletus projectellus
(Murrill) Halling

SYNONYMS *Ceriomyces projectellus* Murrill, *Boletus projectellus* (Murrill) Murrill, *Boletellus projectellus* (Murrill) Singer

Cap reddish-brown with well-developed projecting sterile flap over margin; flesh white with rose tints; pores yellow, not bluing; stem long, coarsely netted, under pines

CAP 4–20 cm wide, convex, margin projecting well beyond flesh as a sterile flap of tissue, grayish-brown with rose tint becoming reddish-brown, surface smooth becoming cracked, minutely velvety, dry. **FLESH** white with rose tints, thick. **PORES** yellow, becoming olive-yellow then dark brownish-olive, round becoming angular, 1–2 mm wide. **TUBES** same color as pores, 10–25 mm deep. **STEM** 9–24 cm long, 6–18 mm broad, equal or more often enlarged downward, very long, colored as cap but becoming white with yellow tints over upper half, with deep coarse net or longitudinal lines over most of surface, also with minute pinkish-cinnamon dots over lower half and occasionally slimy over base when fresh, flesh white with rose tints or pale yellow in base. **SPORE PRINT** olive-brown. **ODOR** not distinctive. **TASTE** somewhat acidic.

Habit and habitat solitary, scattered, or clustered on the ground under various species of pine, such as red pine, white pine, pitch pine, jack pine, and others. July to September. **RANGE** widespread.

Spores subfusiform, smooth, 20–32 × 6–12 μm.

Comments No other bolete in the Northeast is comparable in height with this tall pine-associated species. The very large spores are also characteristic. This is an edible species that has good flavor and is favored by some mycophagists.

Aureoboletus innixus (Frost)
Halling, A. R. Bessette & A. E. Bessette

SYNONYMS *Boletus innixus* Frost, *Pulveroboletus innixus* (Frost) Singer, *Suillus innixus* (Frost) Kuntze

Cap reddish-brown or yellow-brown, slippery when wet; pores bright yellow at first; stem often with enlarged bulb at base and then sharply pinched, slimy over base when young

CAP 3–8 cm wide, convex, reddish-brown or yellow-brown with grayish tints, surface smooth, often cracked with age, slippery when wet, but soon dry. **FLESH** white or pale yellow. **PORES** bright golden-yellow, becoming dull yellow with age, round becoming angular, up to 2 mm broad. **TUBES** same color, 3–10 mm deep. **STEM** 2–6 cm long, 10–15 mm broad, equal but typically enlarged turniplike over base, swelling up to 25 mm broad, and then sharply tapered or pinched below swelling, dark brown or reddish-brown with yellow streaks, yellow mycelium at base, surface glabrous minutely brown-dotted over upper half, flattened fibrillose below and, when very young, viscid over base of stipe, flesh pale yellow. **SPORE PRINT** olive-brown. **ODOR** pungent, like witch hazel or flesh of the thick-skinned earthball. **TASTE** not distinctive.

Habit and habitat solitary, scattered, or in clusters fused at the stipe bases on the ground under hardwoods with oak present. June to October. **RANGE** widespread.

Spores narrow subfusiform, smooth, 8–11 × 3–5 μm.

Comments The cap produces a green reaction to household ammonia. Frequently mistaken for *Aureoboletus auriporus* or *A. roxanae*, that do not produce this reaction to ammonia and differ by several other features. The slime layer at the base of the stipe was considered a type of universal veil by Singer, and thus he assigned this species to the genus *Pulveroboletus*.

Aureoboletus roxanae (Frost) Klofac

SYNONYMS *Boletus roxanae* Frost, *Xerocomus roxanae* (Frost) Snell

Cap purplish-red or orange-yellow, woolly dotted or granulate; pores white then pale golden-yellow; stem yellow or orange-yellow with lines over apex and a pinched bulb at base

CAP 3–9 cm wide, convex becoming plane, purplish-red or yellowish-brown with red tints or orange-yellow, surface minutely woolly dotted or granulate, becoming smooth, dry. **FLESH** pale yellow. **PORES** white at first, soon pale yellow or pale golden-yellow, cinnamon where bruised, round becoming somewhat angular, 1–3 per mm. **TUBES** same color as pores, 6–10 mm deep. **STEM** 2–9 cm long, 5–15 mm broad, usually tapered upward from a swollen base that is pinched below, yellow or pale orange-yellow or with rusty-orange tints, white mycelium at base, surface faintly dotted or with longitudinal lines over apex, glabrous elsewhere, flesh pale or dark yellow. **SPORE PRINT** olive-brown. **ODOR** not distinctive, or pungent. **TASTE** not pleasant.

Habit and habitat solitary or clustered on the ground under hardwoods, especially oak or in mixed oak and pine forests. June to October. **RANGE** widespread.

Spores narrow ellipsoid or oblong, smooth, 8–13 × 3–5 µm.

Comments *Boletus auripes* has a bright yellow stem but is larger and has a smooth dark brown cap. *Pulveroboletus auriflammeus* is golden-yellow on the cap, pores, and stipe. *Boletus subglabripes* has a smooth cinnamon-brown cap and a pale yellow stipe covered with yellow dots and frequent reddish-brown stains.

Austroboletus gracilis (Peck) Wolfe

GRACEFUL BOLETE

SYNONYM *Tylopilus gracilis* (Peck) P. C. Henning

Cap maroon or cinnamon brown and minutely velvety; pores white then pinkish; stem slender, colored as cap, with silvery fuzz overall and fine, netted, raised lines

CAP 3–10 cm wide, convex, maroon or reddish-brown or cinnamon-brown, at times developing yellowish-brown tints, surface minutely velvety, minutely cracked with age, dry. **FLESH** white or tinted pinkish. **PORES** white, then pinkish or pale flesh-brown, round, 1–2 per mm. **TUBES** same color, 10–20 mm deep. **STEM** 8–18 cm long, 5–9 mm broad, slender and equal or enlarged downward, colored as cap or slightly paler, white mycelium over base, distinctly silvery fuzzy overall, often with raised interwoven lines forming a subtle broad meshed net (reticulum), flesh white. **SPORE PRINT** flesh-brown or reddish-brown. **ODOR** not distinctive. **TASTE** not distinctive, or slightly acidic.

Habit and habitat single, scattered, or clustered on the ground or on decaying stumps and logs, under mixed hardwoods or in conifer woods. June to October. **RANGE** widespread.

Spores subfusiform, ornamented with tiny pits, 10–17 × 5–8 μm.

Comments *Austroboletus gracilis* var. *flavipes* T. J. Baroni, Halling & Both differs by having yellow colors on the stem and cap, lacking the raised net on the stem, and having shorter and narrower spores. *Austroboletus gracilis* var. *pulcherripes* Both & Bessette differs by its stout and heavily netted stem.

Boletus edulis Bulliard

PORCINI, KING BOLETE, CEP, STEINPILZ

SYNONYMS *Boletus clavipes* (Peck) Pilát & Dermek, *Boletus edulis* var. *clavipes* Peck, *Boletus edulis* subsp. *aurantioruber* E. A. Dick & Snell, *Boletus aurantioruber* (E. A. Dick & Snell) Both, Bessette & W. J. Neill

Cap bun-brown or cinnamon-brown; flesh white, unchanging, thick; pores white and filled when young, soon yellow-green, very small; stem white or brown with fine white netting

CAP 8–25 cm wide, convex, pale brown, reddish-brown, orange-brown or cinnamon brown, surface smooth or somewhat bumpy, occasionally irregularly wrinkled, slippery when wet but quickly drying, sometimes cracked. **FLESH** white, thick, to 2 cm. **PORES** white and filled when young, becoming yellow-green then dull brown with age, bruising dark olive or yellowish-brown, round, small, 2–3 per mm. **TUBES** same color, to 20 mm deep. **STEM** 10–25 cm long, 20–40 mm broad, narrowly or broadly club-shaped, sometimes bulbous at base, white above but often brown toward base, with a fine white netted mesh (reticulate) either partly over apex or covering stem entirely, flesh white. **SPORE PRINT** olive-brown. **ODOR** not distinctive. **TASTE** sweet and nutty.

Habit and habitat solitary or clustered on the ground under conifers, especially Norway spruce and pine. June to November. **RANGE** widespread.

Spores subfusiform, smooth, 13–19 × 4–6 µm.

Comments Molecular studies indicate a broad variation in form and color for the North American *Boletus edulis*, thus our club-stemmed "clavipes" and the orange-brown or reddish-brown capped "aurantioruber" are genetically the same as the large bulbous stemmed "edulis" of Europe. *Tylopilus felleus* looks somewhat similar but has a brown reticulum, a flesh-pink spore deposit, and a strongly bitter taste making it inedible. The porcini is edible with excellent flavor. *Boletus chippewaensis* A. H. Smith & Thiers differs by the yellow cap splashed with reddish-brown stains and pores that stain vinaceous when bruised. See additional comments under *B. variipes*.

Boletus variipes Peck

Cap grayish or yellowish-tan, smooth becoming cracked; pores white becoming yellow green, not bruising; stem grayish-brown and finely white or grayish-brown netted

CAP 6–20(–30) cm wide, convex becoming plane, grayish-tan, yellowish-tan, or pale brown, surface smooth, slightly velvety, with expansion often cracked like a dried mudflat. **FLESH** white, unchanging. **PORES** white at first and filled, soon pale yellow green, becoming ochre-yellow with age, round, small, 1–2 per mm. **TUBES** colored as pores, 10–30 mm deep. **STEM** 8–15 cm long, 10–35 mm broad, equal or club-shaped, grayish-brown with white mycelium over base, fine mesh network white or same color as stem, covering entire stem or only over upper half, flesh white unchanging. **SPORE PRINT** olive-brown. **ODOR** and **TASTE** not distinctive.

Habit and habitat In clusters on the ground under beech and oaks or in stands of aspen, also reported from under conifers such as pines, hemlock and Norway spruce. (May) July to September. **RANGE** widespread.

Spores subfusiform, smooth, 12–18 × 3.5–6 μm.

Comments Usually appearing after heavy rainfall. *Boletus variipes* var. *fagicola* A. H. Smith & Thiers has a very dark brown cap and stem. The large *B. nobilis* Peck has a white or pale orange-brown stem with a delicate easy-to-miss white net on the stem. *Boletus edulis* differs by its reddish-brown smooth cap that does not become cracked and by the much finer completely white network on the stem. *Boletus separans* is somewhat similar, but can be distinguished by lilac colors on the cap and stem when fresh.

Boletus separans Peck

SYNONYMS *Xanthoconium separans* (Peck) Halling & Both, *Boletus edulis* subsp. *separans* (Peck) Singer

Cap purplish, lilac or reddish-brown, yellow-brown with age; flesh white; pores white then yellow-ochre; stem colored as cap or paler, with lilac colors and fine white net, lilac pigments blue in ammonia

CAP 6–20 cm wide, convex, color highly variable ranging from dark purplish or lilac-brown or reddish-brown or purplish-wine or pinkish-brown, becoming yellow-brown with age, surface often wrinkled-pitted or corrugated, glabrous, dry. **FLESH** white. **PORES** white, becoming yellow or yellow-ochre and then ochre-brown, round, small, 1–2 per mm. **TUBES** same color, up to 10–30 mm deep. **STEM** 6–15 cm long, 10–30 mm broad, equal or enlarged downward, colored as cap or often paler, with wine or pinkish-lilac colors prominent, surface finely white netted, at least over apex, flesh white. **SPORE PRINT** brownish-orange or pale reddish-brown. **ODOR** not distinctive, or disagreeable when dried. **TASTE** not distinctive, or mild, sweet, nutty.

Habit and habitat solitary or clustered on the ground under hardwoods and seems to be associated with red oak in particular. July to September. **RANGE** widespread, but not common.

Spores narrow subfusiform, smooth, 12–16 × 3.5–5 µm.

Comments The purplish, reddish, and lilac areas on the cap and stem turn bright blue with household ammonia dropped on them. Although not commonly collected, it is considered an edible, choice bolete. Also compare *Xanthoconium purpureum* that differs by its uniformly dark maroon-brown cap and shorter stem that lacks a net. Recent phylogenetic studies clearly place *X. separans* back into *Boletus* and not closely allied with other species of *Xanthoconium*.

Boletus auripes Peck

BUTTER-FOOT BOLETE

Cap chestnut brown becoming olivaceous yellow-brown, velvety smooth; pores small, cream yellow; stem golden-yellow with compact reticulum over apex, flesh yellow

CAP 7–15 cm wide, convex becoming plane, chestnut brown becoming yellow-brown with olivaceous tints, surface smooth, velvety, becoming like kid leather, dry. **FLESH** yellow or cream color, unchanging. **PORES** same color as tubes, filled with yellow tissue at first (stuffed), round, small, 2–3 per mm. **TUBES** cream yellow at first, becoming orange yellow then olive-yellow, to 30 mm deep. **STEM** 5–12 cm long, 20–40 mm broad, equal but often flattened, with a tapered base or on some enlarged at base, bright golden-yellow, with compact reticulum over apex to middle, glabrous downward, flesh yellow, unchanging. **SPORE PRINT** yellowish-brown with olive tint. **ODOR** pleasant. **TASTE** mild.

Habit and habitat solitary or scattered on the ground in grassy areas with trees and also in mixed hardwood forests, typically with oak present. June to September. **RANGE** widespread.

Spores subfusiform, smooth, 10–15 × 3.5–5 μm.

Comments A robust, compact bolete that is not common. Compare with the more frequently encountered *Retiboletus ornatipes* that has different cap colors, a very strongly reticulate stem overall, and a bitter taste. Also compare with *Xanthoconium affine*. *Boletus auripes* is considered edible.

Caloboletus inedulis (Murrill)
Vizzini

SYNONYMS *Ceriomyces inedulis* Murrill, *Boletus inedulis* (Murrill) Murrill

Cap white or pale gray and woolly; flesh yellow instantly blue; pores yellow instantly deep blue; stem yellow with red or pinkish-red, turning brown or black over base; bitter tasting

CAP 4–12 cm wide, convex becoming plane, white or pale gray with brown over the disc, surface smooth, densely woolly, often cracking into irregular patches on disc showing white between brown patches, dry. **FLESH** white, instantly blue. **PORES** pale yellow, becoming yellow-green, bruising instantly deep blue, round, small, 1–3 per mm. **TUBES** same colors and reaction to damage. **STEM** 5–12 cm long, 10–20 mm broad, equal, pale or bright yellow over apex and also sometimes in middle, red or pinkish-red elsewhere, often with fine red net over upper stipe or to near base with cream or yellow under reticulum or net lacking, also with fine dense reddish dots covering lower stipe, turning olive-brown or black where handled, flesh yellow turning deep blue. **SPORE PRINT** olive-brown. **ODOR** not distinctive. **TASTE** bitter.

Habit and habitat solitary or clustered on the ground under oaks. July to September. **RANGE** widespread.

Spores subfusiform, smooth, 9–12 × 3–4.5 µm.

Comments The stockier and typically western *Caloboletus calopus* (Persoon) Vizzini is occasionally found in some parts of the Northeast (Michigan, New Hampshire, New York). *Caloboletus calopus* is a bitter-tasting bolete with similar-colored yellow, then rapidly bluing pores and red colors in the stem, but the larger dark yellow-brown cap, ranging up to 20 cm, and the larger spores (13–19 × 5–6 µm) distinguish this rarely collected species in the Northeast.

Boletus pallidus Frost

Cap white or pale grayish-tan; pores white then yellow, often bruising blue; stem long, slender, glabrous, white staining brown when handled

CAP 4–15 cm wide, convex becoming plane, off-white or buff or pale grayish-tan with rose tints, surface smooth or becoming cracked with age, slippery when moist, finely velvety, soft to touch, becoming dry. **FLESH** white or pale yellow, erratically and slowly turning bluish, sometimes pinkish. **PORES** very pale creamy-white at first, soon greenish-yellow, turning blue when bruised or not, angular, 1–2 per mm. **TUBES** same color and reactions as pores, 10–20 mm deep. **STEM** 5–15 cm long, 10–25 mm broad, usually slender and equal, sometimes enlarged gradually over base, white or with yellowish tints over the apex and reddish tints over the base, staining slowly brown when handled, white mycelium at base, surface glabrous or minutely dotted over upper portions, flesh white with brown or reddish-brown discolorations especially in the base. **SPORE PRINT** olive-brown. **ODOR** not distinctive. **TASTE** mild or slightly bitter.

Habit and habitat solitary, scattered or clustered on the ground under hardwoods with oak present. July to September. **RANGE** widespread.

Spores subfusiform, smooth, 9–15 × 3–5 µm.

Comments Considered edible. Compare this species with *Caloboletus inedulis* that also has a pale-colored cap. The more highly colored reddish stem and strongly bitter taste of *C. inedulis* clearly separate it from *Boletus pallidus*.

Xerocomus ferrugineus
(Schaeffer) Alessio

SYNONYMS *Boletus ferrugineus* Schaeffer, *Boletus spadiceus* Schaeffer, *Xerocomus spadiceus* (Schaeffer) Quélet

Cap olive to reddish-brown with delicate yellow bloom, woolly; flesh pale yellow, turning slowly blue; pores yellow, slowly blue, angular; stem yellowish with reddish-brown dots, often forming a coarse net

CAP 5–11 cm wide, convex, quite variable, dark olive with red tints or yellow-brown, reddish-brown or date-brown, with a delicate fine yellow bloom at first, surface smooth, woolly, some becoming cracked, dry. **FLESH** pale yellow with red line near cap skin, usually turning blue, if only sparingly. **PORES** yellow, becoming dingy olive-yellow, slowly blue or blue-green when bruised, angular, 1–2 mm wide. **TUBES** same color as pores, 8–10 mm deep.

STEM 4–10 cm long, 10–25 mm broad, equal, pale sordid yellow or yellowish-tan, white at base but with yellow mycelium radiating into substrate, surface with minute brown or reddish-brown dots overall and frequently with low coarse net or merely longitudinally lined, flesh off-white or pale yellow. **SPORE PRINT** olive-brown. **ODOR** and **TASTE** not distinctive.

Habit and habitat solitary, scattered, or clustered on the ground under conifers. July to September. **RANGE** widespread.

Spores subfusiform, smooth, 8–13 × 3.5–5.5 μm.

Comments The cap surface turns fleetingly green then dark reddish-brown in a drop of household ammonia. See *Xerocomus subtomentosus* and *X. illudens* for further comparisons.

Xerocomus illudens (Peck)
Singer

SYNONYM *Boletus illudens* Peck

Cap cinnamon-brown or pinkish-cinnamon; flesh pale yellow, not turning blue; pores lemon-yellow, not turning blue; stem yellow and mustard yellow, with coarse net

CAP 3–9 cm wide, convex becoming plane, cinnamon-brown or pinkish-cinnamon, surface smooth, velvety, dry. **FLESH** pallid, then yellow. **PORES** lemon-yellow, becoming dingy yellow, angular, 1–2 mm wide radially. **TUBES** same color as pores or darker honey yellow, 10–15 mm deep. **STEM** 3–9 cm long, 6–15 mm broad, equal and often tapered downward, yellow or mustard yellow over apex to mid, white over base, with yellow-brown or mustard-yellow low coarse net or longitudinal lines over most of surface, flesh off-white or pale yellow. **SPORE PRINT** olive-brown. **ODOR** and **TASTE** not distinctive.

Habit and habitat solitary, scattered, or clustered on the ground under oak or in oak-pine forests. July to October. **RANGE** widespread.

Spores subfusiform, smooth, 10–14 × 4–5 μm.

Comments Another helpful feature to identify this species is that the cap surface turns fleetingly bright green then almost black when a drop of household ammonia is applied. Compare with *Xerocomus ferrugineus* that has pores and flesh that turn blue when bruised or exposed and that lack a coarse, well-formed net on the stem. For *X. subtomentosus* the cap surface turns reddish-brown when a drop of household ammonia is applied, and it lacks a well-developed coarse net on the stem.

Xerocomus subtomentosus
(Linnaeus) Quélet

YELLOW-CRACKED BOLETE

SYNONYM *Boletus subtomentosus* Linnaeus

Cap olive-brown or yellow-brown, turning red-brown when ammonia applied, velvety becoming cracked; pores yellow and only slowly faint green when bruised; stem pale yellow, staining brown, sometimes with weak net over apex

CAP 5–20 cm wide, convex, olive-brown or yellow-brown, surface smooth, cracked with age, minutely velvety, dry. **FLESH** white or pale yellow, unchanging. **PORES** yellow, becoming olive-yellow, staining slowly faint green, changing to brown when bruised, mostly angular, 1–2.5 mm broad. **TUBES** same color and slowly greenish when cut, up to 10–25 mm deep. **STEM** 4–10 cm long, 10–30 mm broad, equal or tapered somewhat downward, pale yellowish or ochre-yellow with reddish-brown or brownish streaks from bruising, darker sulfur yellow mycelium at base, surface minutely dotted and also at times with weak widely netted or riblike longitudinal lines over apex, flesh pale yellow or darker. **SPORE PRINT** olive-brown. **ODOR** and **TASTE** not distinctive.

Habit and habitat solitary or clustered on the ground under hardwoods or mixed coniferous forests. June to October. **RANGE** widespread.

Spores narrow subfusiform, smooth, 10–15 × 3.5–5 µm.

Comments Compare with *Xerocomus ferrugineus* that produces a fleeting green reaction to ammonia on the cap surface, and the pores stain quickly and intensely greenish-blue when bruised or exposed. These small to medium-sized velvety-woolly capped boletes with yellow pores and tubes have in the past been placed in either *Boletus* or *Xerocomus*. Recent molecular investigations indicate *Xerocomus* is the best name to use from a phylogenetic perspective.

Boletus badius (Fries) Fries

BAY BOLETE

SYNONYMS *Boletus castaneus* var. *badius*
Fries, *Suillus badius* (Fries) Kuntze,
Xerocomus badius (Fries) E.-J. Gilbert,
Imleria badia (Fries) Vizzini

Cap reddish-brown or chestnut-brown
and slippery when wet; tubes olive-yellow,
bluing slowly; stem brown-colored as cap,
smooth

CAP 3–10 cm wide, convex, becoming flat,
dark reddish-brown or chestnut-brown,
surface smooth, slippery when wet,
minutely woolly when dry. **FLESH** white,
soft, pinkish under cap skin, yellowish
near tubes, rarely turning pale blue. **PORES**
pale yellow becoming pale olive-yellow,
round and small, 1–2 per mm, bluing rapidly or sometimes slowly when bruised,
eventually reddish. **TUBES** olive-yellow,
to 15 mm deep. **STEM**, 4–9 cm long,
10–20 mm broad, equal or enlarged downward, more or less colored as cap, but base
and mycelium white, glabrous or with fine
brown dots, flesh white. **SPORE PRINT** olive-
brown. **ODOR** and **TASTE** not distinctive.

Habit and habitat solitary or scattered on
the ground or on rotted wood in mixed
coniferous forests with pine or spruce, also
reported from under hardwoods. June to
November. **RANGE** widespread throughout
the region.

Spores subfusiform, smooth, 10–14 ×
4–5 µm.

Comments The bay bolete gets its name
from the color of the cap that resembles
the hair coat color of a bay horse. The cap
becomes dry if the weather is dry for a few
days and as the fruit body ages. Wet your
finger and run it across the cap—it should
be slightly slippery. A recent new combination into a new genus *Imleria* may or
may not be supported with future molecular investigations.

Boletus speciosus Frost

SYNONYMS *Ceriomyces speciosus* (Frost) Murrill, *Suillus speciosus* (Frost) Kuntze
Cap grayish-rose, felted-velvety; pores small, bright yellow and rapidly blue when touched; stem yellow above, rhubarb red below, covered with fine yellow reticulum, flesh yellow bluing rapidly

CAP 7–15 cm wide, convex, grayish- or purplish-rose at first and retaining these colors, or becoming paler with age, surface smooth, felted-velvety, dry. **FLESH** pale yellow then quickly blue when exposed. **PORES** bright yellow becoming olive-yellow, rapidly blue when touched, round, small, 2–3 per mm. **TUBES** same color and bluing reaction, to 15 mm deep. **STEM** 5–12 cm long, 15–40 mm broad, equal, with abruptly pinched base, bright yellow above, deep rhubarb red over base, finely yellow reticulate overall or at least to middle, red-spotted over lower half, flesh yellow, instantly blue when exposed. **SPORE PRINT** olive-brown. **ODOR** not distinctive. **TASTE** mild.

Habit and habitat solitary or scattered on the ground in mixed hardwoods often with conifers present. June to October. **RANGE** widespread throughout the region.

Spores subfusiform, smooth, 11–15 × 3–5 μm.

Comments *Boletus speciosus* var. *brunneus* Peck, recently resituated in *Butyriboletus brunneus* (Peck) D. Arora & J. L. Frank, is similar in form and color except the cap is reddish-brown or yellowish-brown. *Boletus speciosus* apparently belongs in this new genus as well, but the new combination has not been formally made.

Boletus miniato-olivaceus
Frost

Cap orange-red slowly becoming olive in spots then more widespread; flesh yellow turning blue; pores lemon-yellow or olive-green, turning blue when injured; stem pale yellow with some red streaks over base, not bluing

CAP 5–15 cm wide, convex, becoming plane, orange-red at first fading to olive or olive-brown in large patches, surface smooth and spongy, minutely velvety becoming glabrous, dry. **FLESH** white to pale yellow, turning blue. **PORES** lemon-yellow becoming olive-green then dull reddish-brown with age, rapidly blue when injured, round, small, 1–2 per mm. **TUBES** same color, turning bluing when exposed, to 16 mm deep. **STEM** 6–12 cm long, 10–20 mm broad, equal or enlarged upward into cap, pale yellow but with red or reddish-brown streaks over the base, glabrous, not reticulate, flesh yellow turning blue when exposed. **SPORE PRINT** olive-brown. **ODOR** and **TASTE** not distinctive.

Habit and habitat solitary or clustered on the ground in mixed hardwoods with beech or oak, but also known from mixed conifers, with hemlock present. June to October. **RANGE** widespread throughout the region.

Spores oblong-elliptical or subfusiform, smooth, 11–14(–16) × 3.5–5 µm, hymenial cystidia inflated over apex (bladderlike or vesiculose).

Comments *Boletus sensibilis* Peck turns instantly blue in all parts when touched or exposed and does not develop the olive discolorations on the cap surface. Additionally, *B. sensibilis* lacks bladder-like hymenial cystidia. *Boletus miniato-olivaceus* is toxic like *B. sensibilis*, causing gastrointestinal irritation when eaten.

Baorangia bicolor (Kuntze)
G. Wu, Halling & Zhu L. Wang

SYNONYMS *Suillus bicolor* Kuntze, *Xerocomus bicolor* (Kuntze) Cetto, *Boletus rubellus* subsp. *bicolor* (Kuntze) Singer, *Boletus bicolor* Peck

Cap rose or apple-red; flesh yellow turning blue, pores small yellow and rapidly blue when touched, tubes very shallow; stem yellow at apex, densely red-velvety-dotted elsewhere

CAP 8–15 cm wide, convex, deep rose to purplish or apple-red at first, becoming duller and with brown colors with age, smooth, felted-woolly, dry. **FLESH** pale yellow, slowly blue, much thicker than tubes. **PORES** yellow becoming olive-yellow, sometimes reddish-tinted with age, turning greenish-blue when injured, round, small, 1–2 per mm. **TUBES** same color and bluing reaction, short compared to depth of flesh, 3–8 mm. **STEM** 5–10 cm long, 10–30 mm broad, equal or narrowly club-shaped, yellow at apex but red or rose red over most of the stem from crowded tiny velvety dots, flesh yellow, erratically blue when exposed. **SPORE PRINT** olive-brown. **ODOR** and **TASTE** not distinctive.

Habit and habitat solitary or clustered on the ground in mixed hardwoods with oak present. June to October. **RANGE** widespread.

Spores oblong-elliptical or subfusiform, smooth, 8–12 × 3–5 µm.

Comments *Boletus bicolor* Peck, described in 1872 from North America, is the one found in guidebooks up until 2014. However, that name cannot be used due to rules of the International Code of Botanical Nomenclature, since the name *Boletus bicolor* Raddi represents a different species in Italy described in 1806. It seems however, that *Suillus bicolor* Kuntze of Europe is the same species Peck described from North America, so that specific epithet can be used in a genus different than *Boletus*. Thus *Baorangia* was proposed as a new genus for several different, but phylogenetically related species found in North America, Europe, and Asia that are only distantly related to *Boletus*. In any case, our *Baorangia bicolor* is edible and quite good.

Boletellus chrysenteroides
(Snell) Snell

SYNONYM *Boletus chrysenteroides* Snell

Cap dark brown, smooth tomentose becoming cracked; flesh pale cream turning blue; pores bright yellow bruising blue; stipe brown with reddish-brown dots

CAP 3–10 cm wide, convex becoming broadly convex, blackish-brown, becoming chocolate or chestnut brown or paler brown with expansion, surface tomentose or velvety, becoming finely cracked with age. **FLESH** pale cream, slowly turning blue when exposed. **PORES** pale or bright yellow becoming olive-yellow, bruising slowly blue then turning brown, angular, 0.5–1 mm broad. **TUBES** same color as pores, bluing, 10 mm deep. **STEM** 3–6(–13) cm long, 6–12 mm broad, equal or enlarged slightly downward, reddish-brown or deep dark-brown but paler and some yellowish over apex, covered with reddish-brown dots, staining slowly brownish when handled, flesh pallid at apex and slowly blue when exposed, with reddish hues in mid, brownish in base and unchanging. **SPORE PRINT** dark olive-brown. **ODOR** not distinctive. **TASTE** slightly acidulous or not distinctive.

Habit and habitat typically single, occasionally in clusters on old well-decayed wood or on the ground around old stumps and logs in oak forests or mixed forests with oaks. July to September. **RANGE** widespread.

Spores elliptical-subfusiform, longitudinally striate, 10–16 × 5–8 μm.

Comments One can learn to tell the different species of *Boletellus* in the field because of their distinctive morphologies, but a microscopic examination of the basidiospores is the best diagnostic method. *Boletellus chrysenteroides* is only vaguely similar to *Xerocomellus chrysenteron* and grows on decaying wood.

Boletellus intermedius A. H. Smith & Thiers may be mistaken as a large form of *Xerocomellus chrysenteron*, but the rose-red colors on the young caps and the longitudinally striate spores are distinctive. The bluing reaction of the tubes is more rapid and more intense than in *X. chrysenteron* as well. Edibility unknown.

Boletellus intermedius

Boletellus russellii (Frost)
E.-J. Gilbert

RUSSELL'S BOLETE

SYNONYMS *Boletus russellii* Frost, *Ceriomyces russellii* (Frost) Murrill, *Frostiella russellii* (Frost) Murrill, *Suillus russellii* (Frost) Kuntze

Cap pinkish-buff or reddish-brown or yellow-brown, woolly, cracked; flesh not bluing; pores olive-yellow not bluing; stem flesh brown, deeply lacerate-netted

CAP 3–13 cm wide, hemispheric at first, becoming broadly convex, pinkish-buff, cinnamon to pale reddish-brown or yellow-brown, or with olive-gray hues, surface woolly and strongly cracked like a mudflat with age. **FLESH** pale yellow, not turning blue. **PORES** yellow, becoming olive-yellow, angular, 1–2 mm broad. **TUBES** 7–12 mm deep. **STEM** 10–20 cm long, 10–20 mm broad, equal or enlarged slightly downward, darker colors than pileus, reddish or deep fleshy-brown, strongly lacerate-netted, often slimy over base when fresh. **SPORE PRINT** dark olive-brown. **ODOR** not distinctive. **TASTE** mild.

Habit and habitat single or clustered on the ground in oak forests or mixed forests with oaks. July to September. **RANGE** widespread.

Spores elliptical-subfusiform, deeply longitudinally winged, the pronounced wings not touching at the apex leaving a space or "cleft," 15–20 × 7–11 μm.

Comments A tall, stately bolete that is easy to recognize because of the long, deeply lacerate-netted stem. Not common. Edible.

Xerocomellus chrysenteron
(Bulliard) Ŝutara

RED-CRACKED BOLETE

SYNONYMS *Xerocomus chrysenteron* (Bulliard) Quélet, *Boletus chrysenteron* Bulliard

Cap olive-brown with reddish tints in the cracked areas, soft; pores angular, yellow, bruising blue; stem, yellow above, red elsewhere, densely red-dotted

CAP 2–11 cm wide, convex becoming plane, dark olive or olive-brown, rarely reddish overall, with reddish or pinkish tints in cracks, velvety, cracked with age, soft, dry. **FLESH** white then yellow, sometimes with red line below cap skin, turning blue slowly near tubes. **PORES** yellow becoming olive-yellow, turning blue when bruised, angular, about 1 mm broad. **TUBES** same color and reactions as pores, 2–10 mm deep. **STEM** 4–9 cm long, 5–16 mm broad, equal, yellow over apex otherwise red or purple-red, with white or pale yellow mycelium over the base, surface rhubarb-red-dotted overall, flesh pale yellow, turning blue in apex. **SPORE PRINT** olive-brown. **ODOR** and **TASTE** not distinctive.

Habit and habitat solitary, scattered or clustered on the ground, especially along moss covered banks of tree covered seasonal roads under hardwoods with oak and beech most frequently. July to September. **RANGE** widespread.

Spores subfusiform, smooth, 9–15 × 3–5 µm.

Comments This species is difficult to separate from *Xerocomellus truncatus* and *Boletellus intermedius* unless one has a microscope, as *X. truncatus* has truncate spores and *B. intermedius* has striate spores with a small germ pore. Reasonably useful field characters are that *X. truncatus* is typically slenderer in stature than *X. chrysenteron*, and *B. intermedius* has rosered colors on the cap and turns blue on the tubes and flesh very rapidly.

Xerocomellus truncatus
(Singer, Snell & E. A. Dick) Klofac

SYNONYMS *Xerocomus truncatus* Singer, Snell & E. A. Dick, *Boletus truncatus* (Singer, Snell & E. A. Dick) Pouzar

Cap olive-brown with reddish tints, deeply cracked, soft; pores angular, yellow, bruising intensely blue; stem slender, yellow above, red elsewhere, densely dotted

CAP 2–7 cm wide, convex becoming plane, dark olive or olive-brown developing wine or reddish-wine tints, velvety, cracked with age, sometimes reddish showing in cracks but mostly pale creamy-yellow, soft, dry. **FLESH** white or pale yellow, turning blue. **PORES** yellow soon greenish-yellow, turning quickly blue when bruised, angular, about 1 mm broad. **TUBES** same color and reactions as pores, 10–15 mm deep. **STEM** 3–6 cm long, 4–10 mm broad, slender, equal, yellow over apex otherwise red, with grayish-buff mycelium over the base, surface minutely dotted overall, flesh pale yellow but reddish centrally. **SPORE PRINT** olive-brown. **ODOR** and **TASTE** not distinctive.

Habit and habitat solitary, scattered or clustered on the ground or on well-decayed wood, often on road banks, under hardwoods and in mixed coniferous forests. July to September. **RANGE** widespread.

Spores subfusiform, smooth, 10–17 × 4.5–6.5 µm, apex flattened and walls thickened on the sides, resembling an apical pore.

Comments This species, described originally from the eastern United States is typically a slender species. See the comments under *Xerocomellus chrysenteron* for comparison. In the warm wet weather, *X. truncatus* is often attacked by the ascomycete fungus *Hypomyces chrysospermus* that covers the bolete with fluffy white mycelium, making it unrecognizable. When the ascomycete mycelium turns yellow the asexual spores are being produced. Eventually very tiny flesh-pink flasks (perithecia) will form and produce sexual spores, completing the life cycle (see *Hypomyces chrysospermus*).

Pseudoboletus parasiticus
(Bulliard) Šutara

PARASITIC BOLETE
SYNONYMS *Boletus parasiticus* Bulliard, *Xerocomus parasiticus* (Bulliard) Quélet

A small yellowish-brown bolete growing from earthballs

CAP 2–8 cm wide, convex, yellow-brown, olive-brown or grayish-brown, surface smooth, dry. **FLESH** pale yellow, unchanging. **PORES** honey-yellow or greenish-yellow, at times staining slowly ochre-reddish when bruised, angular, 1–2 mm broad. **TUBES** same color, 3–10 mm deep, producing lines down stem apex. **STEM** 3–6 cm long, 6–13 mm broad, equal, mostly curved, colored as but paler than cap, surface dotted and streaked with yellow-brown matted fibrils and scales, these turning deep reddish-orange in KOH, flesh pale yellow, unchanging. **SPORE PRINT** olive-brown. **ODOR** and **TASTE** not distinctive.

Habit and habitat solitary or clustered on the thick-skinned earthball, *Scleroderma citrinum* (also called pig-skinned poison puffball), under hardwoods or in mixed coniferous forests, especially with hemlock present. June to October. **RANGE** widespread.

Spores narrow subfusiform, smooth, 10–15 × 3.5–5 μm.

Comments Only found growing on earthballs and claimed to be edible, with caution. Not common. Ammonia and KOH produce a red color reaction on the cap, while KOH produces a dark reddish-orange reaction on the stem. I have collected specimens in the southern states that have slimy caps, which could be a completely different species.

Boletus frostii J. L. Russell

FROST'S BOLETE

SYNONYMS *Suillellus frostii* (J. L. Russell) Murrill, *Suillus frostii* (J. L. Russell) Kuntze, *Exsudoporus frostii* (J. L. Russell) Vizzini, Simonini & Gelardi

Cap dark red, slippery; pores dark red, when young with golden liquid droplets, instantly deep blue when bruised; stem coarsely netted, red but becoming yellow or olive-brown on some

CAP 5–15 cm wide, convex becoming plane, dark red or bright delicious-apple-red, white or yellowish on very margin, sometimes with broad areas becoming yellow, surface smooth, slippery or tacky when wet. **FLESH** lemon-yellow, instantly blue. **PORES** dark red as cap, when young beaded with golden liquid droplets, bruising quickly sordid blue, round, small, 2–3 per mm. **TUBES** yellow or olive-yellow, turning sordid blue. **STEM** 4–12 cm long, 10–25 mm broad, equal or narrowly club-shaped, dark red or with yellow in base or apex between raised network, mesh of network deeply and coarsely covering entire stem, dark red or paler and even yellow or becoming olive-brown on some, with white or pale yellow mycelium over base, slightly turning sordid blue where handled, flesh white over apex, yellow or reddish-streaked in base. **SPORE PRINT** olive-brown. **ODOR** not distinctive. **TASTE** mild, although cap surface slightly acidic.

Habit and habitat solitary or clustered on the ground under oaks. July to October. **RANGE** widespread.

Spores subfusiform, smooth, 11–15 × 4–5 µm.

Comments A very beautiful and distinctive bolete. Occasional, although in some years it can be common. Edible.

Boletus subluridellus A. H. Smith & Thiers

Cap red or orange-red, instantly dark violet when touched; flesh bright yellow instantly blue; pores dark red, blue when bruised; stem lemon-yellow with small yellow dots

CAP 5–10 cm wide, convex, red or orange-red or bay brown, instantly dark violet where touched, surface smooth, velvety and resinous to the touch at first, then like smooth leather. **FLESH** bright yellow, quickly blue becoming white. **PORES** dark red or orange-red with age, bruising quickly blue or blue-black, round, small, 2–3 per mm. **TUBES** yellow, turning blue, up to 10 mm deep. **STEM** 4–10 cm long, 15–25 mm broad, equal, lemon-yellow or paler, yellow-dotted over surface, darkening somewhat when handled, flesh yellow becoming greenish-blue when exposed. **SPORE PRINT** olive-brown. **ODOR** and **TASTE** not distinctive.

Habit and habitat solitary or clustered on the ground under hardwoods, especially oak, often in grassy areas with oak. July to October. **RANGE** widespread, but not common.

Spores subfusiform, smooth, 11–15 × 4–5.5 µm.

Comments Compare with *Boletus subvelutipes*. Edibility unknown, but blue-staining boletes are best avoided, because many cause gastrointestinal upsets.

Boletus subvelutipes Peck

RED-MOUTH BOLETE

Cap yellowish to reddish-brown, blue when touched; flesh instantly blue; pores orange-red, blue when bruised; stem yellowish with small reddish-brown dots, bruising dark blue, dark red tufted hairs at base

CAP 5–18 cm wide, convex, variable from cinnamon-brown, yellow-brown, orange-brown or reddish-brown, surface smooth, woolly and soft, becoming dark blue when bruised. **FLESH** yellow, but so quickly blue, yellow color difficult to catch. **PORES** orange-red, red or brownish-red, becoming dingy reddish or paler orange-red with expansion, bruising quickly blue or blue-black, round, small, 2 per mm. **TUBES** yellow or olive-yellow, turning blue, 10–25 mm deep. **STEM** 3–10 cm long, 10–20 mm broad, equal, red- or red-brown-dotted over yellow surface in apex, mixtures of red and yellow over base, quickly dark blue-black when handled, with yellowish then soon dark burgundy red tufted hairs over base, flesh yellow but quickly blue when exposed. **SPORE PRINT** olive-brown. **ODOR** not distinctive. **TASTE** mild or slightly acidic.

Habit and habitat solitary or clustered on the ground under hardwoods. June to September. **RANGE** widespread.

Spores subfusiform, smooth, 13–18(–20) × 4–6.5 µm.

Comments Toxic, causing gastrointestinal upset. This is a highly variable species as currently conceived, probably representing several different variants or species. *Boletus subvelutipes* is the most commonly collected red-pored bolete in the Northeast, that lacks a net on the stem. If yellow colors are fairly bright and dominate the cap and stem, and the stem is dotted with red-brown, you may have what is being called *B. discolor* (Quélet) Boudier in North America. The species called *B. discolor* in North America may not be the same as this European bolete.

Retiboletus griseus (Frost)
Manfred Binder & Bresinsky

SYNONYMS *Boletus griseus* Frost, *Xerocomus griseus* (Frost) Singer

Cap dark gray; flesh off-white; pores off-white or grayish, turning brown when damaged; stem with a pale yellow then brown net; taste not distinctive

CAP 5–15 cm wide, convex or plane, dark gray, grayish-brown, with some ochre-yellow tints in age, surface smooth, matted fibrillose, dry. **FLESH** off-white, staining dark yellow-brown around insect tunnels. **PORES** off-white, pale grayish or gray-brown, turning brown when bruised, round, 1–3 per mm. **TUBES** same color as pores, 8–20 mm deep. **STEM** 4–15 cm long, 10–35 mm broad, equal or tapered downward, off-white or grayish overall, becoming yellow from the base upward, occasionally with red tints, obvious pale yellow coarse net over most of surface that becomes brown or black eventually, flesh as in cap or dark brown in base. **SPORE PRINT** dark yellow-brown. **ODOR** and **TASTE** not distinctive.

Habit and habitat solitary or scattered on the ground under hardwoods, especially oak. June to September. **RANGE** widespread.

Spores subfusiform, smooth, 9–13 × 3–5 μm.

Comments Compare with *Retiboletus ornatipes* that differs by the bitter taste, the yellow pores, and the coarser yellow or mustard-yellow net on the stipe.

Retiboletus ornatipes (Peck)
Manfred Binder & Bresinsky

ORNATE-STALKED BOLETE
SYNONYM *Boletus ornatipes* Peck

Cap gray or olive or yellow-brown; flesh bright yellow, unchanging; pores bright yellow; stem coarsely lemon-yellow or mustard-yellow netted; taste bitter

CAP 4–16 cm wide, convex, usually pale or dark gray, but also olive or yellow-brown or even yellow, surface smooth, velvety, when wet slippery but soon dry. **FLESH** bright yellow, unchanging. **PORES** bright yellow, turning orange-yellow or orange-brown when bruised, round, 1–2 per mm. **TUBES** same color as pores, 4–15 mm deep. **STEM** 8–15 cm long, 10–20 mm broad, equal or enlarged downward then tapered at very base, bright yellow overall or sordid white with lemon-yellow or mustard-yellow net producing yellow colors, some turning darker orange-yellow where handled, deep coarse net over most of the surface, flesh bright yellow. **SPORE PRINT** dark yellowish-brown. **ODOR** not distinctive. **TASTE** usually quite bitter.

Habit and habitat solitary, scattered, or clustered on the ground under hardwoods such as beech, birch, oak. June to September. **RANGE** widespread.

Spores subfusiform, smooth, 9–13 × 3–4 µm.

Comments Compare with *Retiboletus griseus*. *Retiboletus retipes* (Berkeley & M. A. Curtis) Manfred Binder & Bresinsky, with a yellow cap and described from the southeastern United States, is similar in form but when comparing molecular profiles is phylogenetically distinct.

Xanthoconium affine (Peck) Singer

SYNONYM *Boletus affinis* Peck

Cap reddish-brown becoming yellowish-brown, woolly, smooth and soft; tubes white becoming pale yellow, not bluing; stem paler than cap, smooth; spore print bright rusty-yellow-brown

CAP 2–10 cm wide, convex, reddish-brown becoming yellowish-brown, surface smooth, woolly or velvety, soft. **FLESH** white. **PORES** white, soon pale yellow or yellowish-tan with age, round and small, 1–2 per mm, darker yellow when bruised. **TUBES** colored as pores. **STEM** 3–10 cm long, 10–20 mm broad, equal but with pointed base frequently, paler than cap with white tapered base, glabrous, flesh white, sometimes yellowish-brown slowly or around insect larva tunnels. **SPORE PRINT** bright rusty-yellow-brown. **ODOR** not distinctive when fresh, rather foul in old rotting specimens. **TASTE** not distinctive.

Habit and habitat single, scattered, or clustered on the ground near beech and oak trees in mixed forests. July to October. **RANGE** widespread.

Spores subfusiform, smooth, 9–14 × 3–5 µm.

Comments Compare with species of *Xerocomus* that produce olive-brown spore deposits and have pores and tubes that are darker yellow or yellow-green. *Xanthoconium affine* var. *maculosus* (Peck) Singer, also frequently encountered, has a darker brown, white-spotted (maculate) cap and a darker brown stem. Another variety, *X. affine* var. *reticulatum* (A. H. Smith) Wolfe, is also darker but with a strongly netted stem. This latter variety is less commonly encountered.

Xanthoconium purpureum Snell & E. A. Dick has a maroon or reddish-brown cap, white flesh, and white pores that turn yellow-ochre. The stem is pale yellowish-brown with a delicate reddish net over the apex. Maroon pigments of the cap tun greenish-blue with a drop of household ammonia. This species may be confused with dark brown forms of *X. affine*, but the ammonia reaction of the cap for *X. affine* is rusty-tan, not greenish-blue. When the fruit bodies begin to decay, they take on a very foul-smelling odor.

Xanthoconium purpureum

Bothia castanella (Peck)
Halling, T. J. Baroni & Manfred Binder

SYNONYMS *Boletinus castanellus* Peck, *Xerocomus castanellus* (Peck) Snell & E. A. Dick, *Suillus castanellus* (Peck) A. H. Smith & Thiers, *Boletinellus castanellus* (Peck) Murrill, *Gyrodon castanellus* (Peck) Singer

Cap reddish-brown or chocolate-brown fading to yellow-brown; tubes decurrent, pores compound, elongate-angular; stem central or off-center

CAP 3–10 cm wide, convex becoming plane, bay-red or dark reddish-brown or dark chocolate-brown, fading to yellowish-brown, surface coarsely woolly or granular, dry. **FLESH** off-white, often slowly rust-brown. **PORES** pale brown or pale yellow-brown or cinnamon-brown with pinkish-cinnamon hues on some, staining slowly rusty-brown when bruised, radially elongate angular, usually compound (pores within pores), 3–4 mm broad, forming gills at stem. **TUBES** decurrent, same color as pores, 4–5(–10) mm deep. **STEM** 2–6 cm long, 4–18 mm broad, central or off-center, equal or tapered downward, paler than cap or near wine-brown or dark dingy brown, white mycelium at base, surface with fine broad dark brown net or longitudinal lines over apex, minutely woolly or dotted elsewhere, flesh brown or cinnamon-brown, especially in base. **SPORE PRINT** yellow-brown. **ODOR** not distinctive. **TASTE** mild or acidic.

Habit and habitat solitary or clustered on the ground under hardwoods such as oak, beech, birch or in mixed coniferous forests with white pine or hemlock. July to September. **RANGE** widespread.

Spores narrow ellipsoid or oblong, smooth, 8.5–10.5 × 4–5 µm.

Comments Compare with the more commonly collected *Boletinellus meruli-oides* that also has decurrent tubes with compound, radially elongate pores, but a constantly off-center stem and pale yellow tubes.

Boletinellus merulioides
(Schweinitz) Murrill

ASH-TREE BOLETE

SYNONYMS *Daedalea merulloides*
Schweinitz, *Gyrodon merulioides*
(Schweinitz) Singer, *Boletus merulioides*
(Schweinitz) Murrill, *Boletinus merulioides*
(Schweinitz) Coker & Beers

Cap fan-shaped, brown, smooth; flesh
cream sometimes turning blue-green;
tubes decurrent, shallow, pores elongate
radially, compound; stem off-center or
originating on cap margin; on soil under
ash trees

CAP 5–12 cm wide, convex, becoming flat
and broadly depressed or undulate, more
or less fan-shaped in outline, deep olive-
brown, reddish-brown or yellowish-brown,
surface smooth, glabrous or flattened
fibrillose, soft. **FLESH** pale cream, erratically turning blue-green slowly. **PORES** pale
yellow, ochre-golden or deep olive-yellow
with age, sometimes bruising very slowly
blue, then turning brown, elongate and
radially arranged with raised edges resembling lamellae, large, angular, becoming
compound (pores within pores). **TUBES**
decurrent, very shallow, not separating
from cap easily, same color as pores. **STEM**
2–4 cm long, 5–25 mm broad, off-center
or on cap margin, tapered downward,
yellow over apex, brown to base, bruising
reddish-brown, glabrous, flesh pallid,
sometimes faintly and slowly blue when
exposed. **SPORE PRINT** dark olive-brown.
ODOR not distinctive. **TASTE** not distinctive,
or of potato.

Habit and habitat typically in clusters on the ground around or near ash
trees in mixed forests. July to October.
RANGE widespread.

Spores elliptical, smooth, 7–10 × 6–8 µm.

Comments The ash-tree bolete is very common and easy to identify in the field. Its
biology is interesting, as it is protecting
and receiving food from the woolly ash
aphid (*Prociphilus fraxinifolii*) in the soil. If
you dig up the soil and mycelium around
the fungus and tree roots, you should be
able to find small black "peas" in the soil
around the ash roots. Have a look inside
one of these peas and you should be able to
find the aphids.

Chalciporus piperatoides
(A. H. Smith & Thiers) T. J. Baroni & Both

SYNONYM *Boletus piperatoides* A. H. Smith & Thiers

Cap cinnamon or orange-brown, very slippery when wet; pores yellow becoming orange-brown, turning dark blue when injured; stem colored as cap with bright yellow base; taste slowly chili pepper hot

CAP 3–8 cm wide, convex or conical-convex, becoming plane, dull reddish-brown, rusty-cinnamon, orange-brown, fading to tan or yellow-brown, surface smooth sometimes coarsely cracked with age, velvety and streaked with fibrils and scales when young, small scales remaining on the center, otherwise glabrous, distinctly slimy when wet, becoming dry. **FLESH** pale rusty-brown or pinkish-cinnamon, becoming blue near tubes. **PORES** dingy yellow-golden at first becoming reddish or orange-brown or dull brown with an olive tint, turning dark blue when bruised, elongate angular, often compound (pores in pores), 1.5 mm wide. **TUBES** same color and bluing reaction as pores, 4–10 mm deep. **STEM** 3.5–8 cm long, 5–16 mm broad, mostly colored as cap or paler, with bright yellow mycelium over base, flesh color as cap in apex and erratically blue when exposed, deep yellow in base. **SPORE PRINT** smoky olive when fresh, drying rusty-cinnamon. **ODOR** not distinctive. **TASTE** slowly peppery or sharp and burning.

Habit and habitat solitary, scattered or clustered on the ground under conifers and hardwoods. July to September. **RANGE** widespread.

Spores subfusiform, smooth, 7–9 × 3–4 μm.

Comments *Chalciporus piperatoides* is not as frequently collected as the peppery bolete, *C. piperatus* (Bulliard) Bataille. They are very similar in looks and taste, except the pores, tubes, and flesh of *C. piperatus* do not turn blue when injured or exposed and the cap and tubes are more highly pigmented with orange-brown or reddish-orange colors.

Chalciporus rubinellus (Peck) Singer

SYNONYM *Boletus rubinellus* Peck
Cap bright red, conical; flesh bright yellow; pores red or pink, angular; stem colored as cap; taste mild

CAP 1.5–5 cm wide, broadly conical-convex, becoming convex, red or purplish-red at first, developing yellow tints under the red, becoming reddish-brown, surface smooth, densely matted or tufted-hairy, distinctly slimy when wet, becoming dry. **FLESH** bright yellow. **PORES** bright red or rose-red, becoming pinkish in age, elongate-angular, 1.5 mm wide. **TUBES** same color as pores at first, becoming salmon or yellowish, finally brownish, 6–10 mm deep. **STEM** 1–6 cm long, 3–10 mm broad, mostly colored as cap but can be paler than or darker than cap, sometimes mixed with bright yellow, surface dotted with minute scales and fibrils, flesh bright yellow. **SPORE PRINT** brown. **ODOR** and **TASTE** not distinctive.

Habit and habitat solitary, scattered or clustered on the ground under conifers such as spruce, hemlock, white pine, more rarely in mixed stands of pitch pine and oak. July to September. **RANGE** widespread.

Spores subfusiform, smooth, 12–15 × 3–5 µm.

Comments *Chalciporus rubinellus* has a mild taste and is a more brightly colored than its frequently encountered sister species, *C. piperatus*. Compare also with *C. piperatoides*.

Gyroporus castaneus (Bulliard) Quélet

CHESTNUT BOLETE

SYNONYM *Boletus castaneus* Bulliard

Cap chestnut-brown, velvety matted; pores white becoming pale yellow; stem colored as cap, glabrous, brittle, hollow

CAP 3–10 cm wide, convex, becoming plane, chestnut-brown, yellow-brown or orange-brown, surface somewhat velvety matted, becoming glabrous, dry. **FLESH** white. **PORES** white becoming pale yellow, round becoming angular, 1–2 per mm. **TUBES** same color as pores, 5–8 mm deep. **STEM** 3–9 cm long, 6–15 mm broad, equal or enlarged downward or swollen in the middle, colored as cap or paler over apex, glabrous, very brittle and breaking cleanly, hollow or cottony filled, flesh white turning brown, especially in the base. **SPORE PRINT** pale yellow. **ODOR** and **TASTE** not distinctive.

Habit and habitat single, scattered, or clustered on the ground, under mixed hardwoods and conifers, often with oak, also on road banks and in grassy disturbed areas sometimes distant from trees. June to October. **RANGE** widespread.

Spores elliptically bean-shaped, smooth, 8–13 × 5–6 μm or 6–8 × 4–5 μm.

Comments More than one species appears to be associated with this name, as evidenced by the different reports of spore sizes in the literature. All forms are edible. *Gyroporus* is separated from other boletes by the hollow stem and pale yellow spore print.

Gyroporus cyanescens
(Bulliard) Quélet

BLUING BOLETE

SYNONYMS *Boletus cyanescens* Bulliard, *Gyroporus cyanescens* var. *violaceotinctus* Watling

Cap buff or pale olive, woolly, instantly blue when injured; pores white becoming pale yellow, bluing; stem colored as pileus, brittle, hollow and instantly blue when injured

CAP 4–12 cm wide, convex, becoming plane, pallid buff or straw-yellow or pale olive or olive-tan, rapidly blue when touched, coarsely woolly matted, dry. **FLESH** white or pale yellow, instantly blue when exposed. **PORES** white or pale yellow, turning instantly deep blue when injured, round, 1–2 per mm. **TUBES** same color and reaction as pores, 5–10 mm deep. **STEM** 4–10 cm long, 10–25 mm broad, equal or enlarged downward or swollen in middle, colored as cap or paler, turning blue instantly when touched, coarsely woolly matted over lower half, white and smooth over upper half and marking an annular zone, flesh very brittle, hollow or with cottony filled off-white center. **SPORE PRINT** pale yellow. **ODOR** and **TASTE** not distinctive.

Habit and habitat single, scattered, or clustered on the ground, especially road banks with sandy soils, under mixed hardwoods including birch and poplar. August to October. **RANGE** widespread.

Spores elliptical-bean-shaped, smooth, 8–10 × 5–6 μm.

Comments Some recognize the American *Gyroporus cyanescens* var. *violaceotinctus* based on the dark violet-blue reaction occurring instantly, considering it different from *G. cyanescens* var. *cyanescens* that stains greenish at first then turns blue. Also, different-colored forms of this "species" seem to occur as well in North America, ranging from very pale to uniformly straw-yellow. Edible.

Strobilomyces strobilaceus
(Scopoli) Berkeley

OLD MAN OF THE WOODS
SYNONYMS *Boletus strobilaceus* Scopoli, *Strobilomyces floccopus* (Vahl) P. Karsten, *Strobilomyces strobiliformis* Beck

Cap gray or black, shaggy scaly, margin with gray projecting patches; pores turning reddish when injured; stem colored and adorned as cap, with a gray cottony ring, flesh rapidly turning reddish

CAP 3–15 cm wide, convex, becoming plane, dark grayish or black, surface covered with coarse soft erect or flattened shaggy scales, typically forming cracked pattern of pale gray or dull white between scales, often with gray patches from ring hanging off margin, dry. **FLESH** white, turning rapidly orange-red, then black. **PORES** white, then gray, eventually black, turning reddish, then black when injured, elliptical or slight angular, 1–2 mm wide. **TUBES** same color and staining reactions, 10–15 mm deep. **STEM** 4–12 cm long, 10–25 mm broad, mostly equal, colored and adorned as cap, with a soft cottony ring near apex, flesh rapidly orange-red, then black. **RING** gray, cottony, mostly adhering to cap margin, ring zone may be present. **SPORE PRINT** black. **ODOR** and **TASTE** not distinctive.

Habit and habitat single, scattered, or clustered on the ground or on decaying wood, under mixed hardwoods of beech or oak, or both, or in conifer woods with pine. June to October. **RANGE** widespread.

Spores globose or subglobose, reticulate ornamented, 9.5–15 × 8.5–12 µm.

Comments *Strobilomyces confusus* Singer differs by the spores that are not reticulate but display irregular ridges and projections (like the fruit bodies of *Sparassis*). The scales on the cap of *S. confusus* also tend to be smaller, more acute, and compact. Edible, but the stem is tough.

Pulveroboletus ravenelii
(Berkeley & M. A. Curtis) Murrill

POWDERY SULFUR BOLETE

SYNONYM *Boletus ravenelii* Berkeley & M. A. Curtis

Cap bright yellow, becoming reddish or chestnut-brown, cracked with age, margin with yellow fragile ring patches hanging down; pores yellow, rapidly greenish-blue when injured; stem bright yellow with torn ring, soft cottony-powdery below ring

CAP 1–10 cm wide, convex, becoming plane, bright sulfur-yellow, becoming orange-red or brownish-red from center outward, soft powdery then matted hairy or scaly, cracked in age, with yellow ring patches hanging off margin, dry, but slippery when wet. **FLESH** white or pale yellow, turning pale blue. **PORES** bright yellow then dingy-yellow, eventually pale grayish-olive, turning blue when injured, round or angular, 1–3 per mm. **TUBES** same color, turning blue. **STEM** 4–16 cm long, 6–16 mm broad, equal, bright sulfur-yellow, white or pale yellow mycelium at base, sheathed with soft cottony layer to near apex, leaving a thin tattered ring on the stem, flesh yellow in base. **RING** bright yellow, soft, cottony-powdery. **SPORE PRINT** grayish-olive-brown. **ODOR** and **TASTE** not distinctive.

Habit and habitat single, scattered, or clustered on the ground or on decaying wood, under mixed hardwoods of beech or oak, or both, or in conifer woods with pine. July to October. **RANGE** widespread.

Spores elliptical, smooth, 8–10 × 4–5 μm.

Comments The yellow cottony or powdery material on the stem comes off on your fingers when handling. *Pulveroboletus auriflammeus* (Berkeley & M. A. Curtis) Singer is a smaller, slender, dark orange or orange-yellow species that lacks a distinct ring on the stem, has a broad well-developed net on the stem, and has flesh, pores, and tubes that do not turn blue when injured.

Hemileccinum subglabripes
(Peck) Halling

SYNONYMS *Boletus subglabripes* Peck, *Leccinum subglabripes* (Peck) Singer, *Krombholzia subglabripes* (Peck) Singer, *Pulveroboletus flavipes* E. Horak

Cap chestnut or cinnamon brown, bald and smooth or somewhat wrinkled; pores lemon-yellow; stem pale yellow with yellow or reddish dots and scales

CAP 3–15 cm wide, convex, chestnut-brown, ochre-brown, cinnamon-brown or reddish-brown, surface smooth or irregularly wrinkled, moist or dry. **FLESH** pale yellow becoming white, with reddish-brown discolorations, occasionally faintly blue near tubes, thick. **PORES** lemon-yellow, becoming dull greenish-yellow, turning reddish-brown slowly where injured, round becoming somewhat angular, 1–3 per mm. **TUBES** same color as pores, 5–16 mm deep. **STEM** 5–10 cm long, 10–25 mm broad, equal, yellow with erratic rusty-reddish tints over lower half, white mycelium at base, surface obscurely covered with yellow or faintly reddish dots or scales (use a lens), flesh yellow streaked reddish-brown. **SPORE PRINT** olive-brown. **ODOR** not distinctive. **TASTE** mild or acidic.

Habit and habitat solitary or clustered on the ground under hardwoods, especially birch, beech, aspen or in mixed woods with also hemlock, spruce or balsam fir. July to October. **RANGE** widespread.

Spores subfusiform, smooth, 11–20 × 3–5 μm.

Comments Previously listed as *Boletus subglabripes* in field guides and compared frequently with the corrugated or strongly wrinkly capped *Boletus hortonii* A. H. Smith & Thiers. They do have a somewhat similar aspect, more or less, but they do not appear to be phylogenetically related using molecular comparisons. I find *Hemileccinum subglabripes* to be an excellent edible species.

Leccinum aurantiacum
(Bulliard) Gray

RED-CAPPED SCABER STALK

SYNONYMS *Boletus aurantiacus* Bulliard, *Krombholzia aurantiaca* (Bulliard) E.-J. Gilbert, *Leccinum rufum* (Schaeffer) Kreisel
Cap orange or brick-red, matted hairy at first; flesh white, turning pinkish then black; stem white with orange-brown then black dots, under poplar or aspen and pine

CAP 5–20 cm wide, convex, becoming plane, orange or brick-red or rusty-red or reddish-brown, matted hairy, becoming glabrous, smooth. **FLESH** white, slowly becoming pinkish or wine-red, then blackish. **PORES** white, becoming pale olive-tan or olive-brown, turning darker when bruised, round, 2–3 per mm. **TUBES** same color and reactions as pores, 10–20 mm deep. **STEM** 10–16 cm long, 20–30 mm broad, equal or enlarged downward, white, covered with small rounded knobs (scabers) that are white at first, becoming orange-brown or reddish-brown, then black with age, often with blue-green stains over base as well, flesh white, slowly becoming pinkish or wine-red, then blackish, also some with green discolorations in the base. **SPORE PRINT** brown. **ODOR** and **TASTE** not distinctive.

Habit and habitat scattered or clustered on the ground under poplar or aspen and pine. July to September. **RANGE** widespread.

Spores subfusiform, smooth, 13–18 × 3.5–5 μm.

Comments *Leccinum insigne* is similar in color and form, and found in similar habitats, but the flesh does not turn pink or wine-red at first when exposed, but does become black with time. Both species are listed as edible, but see comments under *Leccinum versipelle*.

Leccinum versipelle (Fries & Hök) Snell

SYNONYMS *Leccinum atrostipitatum* A. H. Smith & Thiers, *Leccinum testaceoscabrum* Secretan ex Singer

Cap dull orange, with cap skin projecting well beyond flesh; flesh white then pink, eventually black; pores round very small, bruising brownish; stem white but very densely black-dotted, under birches

CAP 6–18 cm wide, convex with skin projecting beyond flesh making a roll or flap at the margin, dark dull orange or orange-buff, becoming pale brownish-orange or pinkish-tan with age, dry or slight tacky at first, matted hairy, becoming flattened brown scaly. **FLESH** white turning pink, then purplish-gray, then black. **PORES** pale olive or off-white when young, becoming honey-brown or darker, turning deep olive or brownish when bruised, round, 2–3 per mm. **TUBES** same color and reactions as pores, 10–18 mm deep. **STEM** 8–15 cm long, 10–25 mm broad, equal or enlarged downward, white but densely covered with small black, erect, rounded knobs (scabers), also often with blue or blue-green stains in the base, flesh like cap. **SPORE PRINT** brown. **ODOR** and **TASTE** not distinctive.

Habit and habitat solitary, scattered or clustered on the ground under birches. July to September. **RANGE** widespread.

Spores subfusiform, smooth, 13–17 × 4–5 µm.

Comments A common orange-capped *Leccinum* species found under birches. The dense black knobs, or scabers, on the stem at all stages of development are diagnostic. Compare with *L. aurantiacum*, a symbiont with poplar, that differs in the early stages by the scabers being orange-brown and eventually turning black. Listed as edible in most field guides, but cook this one thoroughly, or dry it for later use. However, recent multiple cases of gastrointestinal distress have been linked to eating orange-capped *Leccinum* species. It may be best to avoid these boletes as food for the table.

Leccinellum albellum (Peck)
Bresinsky & Manfred Binder

SYNONYMS *Boletus albellus* Peck, *Leccinum albellum* (Peck) Singer
Cap white or pale grayish, often developing olive tints, surface shallowly pitted-wrinkled; flesh white, unchanging; stem white with obscure white or pale brown knobs; under oaks

CAP 2–7 cm wide, convex, becoming plane, off-white or pale grayish or pale olive-buff, especially over the center, dry, finely felted or glabrous, smooth or pitted-wrinkled, lacking a sterile flap at the margin. **FLESH** white, unchanging. **PORES** white, becoming olive-buff, round or slight angular, 1–2 per mm. **TUBES** same color and reactions as pores, 10 mm deep. **STEM** 4–9 cm long, 5–12 mm broad, equal or slightly enlarged downward, white or pale olive-buff, small rounded knobs (scabers) obscure, white becoming pale gray or pale brown, flesh white unchanging, mycelium at base white. **SPORE PRINT** olive-brown. **ODOR** and **TASTE** not distinctive.

Habit and habitat solitary, scattered or clustered on the ground under mixed hardwoods with oak. July to October. **RANGE** widespread.

Spores subfusiform, smooth, 10–24 × 4–6.5 µm; cap skin composed of erect, inflated cells, thus the wrinkled surface.

Comments The inflated cells of the cap surface place this species in *Leccinellum*. However, two white *Leccinum* or *Leccinellum* species are found in the Northeast with some frequency. The similar-looking *Leccinum holopus* (Rostkovius) Watling grows with birch in swampy areas. It has a slippery cap that often produces greenish discolorations with age, a longer stem, and cap skin composed of cylindrical and partially erect cells. *Leccinum holopus* var. *americanum*, previously separated based on the flesh turning pinkish, is now considered a synonym of *L. holopus* var. *holopus* based on DNA comparisons. It seems the oxidation reaction is variable in this species.

Leccinellum griseum (Quélet)
Bresinsky & Manfred Binder

SYNONYMS *Gyroporus griseus* Quélet, *Leccinum griseum* (Quélet) Singer
Cap gray-brown or dark brown, strongly wrinkled and pitted, glabrous; flesh white then gray to black; pores and tubes off-white, then grayish-brown; stem slender with gray or brown dots

CAP 3–9 cm wide, convex then plane, margin lacking sterile projecting edge, gray-brown or dark brown or yellowish-brown, with black hues on edges of pits, surface wrinkled and pitted, glabrous, dry. **FLESH** white, turning slowly gray or black. **PORES** off-white, becoming gray or gray-brown, sometimes slowly yellow when bruised, round, 1–2 per mm. **TUBES** same color and reactions as pores, 10–20 mm deep. **STEM** 4–13 cm long, 8–15 mm broad, equal or enlarged downward, white or pale brown beneath dense, small dark gray or brown, erect, rounded knobs (scabers), flesh white then slowly gray, blue-green in the base. **SPORE PRINT** cinnamon-brown. **ODOR** and **TASTE** not distinctive.

Habit and habitat solitary, scattered or clustered on the ground under oak, and quite possibly blue beech. July to September. **RANGE** widespread.

Spores subfusiform, smooth, 11–15 × 4.5–6 µm, also in another source as 10–28 × 4–7 µm, both from North American materials; cap skin end cells spherical inflated and stacked up in layers 2–4 cells deep.

Comments *Leccinum griseum*, a European species name, may not be the same species we find in North America. Our "griseum" fits into *Leccinellum* because of the inflated cells in the cap, but it does not completely match the European species. Some boletologists consider *Leccinellum griseum* to be a synonym of *Leccinum pseudoscabrum* (Kallenbach) Šutara, also known as *Leccinum carpini* (R. Schulz) M. M. Moser ex D. A. Reid. This latter species is mycorrhizal with hornbeam or blue beech (*Carpinus caroliniana*) in Europe. The North American species typically has been cited as growing in older growth stands of hardwoods that include oaks, and perhaps blue beech, but that has not been confirmed. More careful observations are needed.

Leccinum snellii A. H. Smith, Thiers & Watling

Cap dark brown or almost black; flesh white turning pink or red, eventually black; pores round very small, bruising brownish; stem white with black dots, fleshy pink above, blue-green below

CAP 3–9 cm wide, convex, becoming plane, dark brown or black radiating, matted fibrils densely covering pallid surface, dark yellowish-brown with age, dry. **FLESH** white turning pink or red especially near stipe, then very slowly purplish-gray or black. **PORES** off-white, becoming grayish-brown, sometimes turning ochre when bruised, round, 2–3 per mm. **TUBES** same color and reactions as pores, 10–15 mm deep. **STEM** 4–11 cm long, 10–20 mm broad, equal or enlarged downward, white beneath small dark gray or black, erect, rounded scabers, often with blue or blue-green stains in the base, flesh white but becoming pink or red above, blue-green in the base. **SPORE PRINT** brown. **ODOR** and **TASTE** not distinctive.

Habit and habitat solitary, scattered or clustered on the ground under birch, especially yellow birch. June to October. **RANGE** widespread.

Spores subfusiform, smooth, 16–22 × 5–7.5 µm, some of the cap skin end cells inflated.

Comments The common dark brown or almost black-capped *Leccinum* species found under yellow birch. *Leccinum scabrum* is similar, but the flesh does not turn reddish when exposed. Edible.

Leccinum luteum A. H. Smith, Thiers & Watling

Cap yellow or cream-yellow, bumpy or pitted or both, slippery; flesh white turning black; stem white with eventually brown or black dots, flesh white turning black

CAP 3–7 cm wide, convex, pale yellow or cream-yellow, becoming darker yellow or sometimes olive-yellow-brown, slippery when wet, bumpy and pitted usually. **FLESH** white or pale cream, slowly pinkish-gray then dark brown or black. **PORES** white, becoming pale olive-tan, pinkish-gray when bruised, round, 2–3 per mm. **TUBES** same color and reactions as pores, 8–15 mm deep. **STEM** 6–13 cm long, 8–10 mm broad, enlarged downward, white above, may become pale cream over lower portions, covered with small rounded knobs (scabers) that are pale brown over upper stem, darker toward base, eventually black on the tips, especially from handling, flesh white turning black and some with green discolorations in the base. **SPORE PRINT** olive-brown. **ODOR** not distinctive. **TASTE** not distinctive, or acidic.

Habit and habitat solitary, scattered or clustered on the ground under blue beech (*Carpinus caroliniana*) and oaks, not common. June to September. **RANGE** widespread.

Spores subfusiform, smooth, 17–20 × 5–6.5 µm, cap skin with subglobose end cells.

Comments The scabers on the stipe can be very pale at first. Not a commonly collected species.

Sutorius eximius (Peck) Halling, M. Nuhn & Osmundson

LILAC-BROWN BOLETE

SYNONYMS *Tylopilus eximius* (Peck) Singer, *Leccinum eximium* (Peck) Singer

Cap dark chocolate-brown; pores chocolate or purple-brown, small; stem bluish-gray, densely covered with transverse oriented purplish-brown granules

CAP 5–12 cm wide, convex, becoming plane, dark chocolate-brown with purplish tints or becoming reddish-brown, with a silvery bloom when young, surface smooth or becoming pitted, minutely velvety, slippery when wet, soon dry. **FLESH** off-white or pale gray-brown, slowly turning reddish or grayish-purple. **PORES** chocolate or purple-brown or even dark gray or black, round, 1–3 per mm. **TUBES** white at first, then same color and reactions as pores, 9–12 mm deep. **STEM** 5–15 cm long, 10–40 mm broad, equal or tapered at either end, bluish-gray with darker purplish or purple-brown granules densely coating entire stem in transverse lines, flesh white or colored as cap flesh. **SPORE PRINT** reddish-brown. **ODOR** not distinctive. **TASTE** mild or slight bitter.

Habit and habitat single, scattered, or clustered on the ground under conifers, especially hemlock, spruce and balsam fir. July to October. **RANGE** widespread.

Spores subfusiform, smooth, 11–17 × 3.5–5 μm.

Comments A distinctive bolete that has been reported as poisonous for some, causing severe gastrointestinal upsets. Some field guides list this species as edible, but that is clearly not the case.

Harrya chromapes (Frost) Halling, Nuhn, Osmundson & Manfred Binder

CHROME-FOOTED BOLETE
SYNONYMS *Boletus chromapes* Frost, *Tylopilus chromapes* (Frost) A. H. Smith & Thiers, *Leccinum chromapes* (Frost) Singer
Cap pink or pale red; pores white, becoming pinkish-tan; stem white with bright pink dots and a bright yellow base

CAP 3–15 cm wide, convex, becoming plane, pink or pale red, fading to pinkish-tan or paler with age, surface smooth but matted velvety, slightly slippery when wet, then dry. FLESH white. PORES off-white, then pinkish-tan or flesh-brown, angular, 2–3 per mm. TUBES same as pores, 8–15 mm deep. STEM 4–15 cm long, 10–25 mm broad, mostly equal, white under bright pink dots except for bright yellow base, surface evenly dotted, sometimes with a fine net, flesh as cap flesh. SPORE PRINT rosy-brown. ODOR and TASTE not distinctive.

Habit and habitat single, scattered, or clustered on the ground hardwood and coniferous forests, especially with oak, beech and pine. May to October. RANGE widespread.

Spores oblong-elliptical or subfusiform, smooth, 11–17 × 4–5 µm.

Comments This attractive bolete is common. Sometimes the edge or mouth around the pores is bright pink as well. The rosy spore print suggests *Tylopilus*, while the pink-dotted stem surface makes some consider *Leccinum*, but recent DNA phylogenetic studies indicate the species is unique and must be given its own genus. *Harrya chromapes* is known from eastern North America, China, and Costa Rica.

Tylopilus balloui (Peck) Singer

BURNT-ORANGE BOLETE

SYNONYMS *Boletus balloui* Peck, *Gyrodon balloui* (Peck) Snell, *Rubinoboletus balloui* (Peck) Heinemann & Rammeloo

Cap bright orange-red; pores white, bruising brown; stem pallid or rich yellow or orange-yellow, flesh white becoming pinkish-tan when exposed

CAP 5–12 cm wide, convex, becoming plane, bright orange or reddish-orange fading to dull orange or cinnamon-brown with age, surface smooth, dry. **FLESH** white, turning pinkish-buff or violet-brown. **PORES** off-white, then pinkish-tan, bruising brown, angular, 1–2 per mm. **TUBES** white at first, then same color and reactions as pores, 8–10 mm deep. **STEM** 2–12 cm long, 6–25 mm broad, equal or tapered at the base, variable but can be white at least at the apex and base, or deep yellow or orange-yellow, staining brown when handled, surface smooth or finely mealy, sometimes with striations or finely netted, flesh as cap flesh. **SPORE PRINT** reddish-brown or paler brown. **ODOR** not distinctive. **TASTE** mild or slightly bitter.

Habit and habitat single, scattered, or clustered on the ground in grassy areas near trees or in mixed forests, especially with oak, beech and pine. July to October. **RANGE** widespread.

Spores elliptical, smooth, short for a bolete, 5–11 × 3–5 µm in some guides, but we found 5.5–7 × 4–5.5 µm in our Caribbean collections.

Comments A great deal of variation in color and form occurs in this "species," and recent molecular evidence suggests several species are involved on a global scale. The collection shown here has a deeply yellow-colored stem, though many collections have white stems. The Northeast form is considered edible.

Tylopilus alboater (Schweinitz) Murrill

BLACK VELVET BOLETE

SYNONYMS *Boletus alboater* Schweinitz, *Porphyrellus alboater* (Schweinitz) E.-J. Gilbert

Cap dark brown or black; flesh white turning pink; pores white, then pink, staining red; stem colored as cap, but with white apex, glabrous; taste mild

CAP 3–8(–15) cm wide, convex, becoming plane, black to dark grayish-brown, with a silvery bloom when young, surface smooth or minutely velvety, dry. **FLESH** white, turning pinkish. **PORES** white at first, then pinkish, turning red when bruised then becoming black, angular, 1–2 per mm. **TUBES** same color and reactions, 5–10 mm deep. **STEM** 4–10 cm long, 20–40 mm broad, mostly equal, colored as cap or paler, often white at apex, glabrous or with silvery bloom at first, flesh white, turning pink, becoming black in base. **SPORE PRINT** flesh or pinkish-brown. **ODOR** and **TASTE** not distinctive.

Habit and habitat single, scattered, or clustered on the ground under hardwoods, with oaks present. June to October. **RANGE** widespread.

Spores subfusiform, smooth, 7–11 × 3.5–5 µm.

Comments The contrasting blackish cap and stem with the white pores that stain red when bruised are reliable features. A fine edible species with firm flesh and mild taste. If the cap is olive-brown and glabrous, it is most likely *Tylopilus atronicotianus* Both.

Tylopilus felleus (Bulliard)
P. Karsten

SYNONYM *Boletus felleus* Bulliard

Large with brown cap; flesh white turning pink then brown; pores white then pink with age, staining rusty-brown; stem brown, with raised dark brown net; taste bitter

CAP 5–30 cm wide, convex, becoming plane, various shades from dark brown or reddish-brown but paler brown and becoming tan with age, surface smooth or minutely woolly, then bald, slippery or tacky when wet, then dry. **FLESH** white, turning pinkish then brown on some. **PORES** white at first, then pinkish, turning rusty-brown when bruised, round, 1–2 per mm. **TUBES** same color and reactions, 10–20 mm deep. **STEM** 4–20 cm long, 10–30 mm broad, narrowly or broadly club-shaped and bulbous over base, pale brown and with darker brown heavy net to near base, flesh white, becoming dark brown in base or from insect tunnels. **SPORE PRINT** flesh or pinkish-brown. **ODOR** not distinctive, or slightly farinaceous. **TASTE** strongly bitter.

Habit and habitat single, scattered, or clustered on the ground or on decaying wood under conifers, especially hemlock, or in mixed woods. June to October. **RANGE** widespread.

Spores subfusiform, smooth, 11–17 × 3–5 µm.

Comments One of the most commonly collected species of *Tylopilus*. It is separated from the porcini, *Boletus edulis*, with which it can be confused, by the bitter taste, the dark brown net on the stem, the pink pores and tubes that stain rusty-brown, and the flesh-brown spore print.

Tylopilus ferrugineus (Frost) Singer

SYNONYM *Boletus ferrugineus* Frost

Cap reddish-brown; flesh white turning pink then brown; pores pallid staining rusty-brown; stem brown with white base, with faint net over apex; taste mild

CAP 4–14 cm wide, convex, becoming plane, dark reddish-brown or chestnut-brown, surface smooth or minutely felted, then bald, dry. **FLESH** white, slowly turning pink then brown. **PORES** white at first, then pale pinkish-buff, turning rusty-brown when bruised, round or angular, 1–3 per mm. **TUBES** same color and reactions, 6–12 mm deep. **STEM** 4–10 cm long, 15–25 mm broad, equal, brown or reddish-brown but white over base and with white mycelium, faintly netted over apex to mid, surface minutely brown-dotted or glabrous, flesh white. **SPORE PRINT** flesh or pinkish-brown. **ODOR** and **TASTE** not distinctive.

Habit and habitat scattered or clustered on the ground under hardwoods, with especially beech and oak. July to September. **RANGE** widespread.

Spores subfusiform, smooth, 8–13 × 3–5 µm.

Comments There are several brown, mild-tasting species of *Tylopilus* that can be challenging to separate. *Tylopilus badiceps* differs by its maroon or purple-brown cap that frequently has a bent or beveled margin, and *T. indecisus* can be distinguished by the pale almost white stem (before it has been handled since it stains brown quickly). All are edible. See also *T. rubrobrunneus*.

Tylopilus plumbeoviolaceus
(Snell & E. A. Dick) Snell & E. A. Dick

VIOLET-GRAY BOLETE

SYNONYM *Boletus plumbeoviolaceus* Snell & E. A. Dick

Cap purple or purple-brown; pores white then flesh-pink, not changing when bruised; stem dark violet or purple; taste intensely bitter

CAP 4–15 cm wide, convex, dark gray-brown or brown with violet or purple tints in early stages, becoming slate-gray or eventually brown or dull cinnamon or tan, surface velvety or woolly, then bald, dry. **FLESH** white. **PORES** white, then pinkish-flesh, round, 1–2 per mm. **TUBES** same color, 5–20 mm deep. **STEM** 8–12 cm long, 10–20 mm broad, equal or enlarged downward, mostly a deep violet or purple with some white mottling, white mycelium over base and sometimes staining olive, glabrous, flesh white. **SPORE PRINT** flesh or pinkish-brown. **ODOR** not distinctive. **TASTE** intensely bitter.

Habit and habitat scattered or clustered on the ground under hardwoods, with especially oak present. June to September. **RANGE** widespread.

Spores subfusiform, smooth, 10–13 × 3–4 µm.

Comments *Tylopilus violatinctus* T. J. Baroni & Both is similar but has much paler colors, the cap stains rusty-violet when bruised, the tubes stain brown, and the spores are shorter. Of the two, *T. plumbeoviolaceus* is more commonly encountered.

Tylopilus rubrobrunneus Mazzer & A. H. Smith has a purple or purple-brown cap when very young, otherwise it is brown. The pores are pallid, staining brown. While stems of *T. plumbeoviolaceus* and *T. violatinctus* have some shade of violet or purple in them when young, the brown club-shaped stem of *T. rubrobrunneus* does not. *Tylopilus rubrobrunneus* generally produces much larger bitter-tasting fruit bodies as well. Compare also *T. felleus*, another bitter-tasting bolete that is brown, but differs by the stem having an obvious brown reticulum.

Tylopilus rubrobrunneus

Tylopilus porphyrosporus
(Fries & Hök) A. H. Smith & Thiers

DARK BOLETE

SYNONYMS *Boletus porphyrosporus*
Fries & Hök, *Porphyrellus porphyrosporus*
(Fries & Hök) E.-J. Gilbert, *Porphyrellus*
pseudoscaber Secretan ex Singer

Cap dark brown; flesh white turning blue
then red then black; pores dark brown,
turning blue when injured; stem same
color as cap

CAP 5–15 cm wide, convex, becoming
plane, dark chocolate-brown or dark olive-
brown or dark reddish-brown, surface
smooth but matted velvety, dry. **FLESH**
white, slowly turning blue, then reddish
or reddish-brown then black. **PORES** dark
brown or dark reddish-brown, turning
blue when injured, then black, round,
2–3 per mm. **TUBES** same color as pores,
turning blue when injured, 13–20 mm
deep. **STEM** 4–12 cm long, 10–30 mm
broad, equal or enlarged downward, colored as cap but with white base, surface
roughened with erect dots, sometimes
with a fine net, flesh as cap flesh. **SPORE**
PRINT dark reddish-brown. **ODOR** pungent,
strong. **TASTE** not distinctive.

Habit and habitat single, scattered, or clustered on the ground or on well-decayed
logs, in mixed woods and coniferous forests. July to October. **RANGE** widespread.

Spores subfusiform, smooth, 12–18 ×
6–7.5 μm.

Comments Compare with *Sutorius eximius*
that differs by its dotted stem and nonbluing flesh. In many field guides *Tylopilus*
porphyrosporus is called *T. pseudoscaber*
Secretan ex A. H. Smith & Thiers. To
make a long story very short, Secretan's
names are not available for use, thus *T.*
porphyrosporus has to be the accepted
name and is the one used in Europe for
the dark chocolate-brown-colored bolete
with bluing flesh and tubes. Is our North
America material exactly the same as the
European species? That connection has
not been tested using phylogenetic analysis as yet. We may have more than one
species sharing similar colors and form in
North America.

Suillus americanus (Peck) Snell

CHICKEN-FAT SUILLUS, AMERICAN SLIPPERY JACK

SYNONYMS *Boletus americanus* Peck, *Ixocomus sibiricus* Singer, *Suillus sibiricus* (Singer) Singer

Cap yellow, slimy, often with reddish-brown patches, margin with cottony roll; flesh pale yellow turning pinkish-brown; pores yellow, turning reddish-brown when bruised, angular; stem same color as cap

CAP 3–10 cm wide, conical-convex, becoming plane and with or without the conical center, margin with cottony white or yellow roll, bright yellow or ochre-yellow with cinnamon or reddish patches, surface slimy. **FLESH** pale yellow, turning pinkish-brown. **PORES** initially covered by cottony veil that clings to cap margin after expansion, yellow, slowly brown or reddish-brown when injured, angular and radially elongate, 1–2 mm wide.

TUBES short decurrent, same color as pores, 7–10 mm deep. **STEM** 3–10 cm long, 3–10 mm broad, equal but often bent or crooked, yellow with reddish sticky dots and smears that become darker, especially from handling, veil typically not leaving a ring on the stem, flesh as cap flesh, becoming hollow. **SPORE PRINT** dull cinnamon. **ODOR** and **TASTE** not distinctive.

Habit and habitat single, scattered, or clustered on the ground under or near white pine. July to October. **RANGE** widespread.

Spores subfusiform, smooth, 8–11 × 3–4 µm.

Comments This species only grows symbiotically with eastern white pine—no trees, no *Suillus americanus*. Edible but not considered choice. See comments under *S. punctipes* for other species with yellowish-orange colors.

Suillus spraguei (Berkeley & M. A. Curtis) Kuntze

PAINTED SUILLUS

SYNONYM *Suillus pictus* (Peck) Kuntze

Cap red hairy-scaly, dry; pores covered by a cobweb veil at first, yellow, compound, radially arranged; stem red scaly over ochre-yellow ground color, with pale grayish collapsed veil on upper stem

CAP 3–12 cm wide, convex, pale or dark red hairy soft scales over white or pale yellow ground color, margin with pale grayish-pink soft patches or continuous roll of ring material, dry. **FLESH** yellow, turning pinkish-gray. **PORES** yellow becoming ochre-brown, staining reddish-brownish when bruised, angular, radially arranged, compound, 0.5–5 mm wide, covered with grayish-pink cottony veil at first. **TUBES** bluntly attached or decurrent, colored as pores, 4–8 mm deep, not easily removed from cap. **STEM** 4–12 cm long, 10–25 mm broad, equal or club-shaped, ochre-yellow under red scales, flesh white or brownish in the base. **RING** cottony grayish-pink, collapsed on upper stem. **SPORE PRINT** yellow-brown. **ODOR** and **TASTE** not distinctive.

Habit and habitat scattered or clustered on the ground under eastern white pine in mixed forests, on road banks and on lawns. June to October. **RANGE** widespread.

Spores ovate or subfusiform, smooth, 8–11 × 3.5–5 µm.

Comments If you are in the southern reaches of the Northeast, you may find the smaller and paler, orange-brown scaly capped *Suillus decipiens* (Peck) Kuntze. It also differs by its pores that turn only faintly brown when bruised. Compare with *S. cavipes* that is found under larch.

Suillus cavipes (Opatowski) A. H. Smith & Thiers

HOLLOW-STALKED LARCH SUILLUS

SYNONYM *Boletinus cavipes* (Opatowski) Kalchrenner

Cap brown hairy-scaly, dry; stem brown hairy-scaly below ring zone, yellow on upper stem; stem hollow; under larch

CAP 3–10 cm wide, convex becoming plane, reddish-brown or yellowish-brown from densely packed, hairy soft scales, margin with white continuous roll from the veil, dry. **FLESH** white or becoming yellow with age. **PORES** pale yellow turning greenish-yellow, angular, radially arranged, single or compound, 1–1.5 mm long, 0.5–1 mm wide, covered with white cottony veil at first. **TUBES** decurrent, colored as pores, 3–5 mm deep, not easily removed from cap. **STEM** 4–9 cm long, 8–15 mm broad, equal or club-shaped, yellow over apex, same color as cap and hairy below ring zone, with cottony white (veil) collapsed on upper stem, hollow inside. **SPORE PRINT** olive-brown. **ODOR** and **TASTE** not distinctive.

Habit and habitat scattered or clustered on the ground under larch (tamarack). September to October. **RANGE** widespread.

Spores ovate or subfusiform, smooth, 7–10 × 3.5–4 µm.

Comments All other scaly capped *Suillus* species have a solid stem, have different colors, and partner with pines. Compare with *Fuscoboletinus paluster* that also occurs under larch.

Suillus punctipes (Peck) Singer

SYNONYM *Boletus punctipes* Peck

Cap ochre-orange with white cottony margin when young, slimy; pores brown, then ochre-yellow; stem dull orange-brown with sticky brown dots; odor fragrant

CAP 3–10 cm wide, convex, becoming plane with wavy margin, dull yellow or ochre-orange with tufts of white or pale gray or brown cottony patches or hairs especially around the margin, surface soon smooth and glabrous, slippery. **FLESH** pale yellow. **PORES** dingy brown at first, then orange-brown or ochre-yellow, round or angular, 1–2 per mm. **TUBES** bluntly attached or short decurrent, brown then honey-yellow, 4–8 mm deep. **STEM** 4–9 cm long, 9–16 mm broad, equal or club-shaped, often curved, dull orange-brown or ochre-yellow, covered with brown sticky dots and smears, sometimes reddish-streaked over the base, flesh solid, dull ochre-orange, slowly turning brown when exposed. **SPORE PRINT** dark olive-brown. **ODOR** fragrant, of almond extract. **TASTE** not distinctive.

Habit and habitat scattered or clustered on the ground under spruce, balsam fir, eastern white pine and perhaps other conifers. July to October. **RANGE** widespread.

Spores subfusiform, smooth, 7.5–12 × 3–3.5 μm.

Comments The sticky dots on the stem stain your fingers brownish. Compare with the apricot-colored *Suillus subaureus* (Peck) Snell that differs by growing under hardwoods, having a reddish-streaked cap surface from downy patches and scales, and having dingy yellow, not brown pores at first. *Suillus tomentosus* (Kauffman) Singer is somewhat similar but the pores and the cap flesh turn blue, sometimes erratically, when injured or exposed, and it occurs under 2-needled pines.

Suillus grevillei (Klotzsch) Singer

LARCH SUILLUS

SYNONYMS *Boletus grevillei* Klotzsch, *Suillus elegans* (Schumacher) Snell

Cap bright yellow or dark brown, slimy-glutinous; stem thick, yellow, sheathed with cottony white veil with gelatinous yellow coating and lacking dots; veil persistent as cottony collapsed ring; under larch

CAP 5–15 cm wide, convex, becoming plane, dark brown or bright reddish-brown or ochre-yellow, margin sometimes with a thin white roll from veil, surface slimy, shiny when dry. **FLESH** yellowish, turning pinkish-brown. **PORES** cream yellow becoming olive-yellow, turning brown when bruised, angular, 1–3 per mm, covered with two layered veil at first, inner tissue white cottony covered by yellow slimy layer. **TUBES** bluntly attached or short decurrent, colored as pores, 4–12 mm deep. **STEM** 4–10 cm long, 10–30 mm broad, equal, yellow above ring, mottled yellow-brown below ring, white over base, discoloring brownish when handled, with white cottony veil coated with yellow gelatinous layer descending to lower stem making it slimy, veil collapsing as a cottony ring on upper stem, surface glabrous, lacking dots, flesh yellow. **SPORE PRINT** cinnamon-brown. **ODOR** sharp-pungent or not distinctive. **TASTE** not distinctive.

Habit and habitat scattered or clustered on the ground under larch. September to November. **RANGE** widespread.

Spores subfusiform, smooth, 8–10 × 2.5–4.5 μm.

Comments This species is variable in color, odor, taste, and reaction of the stem flesh when exposed. The dark brown-capped form is called *Suillus grevillei* var. *clintonianus* (Peck) Kuntze, while another form showing bright green discoloration in the stipe base is referred to as *Suillus proximus* A. H. Smith & Thiers. All are reported as edible.

Suillus luteus (Linnaeus) Roussel

SLIPPERY JACK
SYNONYM *Boletus luteus* Linnaeus

Cap chocolate-brown, slimy-glutinous; stem thick, white, sheathed with membranous white veil with gelatinous purple-brown coating; veil persistent as a flaring ring; under pine

CAP 5–12 cm wide, conical-convex, becoming plane, dark brown or reddish-brown or ochre or yellowish-brown, margin sometimes with a roll of white tissue from veil, surface slimy-glutinous, shiny when dry. **FLESH** white or pale yellow. **PORES** white or pale yellow becoming dark yellow or olive-yellow, angular, 1–2 per mm, covered with membranous white veil at first. **TUBES** bluntly attached or decurrent, colored as pores, 4–15 mm deep. **STEM** 4–8 cm long, 10–25 mm broad, equal, white, but yellow over apex, discoloring brownish over base, with white sheathing membranous veil coated with gelatinous purple-brown layer over its lower half, this veil making a soft flaring then collapsing skirt on upper stem, surface with sticky pale brownish dots above annulus, flesh white. **SPORE PRINT** olive-brown. **ODOR** and **TASTE** not distinctive.

Habit and habitat scattered or clustered on the ground under pines, especially Scots pine, red pine and eastern white pine, in mixed forests, on road banks and on lawns. June to October. **RANGE** widespread.

Spores subfusiform, smooth, 7–9 × 2.5–3 μm.

Comments The slimy-glutinous cap can be rather off-putting in wet weather. This species is widely collected in many countries and eaten. However, opinions differ on how best to prepare it and on the quality of its texture and flavor. *Suillus brevipes* is similar looking, but lacks the veil.

Suillus brevipes (Peck) Kuntze

SHORT-STALKED SUILLUS

SYNONYMS *Boletus brevipes* Peck, *Suillus pseudogranulatus* (Murrill) Murrill

Cap chocolate-brown or reddish-brown, slimy-glutinous; stem short, thick, white, lacking sticky dots; under 2- and 3-needled pines

CAP 5–10 cm wide, convex becoming plane, dark brown or reddish-brown fading to cinnamon or yellowish-brown, surface slimy-glutinous, shiny when dry. **FLESH** white or pale yellow. **PORES** pale yellow becoming olive-yellow, round, 1–2 per mm. **TUBES** bluntly attached or decurrent, colored as pores, 10 mm deep. **STEM** 2–5 cm long, 10–30 mm broad, equal, white, or pale yellow over apex, surface glabrous, lacking sticky dots at first but may be poorly developed later, flesh white. **SPORE PRINT** olive-brown. **ODOR** and **TASTE** not distinctive.

Habit and habitat scattered or clustered on the ground under 2- and 3-needled pines, in mixed forests, on road banks, in sandy areas, and on lawns. June to November. **RANGE** widespread.

Spores subfusiform, smooth, 7–10 × 3–3.5 μm.

Comments Very similar looking to *Suillus luteus*, but lacking a veil. As *S. luteus*, *S. brevipes* is also widely used as a food source.

Suillus granulatus (Linnaeus) Roussel

DOTTED-STALKED SUILLUS
SYNONYMS *Boletus granulatus* Linnaeus, *Suillus lactifluus* (Withering) A. H. Smith & Thiers

Cap spotted orange-cinnamon or brown over paler ground color, slimy-glutinous; pores to 1 mm wide, irregular elongate; stem white with abundant pinkish-cinnamon sticky dots and smears; often under white pine

CAP 5–15 cm wide, convex, white becoming quickly spotted with cinnamon, orange-cinnamon, tan, or brown over pale buff ground color, appearing finely cracked, surface slippery or slimy. **FLESH** white then pale yellowish. **PORES** off-white or pinkish-buff, becoming dingy yellow, irregular elongate, 1 mm wide, with milky beaded droplets when young. **TUBES** colored as pores, 4–10 mm deep. **STEM** 4–8 cm long, 10–25 mm broad, equal, white becoming bright yellow over apex, cinnamon-brown below, surface covered with pinkish-cinnamon or reddish-brown sticky dots and smears at first, flesh white but bright yellow apically. **SPORE PRINT** cinnamon-brown. **ODOR** and **TASTE** not distinctive.

Habit and habitat single, scattered, or clustered on the ground under pines, especially eastern white pine. June to November. **RANGE** widespread.

Spores subfusiform, smooth, 7–10 × 2.5–3.5 μm.

Comments Compare with *Suillus neoalbidipes* and *S. brevipes*. Like *S. luteus* and *S. brevipes*, *S. granulatus* is widely used as a wild-collected food source.

Suillus neoalbidipes M. E. Palm & E. L. Stewart

Cap orange-yellow or pinkish-cinnamon, slimy-glutinous, with white cottony band on margin; stem white or yellow, lacking dots or these developing in age or on drying; under pine

CAP 4–10 cm wide, convex, mixtures of dark orange-yellow or yellowish-brown or pinkish-cinnamon over cream or tan ground color, darkening with age, margin with cream-white cottony band, not touching the stem, tearing and hanging in patches, surface slippery or glutinous. **FLESH** white then pale yellowish. **PORES** white becoming yellow, round, 2–3 per mm. **TUBES** colored as pores, 3–10 mm deep. **STEM** 3–6 cm long, 7–16 mm broad, equal, white becoming pale yellow over apex, reddish-brown below, surface not obviously dotted at first, but eventually with minute brown sticky dots on lower stem with age, flesh yellow or reddish in middle with age. **SPORE PRINT** cinnamon-brown. **ODOR** and **TASTE** not distinctive.

Habit and habitat single, scattered, or clustered on the ground under pine, typically 2- and 3-needled pines, but also with 5-needled pines. July to October. **RANGE** widespread.

Spores subfusiform, smooth, 6–10 × 2–3 μm.

Comments At one time known as *Suillus albidipes* (Peck) Singer, thus this name will be found in older guides. Unfortunately, the type collection tied to that name was recently determined to be *S. granulatus*. This confusion is not acceptable, thus a new name, *S. neoalbidipes*, and a new type collection were selected. Compare with *S. granulatus* and *S. brevipes*, which both lack the cottony band on the cap margin.

Suillus salmonicolor (Frost) Halling

SLIPPERY JILL

SYNONYMS *Suillus pinorigidus* Snell & E. A. Dick, *Suillus subluteus* (Peck) Snell

Cap variable from ochre-orange to olive-brown, slimy-glutinous; stem with baggy two layered veil, glutinous outside, cottony inside with roll at bottom, flesh salmon-orange

CAP 3–10 cm wide, convex, becoming plane, variable in color, ranging from dull yellow to ochre-salmon or cinnamon-brown to olive-brown, surface smooth, glabrous, slimy-glutinous. **FLESH** pale orange. **PORES** yellow, then orange-salmon becoming brownish, round or angular, 1–2 per mm, covered with a thick baggy two layered ring at first. **TUBES** bluntly attached or short decurrent, colored as pores, 8–10 mm deep. **STEM** 3–10 cm long, 6–16 mm broad, equal, dull white, yellow or pinkish-orange, surface with reddish or dark brown sticky dots, flesh solid, dull salmon-orange. **RING** of two layers, outer layer slimy-glutinous, pale orange, inner layer white, cottony producing a distinctive roll on the bottom of the veil, collapsing on stem and turning black with age. **SPORE PRINT** dark olive-brown. **ODOR** and **TASTE** not distinctive.

Habit and habitat scattered or clustered on soil, often sandy soil, or in moss beds under pines such as pitch pine and jack pine. August to November. **RANGE** widespread.

Spores subfusiform, smooth, 6–11 × 2.5–4 µm.

Comments The species seems to prefer cooler weather and lots of precipitation before it fruits. In older guides it will be found as *Suillus subluteus*. *Suillus acidus* (Peck) Singer and *S. intermedius* (A. H. Smith & Thiers) A. H. Smith & Thiers, both occurring under red and eastern white pine, are somewhat similar, but the stems are long, slender, and covered with blackening sticky dots, the caps are white (*S. acidus*) or yellow (*S. intermedius*), and the rings are thin and collarlike.

Suillus serotinus (Frost) Kretzer & T. D. Bruns

SYNONYMS *Boletus serotinus* Frost, *Fuscoboletinus serotinus* (Frost) A. H. Smith & Thiers

Cap slimy chocolate or date brown; pores ashy-gray, small, tubes short decurrent; stem pale grayish with soft, cottony-membranous annulus; under larch

CAP 4–12 cm wide, convex, becoming plane, chocolate or date brown slime layer over white ground color, slime thick at first, drying down into flattened brown scales, smooth, with small white patches of ring hanging from margin. **FLESH** white or pale yellow, slowly blue then purple-gray then reddish-brown. **PORES** dull white then pale ash-gray, becoming pale reddish-brown, turning cinnamon-brown when injured, angular, 1–3 per mm. **TUBES** same color as pores, short decurrent, 8–15 mm deep. **STEM** 5–10 cm long, 7–16 mm broad, equal, off-white and faintly netted above ring zone, surface pale grayish and matted hairy below ring, occasionally with pinkish or yellowish-brown streaking below ring, white mycelium over base, flesh pallid with yellow in base, slowly entirely dull bluish. **RING** pale grayish, cottony-membranous. **SPORE PRINT** purplish-brown. **ODOR** pungent or not distinctive. **TASTE** not distinctive.

Habit and habitat scattered or clustered on the ground, often covered with *Sphagnum*, under larch. August to October. **RANGE** widespread.

Spores subfusiform or subelliptical, smooth, 8–12 × 4–5 μm.

Comments This bolete only occurs with larch, usually in or around bogs and is one of those not commonly seen unless you frequent boggy areas at the right time of year. Edible.

Suillus paluster (Peck) Kretzer & T. D. Bruns

SYNONYMS *Boletus paluster* Peck, *Fuscoboletinus paluster* (Peck) Pomerleau & A. H. Smith

Cap small and bright purplish-red, densely woolly scaly matted; tubes decurrent, almost gill-like; pores sordid yellow, angular and compound; stem colored as cap; under conifers in cold northern bogs

CAP 2–7 cm wide, conical-convex, becoming plane, often with a central bump, red or purplish-red at first, becoming pinkish-purple, surface woolly matted or flattened scaly, with small patches hanging from margin, dry. **FLESH** yellow, or red under cap skin. **PORES** yellow, becoming ochre-yellow, angular, elongate, compound, 1–2 mm long. **TUBES** same color as pores, decurrent, radially elongate almost gill-like, 2–3 mm deep. **STEM** 2–5 cm long, 4–6 mm broad, equal, colored as cap except for yellow base covered with thick white mycelium, surface with minute scales and fibrils, flesh yellow. **SPORE PRINT** purple-brown drying to pinkish-brown. **ODOR** not distinctive. **TASTE** not distinctive, or slightly acidic.

Habit and habitat solitary, scattered or clustered on the ground or well-rotted wood covered with *Sphagnum* under conifers such as larch, balsam fir, arborvitae in cold northern bogs. August to November. **RANGE** widespread.

Spores subfusiform, smooth, 7–9 × 3–3.5 μm.

Comments A rarely seen little jewel unless you visit bogs. This small bolete has a veil covering the tubes at first, a feature not common in boletes except for the genus *Suillus*. It was previously placed in *Fuscoboletinus* because of the purple-brown spore deposit, but phylogenetic analysis using DNA molecules indicates it belongs in *Suillus*. Compare with *S. spraguei*.

Phylloporus rhodoxanthus
(Schweinitz) Bresadola

GILLED BOLETE

SYNONYM *Agaricus rhodoxanthus*
Schweinitz

Cap variable but reddish or olive-brown
typically; gills decurrent, golden-yellow,
easily removed from cap; stem colored as
cap, but with yellow mycelium over base

CAP 3–10 cm wide, convex, becoming
plane or broadly depressed, dark red or
reddish-yellow or reddish-brown or even
olive-brown, surface somewhat velvety,
often cracked in age, dry. **FLESH** white
or pale yellow. **GILLS** bright yellow then
golden-yellow and eventually ochre-yellow,
decurrent, sometimes forking or with
cross veins, well-spaced and thick, easily
separating from cap. **STEM** 3–9 cm long,
6–15 mm broad, equal or tapered downward, colored as cap or ground color yellowish below red or reddish-brown dots
and fibrils, yellow mycelium over base,
often with raised longitudinal ridges
over upper half, flesh yellow. **SPORE PRINT**
yellow or ochre-yellow. **ODOR** and **TASTE**
not distinctive.

Habit and habitat single, scattered, or clustered on the ground, under mixed hardwoods of beech and oak. June to October.
RANGE widespread.

Spores elliptical or subfusiform, smooth,
8–14 × 3–5 µm.

Comments Variable in cap colors, but the
yellow mycelium at the base of the stem
and the golden-yellow decurrent gills are
distinctive. The cap surface also turns
blue with a drop of household ammonia.
Several different variations have been
described; here are two. *Phylloporus leuco-
mycelinus* Singer is similar except the basal
mycelium is white. *Phylloporus boletinoides*
A. H. Smith & Thiers differs by the poroid,
olive-colored "gills," the erratic bluing
of the gills and flesh when injured, and
the brown color reaction to ammonia on
the cap. The separable gills, spore color,
and shape indicate the relationship to
the boletes.

Polypores

The polypores are a very large group of mainly forest macrofungi characterized by producing poroid or gill-like reproductive surfaces that do not separate easily from the cap (if there is one) and by having mainly white spores (exceptions being *Ganoderma* with brown spores and *Boletopsis* with yellow-brown spores). The polypores are not all closely related to each other, they simply share a similar method of spore production, which is a useful characteristic for identification purposes.

Certainly not all, but most of these fungi have rather tough flesh and do not decay as readily as other macrofungi. If gill-like structures are present, the fruit body is very tough. The fleshier polypores produce poroid reproductive surfaces, called tubes, and the spores are white.

The boletes, also known as the fleshy pore fungi, also have poroid reproductive surfaces, but their tubes easily peel away from the cap and the spores are more highly pigmented, ranging from yellow through fleshy-pink, olive-brown, or black. In general, the spores of boletes are also larger than spores produced by polypores. These two groups are not closely related either.

Polypores are generally wood recyclers, saprobes living on dead or decaying organic matter, but some are parasites of woody plants. A few are mycorrhizal, such as *Albatrellus*, *Neoalbatrellus*, and *Boletopsis*. Some are edible and highly sought after, such as *Laetiporus*, the sulfur shelf or chicken of the woods, and *Grifola*, the maitake. Others have been used not as food, but in various other interesting ways by humans; see comments under *Piptoporus*, *Fomes*, *Ganoderma*, and others.

Gill-like or labyrinthine reproductive surface

Cerrena Thin, densely fuzzy and zoned on white or pale grayish cap, caps often green at attachment from algal growth, pores jagged and mazelike, looks like a species of *Trametes* from the top

Daedalea Cap white, faintly fuzzy or glabrous, pores mostly gill-like, very thick, flesh brown, tough

Daedaleopsis Cap brownish and zoned, pores with thin walls, staining salmon-pink when bruised, fleshy brown, tough

Lenzites Cap brownish, zoned, pores mostly gill-like, well-spaced, flesh white

(Note: see also *Trametes elegans*, *Phaeolus schweinitzii*, and *Abortiporus*)

Toothlike reproductive surface (but pores in early stages)

Irpex White overall, no stem, cap barely projecting, growing on hardwood branches

Trichaptum With purple or lilac colors long remaining on "teeth," densely shelving

Tubes crowded together but free, not fused

Fistulina Cap, and stem when present, blood-red or reddish-orange, flesh producing red juice when cut

Stem and cap present

Albatrellus, *Neoalbatrellus*, and *Boletopsis* Fleshy, boletelike, tubes short, decurrent, not peeling, spores white or pale yellowish-brown

Coltricia Small, tough-fibrous, dark brown overall, cap zoned, on soil

Bondarzewia Large to gigantic, one stem with several caps shelving, at base of trees

Abortiporus Medium to large, fleshy but duplex, on wood or on the ground on buried wood, tubes decurrent, pores often angular or mazelike

Phaeolus Large, dark reddish-brown with bright yellow margin when young, densely woolly on the cap, pores bright greenish-yellow, bruising brown, angular elongate, some labyrinthine in form, on ground at base of conifers mostly, but also on standing living pine trees

Polyporus Small or very large, stem central or eccentric or lateral, on wood, sometimes on the ground but attached to buried decaying wood

Royoporus Medium to large, cap dark reddish-brown smooth, stem black fuzzy, on larger decaying hardwood logs

Grifola Large, multiple grayish caps from fused branching stems, on the ground

Meripilus Large, multiple, fan-shaped, zoned caps, caps and pores turning black when bruised

Cap present, stem absent or stublike at cap margin

Fomes Very hard, hooflike, zoned brown and gray, or brown, gray, and reddish-brown, pores gray or brown, on decaying hardwood trees and logs

Fomitopsis Mostly fan-shaped, hard, pores cream color, on conifers

Ganoderma Small to very large, on hardwoods or conifers, fan-shaped, cap dull brown or shiny varnished reddish-brown, pores white, turning dark brown permanently where marked

Ischnoderma Thick, fleshy, with rounded margin, cap brown velvety, then crusty, pores white, staining brown, on downed decaying logs

Laetiporus Fleshy, bright orange or salmon and lemon-yellow colors on cap, bright lemon-yellow pores or white in one species, on decaying trees or on the ground at the base of living trees

Phaeolus Can be found on standing trees, lacking a stem, see description above

Phellinus Fan- or hoof-shaped, often reddish-brown or almost black and cracked on the cap, pores brown, reddish-brown, or purplish-brown, on hardwoods or conifers

Phlebia Mostly resupinate but with projecting cap, undersurface pitted and wrinkled, lacking distinct pores, peach or orange colored

Piptoporus On dead or dying birch trees, cap with stublike lateral stem and thick rounded margin, soft, leatherlike

Pycnoporus Bright red or orange-red overall, thin but tough, fan-shaped

Trametes Thin but leathery-tough, mostly densely shelving with white minute pores, but also single with stublike stem and pores mazelike

Tyromyces and *Postia* Small to medium-sized, flesh thick, soft, watery, caps and pores white or bluish, one species staining brown when handled

Cap and stem absent, resupinate

Phellinus Pores brown, reddish-brown, or purplish-brown

Phlebia Completely resupinate, with wrinkled, ridged, bumpy peachy-orange or grayish-red colors, on the underside of decaying logs

Schizopora On the top or sides of decaying hardwood logs, white or cream-colored, bloblike, raised, rounded, and covered with round, mazelike pores, or even toothed pores

Albatrellus ovinus (Schaeffer) Kotlaba & Pouzar

SHEEP POLYPORE
SYNONYMS *Boletus ovinus* Schaeffer, *Polyporus ovinus* (Schaeffer) Fries

Cap white becoming tan, cracked with age; pores white, small, circular; tubes decurrent, 1–4 mm deep; stem white or colored as cap; under conifers

CAP 4–20 cm wide, convex becoming plane or shallowly sunken, white becoming pale tan or cream color or crust-brown, smooth, becoming cracked with age, glabrous, dry. **FLESH** white, drying yellow-olive. **PORES** white, becoming yellowish or developing greenish hues, 2–4 per mm, mostly circular. **TUBES** white, decurrent, 1–4 mm deep. **STEM** 3–10 cm long, 10–40 mm broad, central or sometimes off-center, equal or enlarged downward, white or becoming tan, glabrous, dry. **SPORE PRINT** white. **ODOR** and **TASTE** not distinctive.

Habit and habitat single or scattered or gregarious on humus and moist soil under conifers. July to October. **RANGE** widespread.

Spores subglobose or ovoid, smooth, colorless, not amyloid, 3.5–5 × 2.5–3.5 µm; clamp connections absent.

Comments If the cap center is pinkish-buff, pinkish-tan, or pale orange, the stem is off-center, the taste is bitter, and the surface tissues stain purple in KOH (from Michael Kuo, MushroomExpert.com), then you have the less commonly collected *Albatrellus confluens* (Albertini & Schweinitz) Kotlaba & Pouzar. That species also has clamp connections and weakly amyloid spores. Also compare with *Boletopsis grisea* that produces a colored spore print.

Abortiporus biennis (Bulliard) Singer

SYNONYMS *Boletus biennis* Bulliard, *Daedalea biennis* (Fries) Fries, *Heteroporus distortus* (Schweinitz) Bondartsev & Singer Cap large, white, velvety, obscurely zoned; pores white or pale pink, turning reddish-brown when injured, angular and mazelike mixed forms together; on decaying wood, sometimes this wood is buried

CAP 4–15 cm wide, plane or sunken, circular or semicircular or kidney-shaped, white becoming pale tan or pale brown, sometimes obscurely zoned, velvety or flattened fibrillose, becoming glabrous, dry. **FLESH** white or with pale pink tints, duplex with a soft layer over a corky lower layer. **PORES** white or pale pink, discoloring reddish-brown when injured, 3–6 per mm, angular or mazelike. **TUBES** white, decurrent, 6 mm deep. **STEM** 1–5 cm long, 10–15 mm broad when present, central off-center or at the cap edge (lateral), white or colored as cap, velvety, dry. **SPORE PRINT** white. **ODOR** and **TASTE** not distinctive.

Habit and habitat single or scattered or gregarious on woody substrates in hardwood or coniferous forests, also on lawns from decaying roots of stumps. July to October. **RANGE** widespread.

Spores elliptical or ovoid, smooth, colorless, not amyloid, 4–6.5 × 3.5–5 μm; thick-walled chlamydospores, 5–9 μm in diameter, present in flesh; also gloeocystidia present in hymenium.

Comments The fruit bodies may produce a watery red juice when cut or squeezed. Also, the fruit bodies can develop oddly when the cap margin turns strongly upward, covering the surface of the cap with the angular pores.

Boletopsis grisea (Peck)
Bondartsev

SYNONYM *Polyporus griseus* Peck
Cap pale grayish with strongly inrolled margin; pores pale grayish, small; tubes subdecurrent, very short; stem colored as cap, stocky and short

CAP 4–12 cm wide, convex with strongly inrolled margin, becoming plane, white or dingy gray from the center outward, smooth then cracked with age, dry. **FLESH** white, thick. **PORES** white or pale grayish, 2–4 pores per mm, mostly circular. **TUBES** same color, short decurrent, 1–5 mm deep. **STEM** 4–10(–30) cm long, 30–50 mm broad, typically short and stocky, central or sometimes off-center, equal or enlarged downward, same color as cap or darker, dry. **SPORE PRINT** pale yellow-brown. **ODOR** not distinctive, or fragrant. **TASTE** not distinctive.

Habit and habitat single, scattered, or clustered on soil and humus under conifers and hardwoods such as oak. August to October. **RANGE** widespread.

Spores subglobose, nodulose-bumpy, almost colorless, not amyloid, 5–7 × 4–5 µm.

Comments Boletes have tubes that can be separated from the caps, the spore deposits are more darkly colored, and the spores lack bumps. *Boletopsis leucomelaena* (Persoon) Fayod has a dark gray or black cap, the pores bruise pinkish then turn brown, and the flesh is white, turning pinkish-gray to lilac when exposed. However, what is now called *B. leucomelaena* in North America may not be the same species found in Europe, according to recent molecular investigations. Our dark-capped species may need a new name. *Boletopsis subsquamosa* (Linnaeus) Kotlaba & Pouzar, a name used previously in error for *B. grisea*, seems to have varied interpretations, one being that it is the same as *Albatrellus ovinus*. What has been called *B. subsquamosa* in North America is actually *B. grisea*.

Neoalbatrellus caeruleoporus (Peck) Audet

BLUE-PORED POLYPORE

SYNONYMS *Albatrellus caeruleoporus* (Peck) Pouzar, *Polyporus caeruleoporus* Peck

Cap blue or bluish-gray; pores blue, angular; tubes decurrent; stem blue as cap; under hemlock

CAP 2–15 cm wide, convex becoming plane, dark blue or bluish-gray, may develop brown discolorations, smooth or bumpy, glabrous or somewhat hairy, dry. **FLESH** white. **PORES** blue, 1–5 per mm, angular. **TUBES** blue, decurrent, 2–5 mm deep. **STEM** 2.5–7.5 cm long, 5–25 mm broad, equal with tapered base, colored as cap or paler, white mycelioid at base, glabrous, dry. **SPORE PRINT** white. **ODOR** and **TASTE** not distinctive.

Habit and habitat single, scattered or sometimes individual caps fused together on humus and moist soil under hemlock and mixed hardwoods. September to October. **RANGE** widespread.

Spores subglobose or ovoid, smooth, colorless, not amyloid, 4–6 × 3–5 μm.

Comments The tubes do not peel off of the cap easily or cleanly, thus one should not confuse this species with a bolete. Compare *Boletopsis* species, which can be somewhat similar, but they are not blue and the spores are nodulose bumpy. *Albatrellus ovinus* looks similar in shape, but is whitish or pale tan to pinkish-buff.

Bondarzewia berkeleyi (Fries) Singer

BERKELEY'S POLYPORE

SYNONYM *Polyporus berkeleyi* Fries

Very large, funnel-shaped from a single stem, and composed of one or more tan or orange-brown zonate caps; pores angular and exuding a white milky sap (latex) when cut; tubes decurrent

CAP 25–90 cm wide, but may also be composed of 1–5 separate caps with individual caps 5–30 cm wide, funnel-form, individual caps may be fan-shaped, typically described as white or cream or pale tan, but can be orange-brown as well, concentric ringed, also often radially wrinkled or streaked, not scaly but somewhat hairy, dry. **FLESH** white, thick. **PORES** white, up to 2 mm across, mostly angular or somewhat labyrinthine, exuding, when very fresh, a thin white milky sap (latex) when cut. **TUBES** white, long decurrent, 10 mm deep.

STEM 4–10(–30) cm long, 30–50 mm broad, typically short and stocky, central or sometimes off-center, equal or tapered downward, same color as cap or darker, dry, very tough. **SPORE PRINT** white. **ODOR** and **TASTE** not distinctive.

Habit and habitat single on soil at the base of living hardwoods, causing a parasitic rot as it grows into the base of the tree. July to October. **RANGE** widespread.

Spores globose or subglobose, with strongly amyloid ridges and spines in Melzer's reagent, colorless in KOH, 6–9 μm.

Comments When this fungus first appears, it looks like a pale hand with many fingers coming out of the soil. The fruit bodies can take many weeks to mature and last for many days once they are producing spores, more than one month as I have personally observed. Compare with *Grifola frondosa*, *Meripilus giganteus*, and *Laetiporus cincinnatus*.

Polyporus squamosus
(Hudson) Fries

DRYAD'S SADDLE
SYNONYM *Boletus squamosus* Hudson
Large, flat, tan caps with reddish-brown, flattened scales; pores white, angular; tubes white decurrent; smell strong when fresh; in clusters on standing hardwood trees

CAP 6–30 cm wide, single or overlapping, broadly convex, becoming plane with the center sunken, from the top round at first, soon semicircular or fan-shaped or kidney-shaped, mostly tan or creamy-tan with large flat reddish-brown scales overall, fading with age, dry. **FLESH** white, 1–4 cm thick, soft when young, then tough. **PORES** white or creamy-white becoming pale yellow, 1–2 per mm, elongate-angular, edges fraying with age. **TUBES** white, 10 mm deep, decurrent. **STEM** 1–5 cm long, 10–40 mm wide, off-center, eventually lateral, equal, black, minutely hairy or smooth where not covered by tubes. **SPORE PRINT** white. **ODOR** and **TASTE** of watermelon rind or strongly mealy.

Habit and habitat on living or dead, standing or fallen hardwood trees, especially elm, but also maple, willow, poplar, and birch. May to November, but most visible in May and June. **RANGE** widespread.

Spores broadly cylindrical, thin-walled, smooth, colorless, not amyloid, 16–20 × 6–9 μm in the monograph on North American polypores by Gilbertson and Ryvarden, but other works describe it in the range 10–16 × 4–6 μm.

Comments When this polypore fruits, the morels are up or soon will be. The soft tender edges of the caps can be cooked or pickled and eaten. The difference in spore size range suggests there is more to learn about this "common" species of polypore. Maybe we have some cryptic populations that have not yet been studied.

Grifola frondosa (Dickson) Gray
HEN OF THE WOODS, MAITAKE

SYNONYMS *Boletus frondosus* Dickson, *Polyporus frondosus* (Dickson) Fries

Multiple grayish caps on a many branched cluster; pores and tubes white, decurrent; stem stublike; on the ground at the base of living hardwoods, especially oak

Fruit body 40–60 cm wide, with many laterally attached caps on highly branched thickened stem. **CAP** 2–7 cm wide, plane and from the top spathulate, or fan-shaped, typically overlapping, gray or grayish-brown, finely velvety or glabrous. **FLESH** white, firm. **PORES** white or pale cream at first, 2–4 per mm, angular, developing ragged edges. **TUBES** white, long decurrent down stem, often to near the substrate, 5 mm deep. **STEM** stublike, up to 100 mm wide at the base, multibranched, attached laterally to the caps, white.

SPORE PRINT white. **ODOR** pleasant, nutty. **TASTE** mild.

Habit and habitat usually a single massive cluster at the base of living hardwood trees, especially oak. August to November. **RANGE** widespread.

Spores ellipsoid, smooth, colorless, not amyloid, 5–7 × 3–5 µm.

Comments The maitake is considered a mild parasite of hardwood trees. *Polyporus umbellatus* (Persoon) Fries looks somewhat similar but has round caps on well-defined central stalks and the spores are cylindrical and longer (7–10 × 2.5–3.5 µm). *Meripilus sumstinei* has larger, zoned caps that, along with the white pores, stain black when bruised. The maitake is considered a medicinal "mushroom" and a good edible. It is worth reading about on the Internet.

Meripilus sumstinei (Murrill)
M. J. Larsen & Lombard

BLACK-STAINING POLYPORE
SYNONYMS *Grifola sumstinei* Murrill, *Polyporus sumstinei* (Murrill) Saccardo & D. Saccardo

Large, multiple, fan-shaped, zoned white and gray or yellowish-brown finely velvety caps that slowly turn black from injury; pores white turning black, tiny; at bases of hardwood stumps or dead trees

Fruit body 10–40 cm wide, bouquetlike with many laterally attached caps on highly branched thickened stem.**CAP** 5–20 cm wide, spatula-shaped or fan-shaped with thin tapered margins, overlapping, white at first, becoming gray-brown or yellowish-tan, slowly staining black where bruised, finely velvety or radially hairy, zoned. **FLESH** white, firm-tough, 1.5 cm thick. **PORES** white or creamy-white aging to tan, turning black from injury, 3–5 per mm, angular. **TUBES** white, bruising black, decurrent, 3–8 mm deep. **STEM** 2 cm thick, tapered to base, stout, attached laterally to the caps, fibrous. **SPORE PRINT** white. **ODOR** and **TASTE** mild.

Habit and habitat single or clustered on the ground at the base of hardwood stumps. August to November. **RANGE** widespread.

Spores ellipsoid, smooth, colorless, not amyloid, 6–7 × 4.5–6 μm.

Comments Like *Grifola frondosa*, this species fruits in large clusters on the ground at the base of hardwood trees. *Meripilus sumstinei* is edible when young and tender, but *Grifola frondosa* is much better tasting. In older field guides and technical literature, *M. sumstinei* was referred to as *M. giganteus* (Persoon) P. Karsten, a species we now know only occurs in Europe. Murrill recognized our North American species was different in the early 1900s.

Laetiporus sulphureus
(Bulliard) Murrill

CHICKEN MUSHROOM, CHICKEN OF THE WOODS, SULFUR SHELF

SYNONYMS *Boletus sulphureus* Bulliard, *Polyporus sulphureus* (Bulliard) Fries

Bright sulfur-yellow pores and cap margins, cap surface zoned bright salmon-orange-yellow; large and fleshy, with many overlapping, fan-shaped caps; on dead or living hardwoods

CAP 5–30 cm wide, to 20 cm deep, plane or broadly sunken, semicircular or fan-shaped with thick bright sulfur-yellow edges, overlapping, salmon or bright orange with bright yellow concentric zones, turning white with age, smooth or finely wrinkled or undulate, suedelike to touch. **FLESH** yellow, thick, soft, white and punkie with age. **PORES** bright sulfur-yellow, 2–4 per mm, round or angular. **TUBES** sulfur-yellow, 5 mm deep. **STEM** absent or rudimentary. **SPORE PRINT** white. **ODOR** and **TASTE** not distinctive.

Habit and habitat on living and dead wood of standing or downed hardwood trees. May to November. **RANGE** widespread.

Spores ovoid or ellipsoid, smooth, colorless, not amyloid, 5–7 × 3.5–5 µm.

Comments The color of the pores gives it one common name, the taste of it cooked suggests the other common names, although I think it has a lemony taste and only the texture is chickenlike. Most people like this wild polypore. If you drink alcohol, do not eat this fungus if you found it on black locust or a conifer—it is known to cause vomiting. Collected from cherry, maple, oak, or beech, there seems to be no problem. If you do not drink alcohol there also is no problem. If the pores are white and it is growing on soil at the base of trees, it is *Laetiporus cincinnatus* (Morgan) Burdsall, Banik & T. J. Volk.

Fistulina hepatica (Schaeffer) Withering

BEEFSTEAK POLYPORE
SYNONYM *Boletus hepaticus* Schaeffer

Large, fleshy, blood-red or reddish-orange cap; flesh thick, meaty, exuding red juice; tubes produced individually, free from each other, yellowish; stem absent or present, colored as cap; on dead oaks or on the ground from the trunk base

CAP 10–30 cm wide, spatula-shaped or semicircular, occasionally multiple caps from a single lateral stem, blood-red or reddish-orange, roughened with velvety knobs, or becoming smooth, moist or tacky, but eventually dry and gelatinous feeling. **FLESH** pale pink, streaked reddish, exuding a red juice, fleshy, 20–60 mm thick. **PORES** white or pale yellow, becoming reddish-brown slowly or staining dark brown when bruised, 1 mm wide, round. **TUBES** separate and free from each other,

same color as pores, 10–15 mm long. **STEM** when present, 5–10 cm long, 10–30 mm broad, lateral, equal with tapered base when in the ground, same color as cap or darker, dry. **SPORE PRINT** pale salmon. **ODOR** not distinctive but pleasant. **TASTE** sour-acidic.

Habit and habitat single or sometimes clustered on dead oak stumps and trunks of standing trees, or on the ground with a stem, growing at the base of living oaks. July to October. **RANGE** widespread.

Spores ellipsoid or ovoid, smooth, almost colorless, not amyloid, 3.5–4.5 × 2.5–3 μm.

Comments Listed in many books as a "good" edible, but it needs to be prepared properly. Clean, slice into ready-to-cook pieces, and soak these in milk for several hours before frying or grilling. You will eliminate the strongly acidic flavor and enjoy the taste much more.

Phaeolus schweinitzii (Fries)
Patouillard

DYE POLYPORE
SYNONYM *Polyporus schweinitzii* Fries

Large, bouquetlike, woolly, concentrically orange-brown or rust-brown with bright yellow margin; pores greenish-yellow, bruising brown, angular to labyrinthine

CAP 5–35 cm wide, single or composed of several overlapping branches from the stem and bouquetlike, fan-shaped or semicircular, bright yellow or orange-yellow overall at first with thickened margin, becoming orange-brown or rust-brown in concentric color bands and grooves, the margin bright yellow, turning buff with age, woolly or hairy at first, glabrous with age. **FLESH** yellow-brown becoming rusty-brown, to 30 mm thick, fleshy soft at first, becoming leathery. **PORES** bright greenish-yellow, bruising brownish, with age brown or reddish-brown, 0.5–3 mm wide, angular, elongate, becoming labyrinthine.

TUBES brown, 1–15 mm deep, decurrent. **STEM** when present, one or more fused together, 1–6 cm long, 10–40 mm thick, central of off-center, brown or reddish-brown. **SPORE PRINT** white or cream. **ODOR** fragrant or as **TASTE** not distinctive.

Habit and habitat on the ground under conifers and near their trunk, or growing as shelving caps directly from the trunk base, mostly on decaying white pine. June to November. **RANGE** widespread.

Spores ellipsoid, smooth, colorless, not amyloid, 5–8 × 3.5–4.5 μm.

Comments This serious pathogen of old-growth conifers causes a rot at the base of the tree, starting where the roots meet the stem; thus, it is called a butt rot. The fruit bodies will continue to be produced from dead woody material as well. Occasionally *Phaeolus schweinitzii* can be found on the trunks of living pines well above the ground.

Coltricia cinnamomea
(Jacquin) Murrill

SYNONYM *Boletus cinnamomeus* Jacquin
Small, shiny, concentrically zoned, with shallowly funnel-shaped caps; pores cinnamon-brown, angular; stem, dark cinnamon-brown, hairy; on soil or in mosses

CAP 1–5 cm wide, circular and shallowly funnel-shaped, bright cinnamon-brown or orange-brown or darker and concentrically zoned, shiny and radially silky, dry. **FLESH** brown, thin. **PORES** yellow-brown becoming cinnamon-brown, 2–3 per mm, mostly angular. **TUBES** same color as pores, 2–3 mm long. **STEM** 1–4 cm long, 1–3 mm broad, equal, same color as cap or darker, dry, velvety, very tough. **SPORE PRINT** pale brown. **ODOR** and **TASTE** not distinctive.

Habit and habitat single, scattered, or clustered on soil or mosses along trails or unimproved road banks. June to November. **RANGE** widespread.

Spores ellipsoid, smooth, almost colorless, not amyloid, 6–8 × 4.5–6 µm.

Comments The flesh turns black in KOH solutions. The fungus can be used in dry flower arrangements or on toy train settings. They probably would make interesting lapel pins as well. The larger *Coltricia perennis* (Linnaeus) Murrill is paler in color with a nonshiny, velvety cap surface.

Polyporus arcularius (Batsch) Fries

SPRING POLYPORE

SYNONYMS *Boletus arcularius* Batsch, *Favolus arcularius* (Batsch) Fries, *Lentinus arcularius* (Batsch) Zmitrovich

Early spring, small, tough, golden-brown or dark brown fibrillose-scaly caps with projecting golden eyelash hairs around the margin; pores hexagonal; on wood of hardwoods

CAP 1–4 cm wide, convex, becoming plane with the center sharply sunken, circular, golden-brown or dark brown, fibrillose-scaly overall, margin fringed with eyelashlike golden hairs, dry. **FLESH** white, thin, tough. **PORES** white or creamy-white, hexagonal or honeycomb-shaped and radially elongate, 0.5–2 mm long. **TUBES** colored as pores, 1–2 mm deep, bluntly attached or decurrent. **STEM** 2–4 cm long, 2–4 mm wide, central, equal, same color as cap, minutely hairy-scaly, tough. **SPORE PRINT** white. **ODOR** and **TASTE** not distinctive.

Habit and habitat single, scattered, or clustered on decaying wood of hardwoods, sometimes on the ground from buried wood. May to June. **RANGE** widespread.

Spores cylindrical, thin-walled, smooth, colorless, not amyloid, 7–9 × 2.5–3 μm.

Comments Compare with *Neofavolus alveolaris*, another early spring polypore fruiting on wood.

Neofavolus alveolaris
(de Candolle) Sotome & T. Hattori

HEXAGONAL-PORED POLYPORE
SYNONYMS *Merulius alveolaris* de Candolle, *Favolus mori* (Pollini) Fries, *Polyporus alveolaris* (de Candolle) Bondartsev & Singer, *Hexagonia alveolaris* (de Candolle) Murrill

Small to medium-sized, bright orange and fan-shaped on downed hardwood branches, sticks and logs; pores honeycomb-shaped, pale cream, darkening with age; fading to almost white after several months

CAP 1–10 cm wide, broadly convex becoming plane and sunken near attachment, from above kidney-bean-shaped or fan-shaped, bright orange-brown, with darker orange-brown scales over paler yellow-orange ground color, fading over months and eventually white, scaly at first, becoming smooth, dry. **FLESH** white, thin. **PORES** white or cream color or darker with age, radially angular elongate and arranged in rows, up to 3 mm long and 2 mm wide, honeycomb-like. **TUBES** same color as pores, decurrent. **STEM** lateral, stubby, white, tough. **SPORE PRINT** white. **ODOR** and **TASTE** not distinctive.

Habit and habitat single, scattered, or clustered on sticks, branches and logs of hardwoods. June to November. **RANGE** widespread.

Spores cylindrical, smooth, colorless, not amyloid, 9–11 × 3–4 µm.

Comments Also in the literature and field guides as *Polyporus mori* (Pollini) Fries. This species fruits in late spring, at the same time as *P. squamosus*. Both species are good indicators the morels should be fruiting.

Royoporus badius (Persoon)
A. B. De

BLACK-FOOTED POLYPORE
SYNONYMS *Boletus badius* Persoon, *Polyporus badius* (Persoon) Schweinitz

Large, tough, reddish-brown caps or the center darker brown with paler margins; pores tiny, white, difficult to see; stem black and fuzzy over lower portions; on decaying hardwood logs

CAP 5–15(–20) cm wide, convex, becoming plane with shallow sunken center, from the top circular or kidney-shaped or lobed, dark reddish-brown, progressively paler brown or nearly white with center remaining red-brown, eventually turning black from the center outward, glabrous, dry. **FLESH** white, tough. **PORES** white or creamy-white becoming brown when dry, 5–8 per mm, circular or angular. **TUBES** white, 1–2 mm deep, decurrent. **STEM** 1–6 cm long, 5–20 mm wide, central or eccentric, equal, black, fuzzy-pubescent overall, becoming glabrous over upper portions with age. **SPORE PRINT** white. **ODOR** and **TASTE** not distinctive.

Habit and habitat on decaying wood of well-decayed and often bark less hardwoods, also on the ground from buried wood. August to October. **RANGE** widespread.

Spores cylindrical, thin-walled, smooth, colorless, not amyloid, 7.5–9 × 3.5–5 μm.

Comments *Polyporus varius* (Persoon) Fries is smaller with pale brown or tan-colored caps, and it tends to prefer small branches of hardwoods, not the larger well-decayed logs.

Trametes elegans (Sprengel)
Fries

SYNONYMS *Daedalea elegans* Sprengel, *Daedaleopsis elegans* (Sprengel) Domański, *Lenzites elegans* (Sprengel) Patouillard

Large, white, zoned, thick-fleshed, fan-shaped caps; pores a mixture of circular, angular, elongate, and mazelike in appearance; stem absent or stublike; on hardwoods

CAP 2–30 cm wide and deep, single and attached directly to the substrate (sessile) or with a short stemlike base, mostly flat and semicircular, fan-shaped, or kidney-shaped or spatula-shaped, white, pale gray, or pale ochre-buff, sometimes with green algae spreading from the attachment outward along the zoned ridges, surface minutely hairy, becoming glabrous, dry. FLESH white or pale cream, tough, to 15 mm deep. PORES white or pale cream, variable with round or angular pores 1–2 per mm, also short mazelike (daedaloid) or even with thick gills, or with combinations of these all on one fruit body. TUBES white or cream, 6 mm long. STEM absent or short stublike. SPORE PRINT white. ODOR and TASTE not distinctive.

Habit and habitat on decaying wood of hardwoods, stumps, logs, and branches. June to November. RANGE widespread.

Spores cylindrical or elongate and narrowly ellipsoid, smooth, colorless, not amyloid, 5–7 × 2–3 µm.

Comments The fruit bodies can be annual or, in some cases lasting more than one year, perennial, and thus green algae may be growing on the cap surface as shown here. The stublike attachment, besides the variable pore-gill construction, is characteristic for the species when it is present.

Trametes versicolor (Linnaeus) Lloyd

TURKEY-TAIL

SYNONYMS *Boletus versicolor* Linnaeus, *Coriolus versicolor* (Linnaeus) Quélet, *Polyporus versicolor* (Linnaeus) Fries

Shelving or rosette clusters of multicolored, contrasting concentric zones and concentric rings of velvet on the caps; pores white, round and tiny

CAP 2.5–10 cm wide, single or shelving or in rosettes, plane or undulating, from the top circular, semicircular, fan-shaped, or kidney-shaped, multicolored and zoned, white, brown, cinnamon, reddish-brown, orange, blue, green, highly variable, surface densely velvety-hairy, with alternating zones of texture or ridges, dry. **FLESH** thin, tough but flexible when fresh, to 5 mm, with thin black layer below surface skin. **PORES** white or becoming pale gray, 4–5 per mm, round. **TUBES** white, 1–3 mm long. **STEM** absent. **SPORE PRINT** white. **ODOR** and **TASTE** not distinctive.

Habit and habitat on decaying wood of hardwoods, stumps, logs, branches. June to November. **RANGE** widespread.

Spores cylindrical, slightly curved, smooth, colorless, not amyloid, 5–6 × 1.5–2 µm.

Comments This species is the most commonly encountered of the several species of *Trametes*. If the caps are not multicolored in zones, but different shades of the same color and velvety with a pale margin, you probably have *T. pubescens* (Schumacher) Pilát. If the caps are hairy or fuzzy and mostly white or grayish with a yellowish-brown margin, it is *T. hirsuta* (Wulfen) Lloyd. *Trametes ochracea* (Persoon) Gilbertson & Ryvarden has white and brown shades, paler than *T. versicolor* and not as distinctly zoned, lacking the thin black layer below the cap skin, and the caps are firm-tough and not flexible. The nearly weightless colorful fruit bodies make attractive earrings.

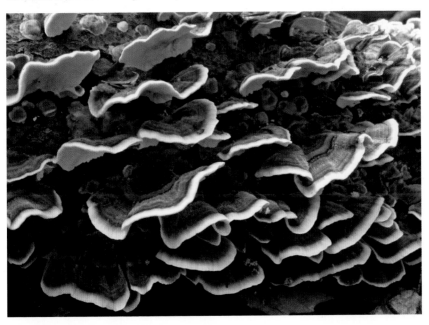

Trichaptum biforme (Fries) Ryvarden

VIOLET TOOTHED POLYPORE

SYNONYMS *Polyporus biformis* Fries, *Hirschioporus pergamenus* (Fries) Bondartsev & Singer

Shelving, fan-shaped with contrasting concentric zones and lilac-purple on the cap margin; pores angular but soon toothlike white, lilac-purple and even after fading always with a hint of lilac; on decaying hardwoods

CAP 1–6 cm wide, single or often densely shelving, plane, semicircular, fan-shaped, zoned white, tan, or pale buff to pale gray with lilac-purple margin at first, surface with fine hairs or smooth, dry. **FLESH** thin, tough, flexible, to 3 mm. **PORES** deep lilac-purple, fading with time to pale buff, 3–5 per mm, angular, elongating unequally or splitting into teeth with age. **TUBES** colored as pores, 1–2 mm long. **STEM** absent. **SPORE PRINT** white. **ODOR** and **TASTE** not distinctive.

Habit and habitat on decaying wood of standing hardwoods, stumps, downed logs, branches. May to December. **RANGE** widespread.

Spores cylindrical, slightly curved, smooth, colorless, not amyloid, 6–8 × 2–2.5 µm.

Comments Easily the most frequently found species in a hardwood forest because of its aggressive saprotrophic nature to recycle dead, decaying hardwoods. The correct ending for this species is *biforme*, not *biformis*, when placed in the genus *Trichaptum*. If similar looking and on conifer trees it is *Trichaptum abietinum* (Dickson) Ryvarden. If the fruit bodies are thick (up to 1 cm), on poplar or aspen in boreal forest areas, and the pores remain mostly circular or angular, then one has *T. subchartaceum* (Murrill) Ryvarden. The green color on some of the fruit bodies in the image of *T. biforme* presented here are from overgrowth by algae.

Tyromyces chioneus (Fries)
P. Karsten

WHITE CHEESE POLYPORE
SYNONYMS *Polyporus chioneus* Fries, *Tyromyces albellus* (Peck) Bondartsev & Singer

White, velvety cap; flesh soft and watery, liquid easily squeezed out like a sponge, fruit body is not damaged; pores small, white; odor fragrant, especially after squeezing; on hardwoods

CAP 1–12 cm wide, 1–8 cm deep, convex, semicircular, white then eventually yellowish or pale tan with age, velvety, becoming glabrous. **FLESH** thick, soft-watery, up to 2 cm thick, liquid easily squeezed out like a sponge. **PORES** white, yellowish with age, 3–5 per mm, circular or slightly angular. **TUBES** colored as pores, up to 8 mm long. **STEM** absent. **SPORE PRINT** white. **ODOR** fragrant, aromatic, pleasant. **TASTE** not distinctive.

Habit and habitat on decaying downed logs and larger branches of hardwoods, common on birch. July to November. **RANGE** widespread.

Spores cylindrical, slightly curved like a sausage, smooth, colorless, not amyloid, 4–5 × 1.5–2 μm.

Comments If the fruit bodies are watery but grayish-blue on the cap and pore surface, then *Postia caesia* (Schrader) P. Karsten is the species. If white but readily staining brown, then it is *P. fragilis* (Fries) Jülich. *Postia* species look like *Tyromyces* species, but differ by producing a brown rot, mainly on conifers. By far, most polypores produce white rots.

Irpex lacteus (Fries) Fries

MILK-WHITE TOOTHED POLYPORE

SYNONYMS *Sistotrema lacteum* Fries, *Steccherinum lacteum* (Fries) Krieglsteiner, *Irpiciporus lacteus* (Fries) Murrill, *Irpex tulipiferae* (Schweinitz) Schweinitz

Crustlike, tough, with small projecting margin, white or pale cream overall; caps when present finely velvety; tube edges growing unequally and becoming flattened and toothlike

Fruit body 1–7 cm deep, tough, crustlike and tightly adhering to branches, eventually producing a short projecting cap (effused-reflexed). **CAP** up to 1 cm deep, plane or incurved, from the top elongated, fan-shaped with irregular margin, white becoming yellowish, finely velvety. **FLESH** white, very thin. **PORES** white or pale cream at first, 2 per mm, angular. **TUBES** white, edges splitting and elongating on one side early in development (irpiciform) producing flattened teeth, 1–5 mm deep. **STEM** absent. **SPORE PRINT** white. **ODOR** pleasant, nutty. **TASTE** mild.

Habit and habitat on branches of hardwoods, in the 5–10 cm diameter ranges most frequently, typically seen when the branch is on the ground. Year-round. **RANGE** widespread.

Spores oblong or cylindrical, smooth, colorless, not amyloid, 5–7 × 2–3 μm; cystidia crystal encrusted, thick-walled, numerous.

Comments This very common resupinate polypore is most often mistaken for a tooth fungus when it is found in mature stages. The pores are more easily observed in early stages of growth or near the cap margin. Compare with *Trichaptum biforme*, another "tooth"-forming polypore that is very common in our forests.

Phlebia tremellosa (Schrader)
Naksone & Burdsall
TREMBLING MERULIUS
SYNONYM *Merulius tremellosus* Schrader
Fruit body coating the woody substrate with small, projecting, fan-shaped, woolly cap; undersurface orange or pinkish, wrinkled and pitted, gelatinous to touch; mostly hardwood logs

Fruit body 3–10 cm wide, with half or more covering the substrate (resupinate), usually forming a cap on the upper edges, many fused laterally and also shelving. **CAP** plane or undulating, from the top fan-shaped or semicircular, whitish, woolly or hairy, also bumpy and scaly. **FLESH** white, 1–2 mm, fleshy or waxy to gelatinous. **UNDERSURFACE** moderately to strongly wrinkled and pitted, not distinct pores, orange or pinkish-orange, gelatinous looking. **STEM** absent. **SPORE PRINT** white. **ODOR** and **TASTE** not distinctive.

Habit and habitat on decaying branches and logs of downed hardwoods like birch and maple, also reported on conifer wood. July to November. **RANGE** widespread.

Spores cylindrical and curved like sausage, smooth, colorless, not amyloid, 3.5–4.5 × 1–2 µm.

Comments Most species of *Phlebia* are entirely resupinate, or growing flattened on the woody substrate, but this jellylike species produces a distinctive cap. In older guides it will be found as *Merulius*, but cultural studies and recent DNA comparisons place this species in *Phlebia*.

Phlebia radiata Fries, radiating phlebia, is crustlike, tough, peachy-orange, strongly wrinkled in a radiating pattern with ridges, folds, and warts covering the surface. It is often on the undersides of downed logs of hardwoods and conifers, especially on beech, from August to November. This brightly colored crust fungus is one of the easier ones to identify, even without a microscope.

Phlebia radiata

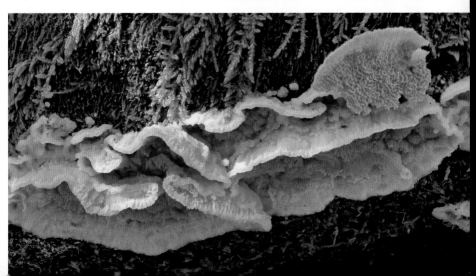

Schizopora paradoxa
(Schrader) Donk

SYNONYMS *Hydnum paradoxum* Schrader, *Irpex paradoxus* (Schrader) Fries, *Irpex obliquus* (Schrader) Fries, *Boletus incertus* (Persoon) Murrill, *Daedalea mollis* Velenovský, *Polyporus versiporus* Persoon, *Poria incerta* (Persoon) Murrill

Bloblike patches, lumpy, white or cream-colored, covered with variably shaped pores or when in vertical positions, toothlike structures; on the top, side, or bottom of decaying hardwoods

Fruit body variable in size, irregular in outline, from the fusion of smaller circular mounded fruit bodies developing on upper surfaces of logs or undersurfaces or vertically, lacking a cap (resupinate), in side view as rounded bumps of differing heights, white, cream, gray becoming grayish-ochre or brownish with age. **FLESH** white or pale cream, firm, 1–5 mm deep. **PORES** colored as above, variable in shape and size, either round and 1–4 per mm or oblong or angular pores, also short maze-like (daedaloid) or pores becoming tooth-like on fruit bodies in vertical orientations. **TUBES** white or cream, 1–5 mm long. **CAP** and **STEM** absent. **SPORE PRINT** white. **ODOR** and **TASTE** not distinctive.

Habit and habitat on decaying wood of hardwoods, stumps, logs, and branches. June to November. **RANGE** widespread but not common.

Spores ellipsoid, smooth, colorless, not amyloid, 5–6.5 × 3.5–4 µm; cystidioles cylindrical with swollen head (capitate) and encrusted with sandy crystals or resinous substances.

Comments The irregular lumpy masses, when found on the sides or undersides of logs and branches, can mimic the look of a tooth fungus. Compare with *Irpex lacteus*, a much more commonly encountered and completely white flat species, which occurs on moderately sized stems, often found on the forest floor.

Pycnoporus cinnabarinus
(Jacquin) P. Karsten

CINNABAR-RED POLYPORE

SYNONYMS *Boletus cinnabarinus* Jacquin, *Polyporus cinnabarinus* (Jacquin) Fries

Bright cinnabar-red cap and pores; cap fading to orange with age, pores retaining bright colors; no stem, directly attached to decaying logs

CAP 2.5–12.5 cm wide, broadly convex, becoming plane, from the top semicircular or fan-shaped, bright cinnabar-red or ochre-salmon or apricot orange, fading to dull orange with age, hairy and rough, becoming smooth and suedelike or developing rough surface in age, dry. **FLESH** red or pale orange, 5–20 mm thick. **PORES** bright cinnabar-red or orange-red, 3–4 per mm, circular or angular. **TUBES** colored as pores, 1–6 mm deep. **STEM** absent. **SPORE PRINT** white. **ODOR** not distinctive. **TASTE** slightly bitter.

Habit and habitat on downed hardwood logs, rarely on conifers. Year-round **RANGE** widespread.

Spores cylindrical and curved, somewhat sausagelike, thin-walled, smooth, colorless, not amyloid, 6–8 × 2.5–3 μm.

Comments As brightly colored as this polypore can be, it is not collected all that frequently. Its tropical counterpart, *Pycnoporus sanguineus* (Linnaeus) Murrill, differs by thinner flesh and the red pigments not fading so quickly; otherwise, they look very similar.

Lenzites betulina (Linnaeus) Fries

MULTICOLOR GILL POLYPORE

SYNONYMS *Agaricus betulinus* Fries, *Daedalea betulina* (Linnaeus) Rebentisch

Fan-shaped with conspicuous zones of mixed white, gray, cream, and brown colors, surface velvety; pores in the form of gills; flesh white, tough-corky

CAP 5–10 cm wide, plane, from the top semicircular or fan-shaped, loosely overlapping, white, gray, cream, and various shades of brown, concentrically zoned, velvety, dry. **FLESH** white, tough-corky, thin (1–2 mm). **PORES** white, then cream or ochre, gill-like or in early phases dichotomously forked mostly near cap margin, well-spaced. **TUBES** white, gill-like, up to 12 mm deep. **STEM** absent. **SPORE PRINT** white. **ODOR** and **TASTE** not distinctive.

Habit and habitat on many different types of decaying hardwoods, birch included, also occasionally on conifers. July to November. **RANGE** widespread.

Spores ellipsoid, smooth, colorless, not amyloid, 5–6 × 2–3 μm.

Comments The image shown here is of a very young fruit body. In mature stages, the pores are distinctly gill-like in form. Compare with *Daedaleopsis confragosa* and *Daedalea quercina* that have dark brown flesh.

Daedaleopsis confragosa
(Bolton) J. Schröter

THIN-MAZE FLAT POLYPORE

SYNONYMS *Boletus confragosus* Bolton,
Daedalea confragosa (Bolton) Persoon,
Daedalea rubescens Albertini & Schweinitz
Circular or fan-shaped with obscure brown
colors and ridge zones on the cap, the surface is also matted with hairs or glabrous;
pores mostly elongate and mazelike, bruising salmon-pink

CAP 5–15 cm wide, plane, from the top
round or fan-shaped, loosely overlapping or in small rosettes, pale buff-gray,
reddish-brown or various shades of brown,
obscurely concentrically zoned with color
and ridges, sparsely matted hairy or glabrous, dry. **FLESH** buff-brown or brown or
pinkish-brown, tough-corky, 10–20 mm.
PORES white, then dingy brown, when
white typically bruising salmon-pink
or reddish, typically elongate, mazelike or poroid or gill-like. **TUBES** white,
up to 12 mm deep. **STEM** mostly absent.
SPORE PRINT white. **ODOR** and **TASTE**
not distinctive.

Habit and habitat on many different types
of decaying hardwoods. July to December.
RANGE widespread.

Spores sausage-shaped, smooth, colorless,
not amyloid, 9–11 × 2–3 µm.

Comments Typically it lacks a stem and the
caps are attached to the logs, making a fan-
shaped fruit body. Also, the salmon-pink
discoloration of the pores injured from
handling occurs fairly rapidly. *Lenzites
betulina*, somewhat similar but found less
frequently, differs by the strongly velvety
zoned cap, the white flesh, and the gills
instead of pores that do not discolor
salmon-pink when bruised.

Daedalea quercina (Linnaeus) Persoon

THIN-MAZE OAK POLYPORE

SYNONYMS *Agaricus quercinus* Linnaeus, *Lenzites quercina* (Linnaeus) P. Karsten
Circular or fan-shaped white or grayish, thick, corky in texture; pores mostly gill-like, very thick

CAP 5–20 cm wide, plane, fan-shaped, white, then gray or brown or even black with age, fuzzy or glabrous, dry. **FLESH** orange-brown or tobacco-brown with indistinct zones, tough, corky or woody, 10 mm. **PORES** white, then dingy yellow or tan, mostly mazelike or gill-like, very thick walls, 1–3 mm wide. **TUBES** colored as pores, up to 40 mm deep. **STEM** absent. **SPORE PRINT** white. **ODOR** and **TASTE** not distinctive.

Habit and habitat on decaying hardwoods, mostly oak, but can be found on ash, beech, aspen, cherry, and elm. Year-round. **RANGE** widespread.

Spores cylindrical, smooth, colorless, not amyloid, 5.5–6 × 2.5–3.5 µm.

Comments Compare with *Lenzites betulina* and *Daedaleopsis confragosa*.

Phellinus gilvus (Schweinitz)
Patouillard

MUSTARD-YELLOW POLYPORE
SYNONYMS *Boletus gilvus* Schweinitz,
Fomes gilvus (Schweinitz) Speggazini,
Polyporus gilvus (Schweinitz) Fries
Fan-shaped, dark rusty-brown, when
young with bright rusty-yellow margin;
pores very small, round, reddish or
purplish-brown at first, becoming yellow-
brown; flesh bright yellow or rusty-brown,
corky, zoned

Upper surface

CAP 2–15 cm wide, single or several overlapping, broadly convex or plane, from
the top fan-shaped or semicircular, ochre-
brown or rusty-yellow, becoming dark
rust-brown or nearly black, sometimes
somewhat velvety at first, glabrous with
age, surface uneven, grooved zonate,
cracking radially. **FLESH** bright yellow
or rusty-brown, to 20 mm thick, zoned,
corky. **PORES** grayish-brown or reddish-
brown or darker purplish-brown at first,
6–8 per mm, round. **TUBES** white stuffed at
first, soon brown, 1–5 mm deep per layer,
sometimes several layers. **STEM** absent,
caps attached directly to substrate. **SPORE
PRINT** white or cream. **ODOR** and taste
TASTE not distinctive.

Undersurface

Habit and habitat single or overlapping on
decaying trunks, or logs of hardwoods. All
months. **RANGE** widespread.

Spores ellipsoid, smooth, colorless, not
amyloid, 4–5 × 3–3.5 µm; thick-walled,
dark brown setae abundant in the hymenium (tube walls).

Comments The rusty-brown colors with a
cracked cap surface, along with the bright
yellow or rusty-brown flesh, and the dark
reddish or purplish pores that become yellowish-brown with age, help to distinguish
this species in the field.

Phellinus igniarius (Linnaeus) Quélet

FLECK-FLESHED POLYPORE

SYNONYMS *Boletus igniarius* Linnaeus, *Fomes igniarius* (Linnaeus) Fries, *Polyporus igniarius* (Linnaeus) Fries

Very large, very hard, hoof-shaped, gray or eventually black, wrinkled, then cracked cap; pores cinnamon or darker rusty-brown, tiny; on various living or dead hardwoods

CAP 5–25 cm wide, to 11 cm deep, broadly convex or hoof-shaped, sometimes plane, from the top fan-shaped or semicircular, gray then nearly black, glabrous, deeply radially wrinkled, becoming cracked. **FLESH** dark reddish-brown with white flecks, to 2 cm thick, zoned, woody. **PORES** pale cinnamon-brown or darker to purplish-brown, 5–6 per mm, round. **TUBES** reddish-brown, and white stuffed, 2–5 mm deep, but in annual layers, up to 10 cm deep. **STEM** absent. **SPORE PRINT** white. **ODOR** and **TASTE** not distinctive.

Habit and habitat on many different species of living hardwoods, usually on the lower trunk of the tree, near the ground, also fruiting from downed dead trunks. Year-round **RANGE** widespread.

Spores subglobose or ovoid, thick-walled, smooth, colorless, not amyloid, 5–7 × 4.5–6 µm; setae thick-walled, dark brown, fusiform, 14–17 × 4–6 µm.

Comments There are many species of *Phellinus* in North America. Not all have a cap, but all have the characteristic brown small circular pores, that on some species may have a purplish-brown-colored phase. *Phellinus igniarius* is one of the more commonly encountered species of *Phellinus*.

Fomes fomentarius (Linnaeus) Fries

TINDER POLYPORE

SYNONYM *Boletus fomentarius* Linnaeus
Very hard and woody, hoof-shaped, brown to grayish zoned; pores brown, tiny; on hardwoods

CAP 5–20 cm wide, horse-hoof-shaped, zoned with brown or grayish-brown or reddish-brown or gray or mixtures of these colors, mostly smooth and hard, may be soft and velvety at first, usually cracked with age. **FLESH** tan or brown, to 30 mm thick. **PORES** grayish-tan to brown, becoming darker with age, only slightly darkening when bruised, 3–4 per mm, round. **TUBES** colored as pores but becoming white-stuffed inside, several obscure layers deep (perennial), 5–60 mm deep when combined. **STEM** absent. **SPORE PRINT** white. **ODOR** not distinctive, or fruity. **TASTE** not distinctive, or acrid.

Habit and habitat single, scattered, or clustered on dead hardwood trees or wounds in living hardwood trees as a weak parasite, frequently on beech and yellow birch, but cherry, maple, poplar, and hickory are not immune. Year-round, slow-decaying tissues. **RANGE** widespread.

Spores cylindrical, smooth, colorless, not amyloid, 12–20 × 4–7 µm.

Comments Ancient humans, thousands of years ago, used this fungus to help start fires, since once lit it burns slowly like punk, hence the common name. The ground up fresh flesh produces a substance known as amadou once it has dried out, and besides being used as tinder, has been used medicinally to dry teeth and staunch blood loss during surgery. Strips from the fruit bodies can be cut, soaked in water, beaten and stretched to loosen the fungal hyphae, and once pliable, assembled into clothing such as gloves, pants, and caps. The fruit bodies attached to decaying trees in nature eventually become nonfunctional, turn hollow and completely black, and can be easily crushed.

Fomitopsis pinicola (Swartz)
P. Karsten

RED-BELTED POLYPORE
SYNONYMS *Boletus pinicola* Swartz, *Fomes pinicola* (Swartz) Fries, *Ungulina pinicola* (Swartz) Singer

Medium to large, hard bodied, stemless, fan-shaped with red-belted, shiny, varnished zone just inside a cream-colored, thickened margin; pores cream, minute; on dead or dying conifers

CAP 5–40 cm wide, to 10 cm deep, hoof-shaped, from the top semicircular or fan-shaped, margin thick, white or creamy and rounded, surface reddish-brown or orange-red or orange-brown overall at first and resinous or appearing shiny-varnished, becoming dark ruby-red or brown to black from the attachment point outward leaving a reddish band near the margin, concentrically undulate or furrowed, often wrinkled, hard. **FLESH** pale cream, to 60 mm thick, tough, corky. **PORES** cream, not changing color dramatically when bruised, 5–6 per mm, round. **TUBES** colored as pores, several, mostly obscure, layers deep (perennial), occasionally running down the substrate (effused). **STEM** absent. **SPORE PRINT** cream. **ODOR** musty-pungent or not distinctive. **TASTE** not distinctive.

Habit and habitat single most often but can be clustered on dead wood of conifers, occasionally on living conifers and some hardwoods such as black cherry, aspen, or birch. Year-round, slow-decaying tissues. **RANGE** widespread.

Spores cylindrical-ellipsoid, smooth, colorless, not amyloid, 6–9 × 3.5–4.5 μm.

Comments This wood-rotting fungus is an important producer of soil stabilizing and moisture absorbing humus in coniferous forests. Compare with *Ganoderma tsugae*, the hemlock varnish shelf. Species of *Fomitopsis* cause brown rots, while *Fomes* causes white rot, the more common form of decay.

Fomitopsis cajanderi
(P. Karsten) Kotlaba & Pouzar
ROSY POLYPORE

SYNONYMS *Fomes cajanderi* P. Karsten, *Fomes subroseus* (Weir) Overholtz, *Fomitopsis subrosea* (Weir) Bondartsev & Singer

On conifers with rosy-pink cap and pores; often several caps fused laterally, also shelving and the tube layer often extending down the woody substrate below the cap

CAP to 20 cm wide, to 10 cm deep, often fused laterally with two or more caps together, also shelving on top of each other, convex becoming plane, semi-circular or fan-shaped, pinkish or rosy overall, then more darkly purple or dark brown to nearly black where attached and spreading outward to pinkish or white margin, mostly hairy, soft, often wrinkled, margin tapered, not thick. **FLESH** rose to pinkish-brown, to 15 mm thick, tough, corky. **PORES** rose-pink, becoming reddish-brown with age, 4–5 per mm, round. **TUBES** colored as pores, several layers deep (perennial), occasionally running down the substrate (effused). **STEM** absent. **SPORE PRINT** white. **ODOR** fragrant. **TASTE** not distinctive, or acrid.

Habit and habitat usually clustered on dead wood of conifers, often spruce, hemlock or larch in the Northeast. Year-round, slow-decaying tissues. **RANGE** widespread.

Spores sausage-shaped, cylindrical and distinctly curved, smooth, colorless, not amyloid, 5–7 × 1.5–2 µm.

Comments *Fomitopsis rosea* (Albertini & Schweinitz) P. Karsten is similar in colors but differs most readily by the hoof-shaped individual fruit bodies and the straight, cylindrical, and wider spores (2–2.5 µm wide).

Ischnoderma resinosum
(Schrader) P. Karsten

RESINOUS POLYPORE

SYNONYMS *Boletus resinosus* Schrader, *Polyporus resinosus* (Schrader) Fries, *Polyporus benzoinus* (Wahlenberg) Fries, *Ischnoderma benzoinum* (Wahlenberg) P. Karsten

Moderately large and fleshy when young, fan-shaped with thick, rounded margin, brown velvety becoming darker and developing a resinous surface; pores white, staining brown; stem absent

CAP 7–25 cm wide, to 12 cm deep, plane, from the top semicircular or fan-shaped with thick, rounded white margin, often overlapping, most of cap pale brown or ochre-brown or becoming dark brown to black near attachment, fine velvety, with age becoming zoned and resinous-crusty, sometimes radially wrinkled with age as well. **FLESH** white becoming cinnamon, thick, soft, becoming hard with age. **PORES** white or pale cream at first, staining brown when bruised, 4–6 per mm, round or angular. **TUBES** white, 10 mm deep. **STEM** absent. **SPORE PRINT** white. **ODOR** and **TASTE** not distinctive.

Habit and habitat on dead wood of hardwood and conifer trees. September to October. **RANGE** widespread.

Spores oblong or cylindrical, slightly curved and sausage-shaped, smooth, colorless, not amyloid, 5–7 × 1.5–2 µm.

Comments Very soft and fleshy for a polypore, at least when young. In wet weather, droplets of liquid may form on the cap margin and pore surfaces.

Ganoderma applanatum

(Persoon) Patouillard

ARTIST'S CONK

SYNONYMS *Boletus applanatus* Persoon, *Fomes applanatus* (Persoon) Fries

Medium to very large, hard bodied, stemless, fan-shaped and zonate, dull brown; pores white, bruising brown; on hardwoods generally

CAP 5–75 cm wide, to 10 cm deep, convex, becoming plane, semicircular or fan-shaped, brown to grayish-brown and white at the very margin, sometimes bright reddish-brown from its own spores falling densely on the cap, not varnished, dull, concentrically undulate or furrowed. **FLESH** brown, to 50 mm thick, woody. **PORES** stark white at first, instantly and dramatically dark brown when bruised or scratched, eventually dirty yellow, 4–6 per mm, round. **TUBES** brown but older layers white stuffed, several layers of tubes with age (perennial). **STEM** absent. **SPORE PRINT** reddish-brown. **ODOR** and **TASTE** not distinctive.

Habit and habitat single or clustered on dead hardwoods, also reported from conifers. Year-round. **RANGE** widespread.

Spores ellipsoid with truncate end, smooth but thick-walled with spines or channels visible inside the walls, brown, not amyloid, $8–12 \times 6.5–8$ µm.

Comments The fresh pore surface can be marked with a sharp instrument, and the markings becomes permanent if not further damaged. Artists use the conks to make elaborate drawings, developing a cottage art industry in upstate New York. My mycology class once received a small artist's conk letter from Kathie Hodge and her Cornell Mycology Class, via the United States Postal Service—it was stamped, addressed on the cap surface with a sharpie, and legal. We had been collecting in the same location a week before their visit and left our calling card, Cortland Mycology Class, etched on a small ganoderma on a log in the forest.

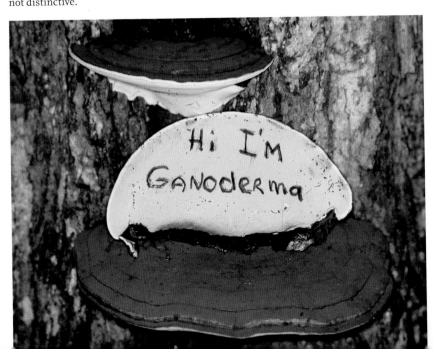

Ganoderma tsugae Murrill

HEMLOCK VARNISH SHELF

SYNONYMS *Polyporus tsugae* (Murrill)
Overholtz, *Fomes tsugae* (Murrill) Saccardo
& D. Saccardo

Cap reddish-brown, fan-shaped and highly
varnished; pores tiny, white, bruising
brown; stem varnished like cap, lateral but
bent vertically; on dead hemlocks

CAP 5–30 cm wide, to 20 cm deep,
broadly convex or plane, from the top at
first spathulate, but soon semicircular
or fan-shaped, shiny reddish-brown or
orange-brown or mahogany and with
white margin early on, surface often dull
cinnamon-brown from spores covering
the cap (see image), highly varnished
overall, smooth or undulate. **FLESH** white,
to 30 mm thick, tough. **PORES** white or
pale cream at first, changing color to
dark brown when bruised or scratched or
overall with age, 4–6 per mm, round or
angular. **TUBES** brown, one layer (annual),
15 mm deep. **STEM** usually present, 9 cm
long, 50 mm wide, lateral and often
upright then curved horizontal, equal,
colored and varnished like cap. **SPORE
PRINT** reddish-brown. **ODOR** and **TASTE**
not distinctive.

Habit and habitat single or clustered
on dead hemlock logs or standing trees
or stumps. May to November. **RANGE**
widespread.

Spores ellipsoid with truncate end, smooth
but thick-walled with spines or pillars visible inside the walls, brown, not amyloid,
9–11 × 6–8 μm.

Comments If you find something similar
on hardwoods, you may have *Ganoderma
lucidum* (Curtis) P. Karsten, called the ling
chih, lingzhi, or reishi, that should have
brownish flesh. The ling chih is a highly
prized medicinal polypore in many cultures and much has been written about
this "divine fungus," some of which can be
accessed on the Internet. The dull brown
color of some of the caps in this image is
due to a heavy deposit of spores being produced by these fruit bodies.

Piptoporus betulinus (Bulliard) P. Karsten

BIRCH POLYPORE
SYNONYMS *Boletus betulinus* Bulliard, *Fomes betulinus* (Bulliard) Fries
Cap white or grayish-tan or brown with thick rounded margin, surface leather soft, smooth mostly; pores white then brown, easily peeling from cap; on dead birch trees

CAP 5–25 cm wide, convex or plane, semicircular or fan-shaped or somewhat hooflike with thick, inrolled-rounded margin, white or pale grayish-tan, becoming darker brown, surface smooth, soft like leather, easily dented, the skin sometimes becoming disrupted and scaly. **FLESH** white, thick, tough. **PORES** white, becoming pale brown with age, 3–5 per mm, round or angular. **TUBES** white, one layer (annual), 10 mm deep, easily separating from cap. **STEM** when present, short, stublike, on the edge of the cap. **SPORE PRINT** white. **ODOR** and **TASTE** not distinctive.

Habit and habitat single or clustered on dead birch logs or standing birch trees or stumps. Year-round. **RANGE** widespread.

Spores sausage-shaped, smooth, colorless, not amyloid, 5–6 × 1.5–2 µm.

Comments Here is another interesting polypore with medicinal properties attributed to it. In 1991 a 5300-year-old human male mummy, given the name Ötzi by the discoverers, was found preserved in ice in the Italian Alps. Among Ötzi's possessions were dried pieces of *Piptoporus betulinus*. The assumption is that he carried these "mushroom" pieces for medicinal use since such use has been documented in early recorded history. Explore the Internet for more information on the several other uses attributed to the birch polypore.

Chanterelles

Chanterelles can look very much like the true gilled mushrooms, especially when they produce well-developed folds, or false gills, on the underside of the cap. Some in this group, for example *Cantharellus flavus* and *C. cibarius*, are highly regarded as choice edible fungi by chefs around the world and are often featured on televised cooking shows. *Cantharellus flavus* is on the top of my list when I go out to forage for wild mushrooms. Some species can also cause gastrointestinal upset, for example *Turbinellus floccosus*, so learn each species thoroughly if you plan to use them for the table.

The following diagnostic key should help identify the different genera we commonly find in the Northeast. For example, I have not included *Gloeocantharellus*, which is mycorrhizal with oaks in the southeastern areas of the United States. We have an abundance of oaks and oak forests in certain areas in the Northeast, but this genus has not been reported as yet from our cooler, temperate regions.

Flesh thick

Cantharellus Cap smooth and glabrous, colors egg-yolk-yellow, pale pinkish, reddish-orange, or yellow-brown, false gills usually well-developed (but edges rounded-blunt) or underside of cap smooth (see discussion under *C. flavus*)

Gomphus Cap smooth, large and often fused as multiples, false gills violet-purple

Turbinellus Cap cottony-scaly, deeply vase-shaped, false gills cream-colored

Polyozellus Caps large, clustered and fused as multiples, purple or blue-black overall, false gills absent, surface veined or wrinkled at most (spores nodulose-angular)

Flesh mostly thin

Craterellus Cap often dark-colored, brown or black, sometimes yellow but then with at least some brown fibrils, typically perforate, false gills present or absent

Cantharellus flavus Foltz & T. J. Volk

AMERICAN GOLDEN CHANTERELLE

Medium to large and entirely egg-yolk-yellow; gills decurrent, forked, edges blunt or rounded; odor faintly of apricots or peaches

CAP 2–10 cm wide, convex with inrolled margin, becoming plane then arching up with broadly sunken center and with a wavy and often scalloped margin, egg-yolk-yellow, becoming paler with age, glabrous, moist or dry. **FLESH** yellow. **GILLS** egg-yolk-yellow, decurrent, subdistant, forking and cross-veined frequently, edges bluntly rounded (false gills). **STEM** 3–8 cm long, 5–20 mm broad, equal or tapered to the base, egg-yolk-yellow or mixed with white, glabrous, solid. **SPORE PRINT** yellow or ochraceous in heavy deposits. **ODOR** faintly of apricots or dried peaches. **TASTE** mild or perhaps slightly peppery after long chewing.

Habit and habitat single, scattered, or clustered on soil under hardwoods or mixed conifer and hardwoods, with oak typically present. July to September. **RANGE** widespread.

Spores elongate ellipsoid or oblong, smooth, colorless, not amyloid, 8–11 × 4.5–6 µm.

Comments This species is among several that have gone under the name *Cantharellus cibarius* in North America, though molecular evidence shows this to be an exclusively European species. Mycologists are methodically describing them using molecular evidence to back up the morphological differences. If the false gills and stipe are white and contrasting with the egg-yolk-yellow cap, and the spore print is pink, it is *C. phasmatis* Foltz & T. J. Volk. This and *C. flavus* are the two most common forms found in the Northeast and both are choice edibles. Also *C. spectaculus* Foltz & T. J. Volk differs by its orange-salmon-colored caps and salmon-pink gills, while *C. roseocanus* (Redhead, Norvell & Danell) Redhead, Norvell & Moncalvo is associated with conifers and displays a pale to dark pink bloom on the cap and stem when young. If gills are lacking or poorly developed, and oaks are present, you have the *C. lateritius* complex, also edible.

Cantharellus cinnabarinus
(Schweinitz) Schweinitz

CINNABAR-RED CHANTERELLE
SYNONYMS *Agaricus cinnabarinus* Schweinitz, *Hygrophorus cinnabarinus* (Schweinitz) Saccardo

Medium-sized with cap and stem vibrant reddish-orange; gills decurrent, conspicuously cross-veined, especially near cap margin; taste slowly burning acrid

CAP 1–4 cm wide, convex, becoming plane then arching up with shallow sunken center and a wavy scalloped margin, vibrant reddish-orange or cinnabar or even dark pink, becoming paler with age, glabrous, moist or dry. **FLESH** white or tinted with cap color. **GILLS** pale pinkish or same color as cap, decurrent, subdistant or close, occasionally forking, conspicuously crossveined, edges thick and rounded (false gills). **STEM** 1–6 cm long, 3–10 mm broad, equal or tapered to the base, same color as cap or paler, glabrous. **SPORE PRINT** pinkish or pinkish-cream. **ODOR** not distinctive. **TASTE** slowly burning acrid.

Habit and habitat single, scattered, or clustered on soil and mosses under hardwoods, with beech, oak, or aspen typically present. June to October. **RANGE** widespread.

Spores ellipsoid or somewhat oblong, smooth, colorless, not amyloid, 6–11 × 3.5–6 µm.

Comments Could be confused with some of the reddish *Hygrocybe* species until you observe the thick, decurrent, cross-veined false gills. *Hygrocybe* species have thin, unforked true gills that lack cross veins. People do eat this chanterelle, but it is not as choice as the golden chanterelle.

Cantharellus minor Peck

Small orange-yellow fruit bodies, with decurrent, blunt-edged, distantly spaced gills; stem long for a typical chanterelle; on soil and mosses under hardwoods with oak present

CAP 0.5–3 cm wide, convex, becoming plane then arching up with shallow sunken center and a wavy and often scalloped margin, vibrant orange-yellow or egg-yolk-yellow, becoming paler with age, glabrous, moist or dry. **FLESH** yellow, thin. **GILLS** same color as cap, decurrent, subdistant or distant, occasionally forking, but not cross-veined, edges rounded (false gills). **STEM** 1–4 cm long, 1–3(–7) mm broad, equal or tapered to the base, long, same color as cap or paler, glabrous, hollow. **SPORE PRINT** pale yellowish-orange. **ODOR** not distinctive. **TASTE** mild.

Habit and habitat single, scattered, or clustered on soil and mosses under hardwoods, with oak typically present. July to September. **RANGE** widespread.

Spores ellipsoid, smooth, colorless, not amyloid, 7–11 × 4.5–6 µm.

Comments A tiny version of *Cantharellus flavus*. It seems to like unimproved muddy roads and is frequently found either in ditches or on banks, often growing with mosses. An even more brightly colored variant can be found associated with hemlock, *C. minor* f. *intensissimus* R. H. Petersen. Compare with the bright yellow, but very slimy, *Gloioxanthomyces nitidus*. The wax caps have true gills that are thin, not blunt-edged false gills.

Cantharellus appalachiensis
R. H. Petersen

Medium-sized chanterelle with brown colors on cap and stem; gills yellow, forked and with cross veins common; under oak

CAP 1–6 cm wide, convex, becoming plane with a wavy and often scalloped margin, brown at first, turning yellow or yellowish-brown over the margin, obscurely zoned, glabrous, moist or dry. **FLESH** yellow, thin. **GILLS** same color as cap, decurrent, subdistant or distant, some forking and most with cross veins, edges rounded (false gills). **STEM** 2–6 cm long, 3–10 mm broad, equal or tapered to the base, pale brown or yellowish-brown, glabrous, white mycelioid over base. **SPORE PRINT** pale cream. **ODOR** fragrant. **TASTE** mild.

Habit and habitat single, scattered, or clustered on humus under hardwoods, with oak typically present. July to September. **RANGE** widespread.

Spores ellipsoid, smooth, colorless, not amyloid, 6–9 × 3.5–5 μm.

Comments The caps start mostly brown, but eventually turn yellowish from the margin inward, leaving the center brown. The blunt-edged gills are also well developed.

Craterellus fallax A. H. Smith

BLACK TRUMPET
Medium to large trumpetlike, vase-shaped, with black cap; gills not present, or the surface barely wrinkled, bluish-gray, becoming salmon-buff; stem very short or absent

CAP 1–8 cm wide, 3–14 cm high, tubular then deeply vase-shaped, margin arching and undulate, dark grayish-brown or black, fading to gray, splitting with age, inside the surface covered with scales or fibrils, moist or becoming dry. **FLESH** same as cap color, thin, brittle. **GILLS** absent, surface smooth or wrinkled, decurrent, dark grayish or bluish-gray, developing salmon-buff or orange colors with spore maturation. **STEM** short, indistinct, colored as the cap, hollow. **SPORE PRINT** ochraceous-buff or ochraceous-orange. **ODOR** not distinctive, or somewhat fragrant. **TASTE** mild.

Habit and habitat single, scattered, or clustered on humus or well-decayed leaf litter under hardwoods. July to October. **RANGE** widespread.

Spores ellipsoid, smooth, colorless, not amyloid, 10–15(–20) × 7–10 µm.

Comments These very dark trumpetlike chanterelles are highly prized for the flavor they produce when dried and then crumbled up for use as a condiment in scrambled eggs and other dishes. *Craterellus dubius* Peck, described from the Adirondack Mountains in New York, looks like a much smaller version of *C. fallax*, but it differs by brown colors in the pileus and smaller spores. There are also a few other larger dark-colored species of *Craterellus*, but for those, the gill surface is obviously veined and ridged, making the stem obvious.

Polyozellus multiplex
(Underwood) Murrill

CLUSTERED BLUE CHANTERELLE
SYNONYMS *Cantharellus multiplex*
Underwood, *Craterellus multiplex*
(Underwood) Shope, *Thelephora multiplex*
(Underwood) Kawamura

Large with many fused at the base, fan-shaped or vase-shaped, deep purple or blue-black; gills absent but surface with veins, wrinkled or poroid, similar color as caps

CAP 2–10 cm wide, with many in clusters reaching 1 m wide, round or fan-shaped individually, also deeply vase-shaped, margin wavy and lobed, deep purple or blue or black, the surface with fine fuzz in concentric zones, then glabrous, moist or becoming dry. **FLESH** same as cap color, thick, but soft and brittle. **GILLS** absent, surface veinlike or wrinkled to almost poroid, decurrent, purple or violet. **STEM** 2–5 cm long, 5–20 mm broad, compound and many fused together, with grooves, colored as the cap. **SPORE PRINT** white. **ODOR** aromatic. **TASTE** mild.

Habit and habitat usually a single cluster on soil or humus among lichens under conifers, especially spruce and fir. July to August. **RANGE** widespread, but only rarely collected in the northern spruce and fir zones.

Spores globose or ellipsoid, nodulose-angular, colorless, not amyloid, 6–8.5 × 5.5–7.5 µm.

Comments It is listed as edible, but it seems to be rarely encountered and then mainly in old-growth spruce and fir forests across North America, from coast to coast. One should probably not be consuming rare species.

Craterellus tubaeformis
(Fries) Quélet

TRUMPET CHANTERELLE
SYNONYMS *Cantharellus tubaeformis* Fries, *Cantharellus xanthopus* (Persoon) Duby

Medium-sized with a trumpetlike, vase-shaped, brown cap; gills violet-gray, thick and decurrent; stem bright orange-yellow; in conifer woods, often on mosses

CAP 1–7 cm wide, convex, becoming vase-shaped, then perforate centrally, margin wavy-undulate and sometimes scalloped, dark brown or dark yellowish-brown, smooth except for the very margin that can become fibrillose-scaly, moist or dry. **FLESH** same as cap color, thin. **GILLS** pale violaceous-gray most often, but may develop yellowish or brownish colors as well, decurrent, distant, often forking, also cross-veined, edges thick and rounded (false gills). **STEM** 2–9 cm long, 3–10 mm broad, equal or enlarged downward, often compressed or furrowed, orange-yellow or dark golden-yellow, becoming duller grayish-orange or even brownish, glabrous. **SPORE PRINT** white or pale cream. **ODOR** not distinctive, or slightly fragrant. **TASTE** mild.

Habit and habitat single, scattered, or clustered on mosses or well-decayed moss covered logs in conifer bogs, often on *Sphagnum*. June to October. **RANGE** widespread.

Spores ellipsoid, smooth, colorless, not amyloid, 9–12 × 6–8 µm.

Comments This very common chanterelle can produce large numbers of fruit bodies. It is edible, but the flesh is very thin. Black-capped forms with violaceous-gray gills and golden-yellow stems may be found in the Adirondack mountain habitats as well. A review of the literature suggests there may be cryptic species occurring in the Northeast that have yet to be sorted out. The color form presented here is mostly shades of brown.

Craterellus ignicolor
(R. H. Petersen) Dahlman, Danell & Spatafora

SYNONYM *Cantharellus ignicolor* R. H. Petersen

Medium-sized with a trumpetlike, vase-shaped, perforate, yellowish-orange cap with scattered brown fibrillose points; gills obvious, veinlike, yellow becoming pinkish-gray; stem bright yellow or orange-yellow; in seepage areas in forests

CAP 1–7 cm wide, convex, becoming perforate centrally, eventually vase-shaped, dull orange fading to pale yellowish-orange, mostly smooth but when young with scattered fibrillose points making the surface look roughened, otherwise glabrous, moist or becoming dry. **FLESH** same as cap color, thin. **GILLS** veinlike but well-developed, forking or branched, or both, and interconnected, decurrent, orange-yellow, developing vinaceous-buff or pinkish-gray tints with age, then with brown hues eventually. **STEM** 2–6 cm long, 2–12 mm broad, equal or base narrowly club-shaped, bright yellow or orange-yellow, becoming orange, glabrous, yellow mycelioid at base, hollow. **SPORE PRINT** ochraceous-salmon or pinkish-yellow. **ODOR** and **TASTE** not distinctive.

Habit and habitat single, scattered, or clustered on humus or well-decayed mossy logs, under hardwoods or conifers, often on mosses, including *Sphagnum*, in damp shady areas. July to October. **RANGE** widespread.

Spores ellipsoid, smooth, colorless, not amyloid, 9–13 × 6–9 μm.

Comments Well-developed forking and branching pinkish-gray gills are present for *Craterellus ignicolor*, while the underside of the cap is yellowish and merely wrinkled for the somewhat similar-looking *C. lutescens* (Fries) Fries. *Cantharellus minor* is also golden-yellow, but the fruiting bodies are smaller, the cap is not perforate, and the gills are yellow like the rest of the fruit body, and not developing vinaceous or pinkish-gray tints.

Gomphus clavatus (Persoon) Gray

PIG'S EAR GOMPHUS

SYNONYMS *Cantharellus clavatus* (Persoon) Fries, *Cantharellus brevipes* Peck

Large, vase-shaped with brown cap; gills violet-purple, wrinkled-ridged, decurrent; flesh thick; under conifers

CAP 2–10 cm wide, fused clumps with multiple caps up to 20 cm high, truncate club-shaped, with two or more caps arising from a common stem, often grown out more on one side than the other, centrally sunken then vase-shaped, margins irregularly wavy and lobed, violet or purplish-brown at first, becoming pale brown or with lilac tints or olive hues as well, eventually sordid yellowish or creamy-tan, smooth and felted, becoming scaly, moist or becoming dry. **FLESH** white or pale pinkish-buff, thick. **GILLS** purplish or violet, becoming violet-brown or buff, wrinkled, ridged or veined, profusely cross-veined, decurrent. **STEM** 1–5 cm long, 10–20 mm wide, short, tapered downward, fused to other stems, sordid buff or pale lilac, smooth but finely hair, white solid inside. **SPORE PRINT** ochre. **ODOR** and **TASTE** not distinctive.

Habit and habitat single, scattered, or clustered on humus under conifers, especially spruce, fir, and hemlock. August to October. **RANGE** widespread.

Spores ellipsoid-cylindrical, bumpy, ochre-yellow in KOH, not amyloid, 10–13 × 5–6 µm.

Comments Considered a choice edible, if you can find it before the insects do. Locally it is not common, but always grows under hemlock.

Turbinellus floccosus
(Schweinitz) Earle ex Giachini & Castellano

SCALY VASE CHANTERELLE

SYNONYMS *Cantharellus floccosus* Schweinitz, *Cantharellus princeps* Berkeley & M. A. Curtis, *Gomphus bonarii* f. *wilsonii* (A. H. Smith & Morse) R. H. Petersen, *Gomphus canadensis* (Klotzsch ex Berkeley) Corner

Large vase-shaped or funnel-shaped, with orange, yellowish-orange or reddish-orange, scaly caps; gills cream or yellow, deeply decurrent, strongly ridged and veined; under conifers

CAP 5–15 cm wide, up to 20 cm high, cylindrical truncate at first, expanding and becoming deeply sunken or funnel-shaped, pale orange or yellowish-orange or reddish-orange, paler with age and more yellowish, densely felted-scaly or woolly with uniformly sized soft scales, dry. **FLESH** white, thick. **GILLS** cream or yellow, ochre tinged with age, wrinkled, ridged or veined, edges thick and rounded, decurrent. **STEM** 5–10 cm long, 15–50 mm wide, tapered downward, mostly imbedded in the substrate, same color as gills, may turn brown when bruised, smooth, white solid inside. **SPORE PRINT** yellowish-brown. **ODOR** and **TASTE** not distinctive.

Habit and habitat scattered or clustered on humus under conifers, in mixed woods, often with fir. August to October. **RANGE** widespread.

Spores ellipsoid, bumpy or somewhat wrinkled looking, pale yellowish in 3% KOH, not amyloid, 11–15 × 7–8 μm.

Comments This scaly chanterelle is known to cause gastrointestinal upset, including nausea, vomiting, and diarrhea, sometimes many hours after ingestion. This one is to admire but not eat.

Tooth Fungi

The tooth fungi are variable in construction, in that they may or may not have a cap and the cap may or may not have a stem, depending on the genus. The majority do have a cap and stem, but *Hericium*, for example, does not produce a cap, just teeth hanging from branches, while *Steccherinum* and *Sarcodontia*, also lacking a cap, have teeth simply attached to a flat crust of tissue that uses a woody branch to provide support and orientation. The toothlike reproductive structures are the only feature the fungi in this group have in common, and molecular studies have now confirmed that they are not all related. For instance, *Hericium* is more closely related to the mushroom genera *Russula* and *Lactarius* and not to the other tooth fungi. The amyloid, ornamented spores of all *Hericium* species are good indicators of this relationship from a morphological perspective.

Lacking a distinct cap and stem

Hericium Large, fleshy, white, cream, or rarely pinkish, often highly branched with well-developed teeth, growing on dead trees and logs

Steccherinum Small or medium, crustlike on the undersides of branches, teeth orange, short

Sarcodontia Medium to large, crustlike on the undersides of dead branches on living apple tree branches, teeth pale cream, long

Having a cap, stem present or absent, growing on wood or pine cones

Auriscalpium Growing on pine cones, small, tough, stem off-center

Climacodon Growing on living trees, usually maples, large, caps thick, shelving, stem absent

Growing on the ground or well-decayed duff

Bankera Fleshy, white, turning brown when injured, spores white (minutely bumpy)

Hydnum Fleshy, white, orange-buff, or orange-brown, turning bright orange on some when injured, spores white (smooth)

Sarcodon Fleshy, brown or more darkly colored, spores brown (warty)

Hydnellum Tough-fibrous, flesh zoned, spores vinaceous-brown (warty)

Auriscalpium vulgare Gray
PINECONE TOOTH
Small, hairy brown caps with lateral or eccentric stems; small crowded, gray teeth; stem rigid, dark brown, densely hairy; arising from buried pine cones

CAP 1–3 cm wide, convex, then plane, from the top kidney-shaped or circular, pale gray-brown at first, soon dark brown or reddish-brown, sometimes black, surface densely hairy. **FLESH** brown, thin, tough. **TEETH** ash-white, turning brown with age, crowded, tapered to a point, up to 3 mm long. **STEM** 2–8 cm long, 0.5–3 mm broad, off-center (eccentric) or attached at edge of cap (lateral), equal or swollen and spongy over base, reddish-brown or dark brown, densely hairy coated, rigid, tough. **SPORE PRINT** white. **ODOR** not distinctive. **TASTE** mild or slightly bitter.

Habit and habitat single, scattered, or clustered on rotting, often buried conifer cones on the ground, sometimes 3–5 fruit bodies per cone, mainly on various species of pine, but also reported on Douglas fir and spruce as well. August to November. **RANGE** widespread.

Spores ovoid or subglobose, minutely spiny, colorless in KOH, but amyloid, 4.5–5.5 × 3.5–4.5 μm.

Comments Nothing else is similar to this stalked, hairy, tough, cone-loving species. Often only the fruit bodies are visible, but when you pick them, the cone comes up from the needle covering since this tooth fungus is very tenacious.

Bankera violascens (Albertini & Schweinitz) Pouzar

SYNONYMS *Hydnum violascens* Albertini & Schweinitz, *Sarcodon violascens* (Albertini & Schweinitz) Quélet, *Bankera carnosa* (Banker) Snell, E. A. Dick & Taussig, *Phellodon carnosus* Banker

White, felted cap and stem that turn dark brown when handled; flesh soft turning dark brown when exposed; teeth white, turning brown when bruised; spores white; under spruce

CAP 2–4 cm wide, plane with center often bumpy, margin sometimes splitting with age, white but quickly turning brownish when bruised, with a fine soft felted coating at first that can be erect-tufted, but felt also readily squashed down from handling, becoming glabrous and smooth with age or surface cracking irregularly and then scaly. **FLESH** white then grayish-brown to dark brown when exposed, subzonate. **TEETH** white or ash-gray, staining brown from bruising, decurrent, moderately crowded, awl-shaped, 2–3 mm long. **STEM** 1–3 cm long, 10–15 mm broad, stout, central, equal with tapered base, white or ash-white as cap, turning brown from handling, finely felted, then glabrous. **SPORE PRINT** white. **ODOR** fragrant, pleasant. **TASTE** mild.

Habit and habitat single, scattered, or clustered on the ground under spruce. July to October. **RANGE** widespread.

Spores globose, spiny, colorless in KOH, not amyloid, 4–5 μm in diameter.

Comments The fruit bodies can grow around small twigs or living mosses, as they develop quickly. This fast growth is an unusual feature attributable to only some of the tooth fungi. The brown discolorations can be grayish-brown or fuscous with a tint of smoky-violet, hence the species name.

Hydnellum cristatum
(Bresadola) Stalpers

SYNONYMS *Hydnum cristatum* Bresadola, *Sarcodon cristatus* (Bresadola) Coker
Medium to large, pale cream becoming cinnamon-brown, hairy and pitted caps; teeth straw-yellow, progressively darker with pale tips, decurrent; stem color of cap, glabrous; odor farinaceous, taste farinaceous becoming peppery; under hardwoods

CAP 3–10(–19) cm wide, convex or plane with broadly sunken and undulating surface, margin irregularly lobed, creamy-white becoming buff-brown or yellowish-brown or cinnamon-brown eventually, surface plush-velvety then densely hairy, eventually pitted areas scattered over hairy surface, opaque. **FLESH** same color as cap, zoned, 5 mm thick. **TEETH** straw-yellow becoming progressively darker brown then black with tips remaining pale, long decurrent, close, round, to 5 mm long. **STEM** 2–6.5 cm long, 7–15 mm broad, central or off-center (eccentric), equal or enlarged downward, same color as cap, glabrous. **SPORE PRINT** brown with vinaceous tints. **ODOR** farinaceous. **TASTE** farinaceous, but distinctly biting peppery.

Habit and habitat single, scattered, or clustered on the ground under hardwoods or mixed hardwood and coniferous forests. July to November. **RANGE** widespread.

Spores subglobose, warty, brown in KOH, 4–5 × 3.8–4.2 µm; hyphae of cap less than 7 µm wide.

Comments A similar-looking species, *Hydnellum mirabile* (Fries) P. Karsten grows under conifers, has olive colors in the mature caps, and the flesh is duplex, having a hard layer next to the teeth and soft layers above that. *Hydnellum cristatum* lacks duplex flesh. Species of *Hydnellum* differ from those of *Sarcodon* by possessing concentric zones in the pileus flesh, and the hyphae are not inflated to 10 µm or more as found in *Sarcodon*. The flesh of *Sarcodon* species is brittle, while it is pliable-elastic for species of *Hydnellum*.

Sarcodon scabrosus (Fries)
P. Karsten

SYNONYM *Hydnum scabrosus* Fries

Medium to large, brown flattened scaly caps; teeth pale brown, decurrent; stem paler brown than cap, with dark green or black colors over and in the base; taste bitter

CAP 3–10 cm wide, convex becoming sunken centrally, margin inrolled, brown becoming reddish-brown or with flesh tints between scales, surface hairy, soon with flattened scales, opaque. **FLESH** white or flesh-brown, dark green to black in stem base. **TEETH** brown becoming darker with age and with white tips, decurrent, crowded, 2–8 mm long. **STEM** 4–10 cm long, 10–35 mm broad, central, mostly equal, paler than cap, base with white fluffy covering but also turning dark green or black over surface and in the flesh, smooth. **SPORE PRINT** brown. **ODOR** not distinctive, or farinaceous. **TASTE** usually slowly bitter.

Habit and habitat single, scattered, or clustered on the ground under conifers or mixed hardwoods. July to November. **RANGE** widespread.

Spores nearly globose or subglobose, nodulose, pale brown in KOH, 5–6 µm; 3–10% KOH on cap surface and flesh turns green, then fades.

Comments *Sarcodon imbricatus* (Linnaeus) P. Karsten has an erect scaly cap, a mild taste, and is typically found with spruce. If the cap is less scaly, the taste is intensely bitter-farinaceous, and the cap surface and flesh do not turn green in KOH, then you may have *S. fennicus* (P. Karsten) P. Karsten.

Hydnum repandum Linnaeus

SWEET TOOTH

SYNONYM *Dentinum repandum* (Linnaeus) Gray

Medium to large buff or orange cap; teeth cream or orange, stout, irregularly short decurrent; stem white or color of cap; teeth, cap flesh and stem surface may slowly stain orange-brown or brownish

CAP 2–17 cm wide, convex becoming plane, margin inrolled, lobed, sometimes deeply so, undulate, variable in color forms, buff or buff-orange, others dark orange or orange-tan, surface smooth like leather, opaque. **FLESH** white, sometimes slowly ochre-yellow when exposed. **TEETH** cream becoming pale orange, may bruise darker orange or yellowish-brown, irregularly short decurrent, moderately spaced, rather stout, short and long awl-shaped types intermingled, up to 2–7 mm long. **STEM** 2–10 cm long, 10–30 mm broad, central or off-center (eccentric), equal or enlarged downward, off-white or cream-buff or color of cap, turning brown when bruised, glabrous. **SPORE PRINT** white. **ODOR** not distinctive. **TASTE** mild or sometimes peppery.

Habit and habitat single, scattered, or clustered on the ground under conifers or mixed hardwoods. July to November. **RANGE** widespread.

Spores nearly globose, smooth, colorless in KOH, not amyloid, 6.5–9 × 6.5–8 μm.

Comments This species is considered a choice edible and easy to positively identify. However, the pure white form of this large fleshy tooth fungus, *Hydnum repandum* var. *album* (Quélet) Rea, has a bitter taste, stains orange where bruised, and is not one to be used for food.

Hydnum albidum Peck has a white cap and is found in seepage areas under mixed hardwoods from July to September. Peck described it from Sandlake, New York, and did not note the orange-staining reaction in his original description, but our collections from near that location and others described in the literature clearly discuss this distinctive feature. The small ovoid spores, the small stature, and white colors separate *H. albidum* from all other species in the genus.

Hydnum albidum

Hydnum rufescens Persoon

SYNONYMS *Dentinum rufescens* (Persoon) Gray, *Hydnum repandum* var. *rufescens* (Persoon) Barla

Orange-brown, subzonate, small to medium-sized cap; spines pale salmon-buff, not decurrent; stem white or pale flesh, bruising yellow, finely woolly covered, then glabrous

CAP 2–5 cm wide, convex, becoming plane with center broadly and shallowly sunken, ochraceous, orange-yellow or orange-brownish, becoming paler when dry, some with concentric zones, surface with a fine woolly coating at first, but becoming glabrous, opaque. **FLESH** pale flesh color, yellowing when exposed. **TEETH** pale salmon-buff, not decurrent, moderately crowded, awl-shaped, 5 mm long. **STEM** 1–4.5 cm long, 1.5–10 mm broad, central or off-center (eccentric),

equal, round or flattened, white or pale flesh pinkish, bruising dark yellow, with fine woolly coating, then glabrous. **SPORE PRINT** white. **ODOR** not distinctive. **TASTE** mild.

Habit and habitat single, scattered, or clustered on the ground in wet areas, also in *Sphagnum*, with conifers and mixed hardwoods. July to October. **RANGE** widespread.

Spores broadly ellipsoid or subglobose, smooth, colorless in KOH, not amyloid, 6.5–8 × 5.5–7 μm.

Comments Similar to *Hydnum repandum* that is usually larger and much fleshier, and has paler ochraceous or yellowish cap colors and at least partly decurrent spines on the stem. Also similar in color to *H. umbilicatum*, but that species has a sharply sunken center on the cap. *Hydnum rufescens* has clearly smaller basidiospores as well when compared to *H. umbilicatum*.

Hydnum umbilicatum Peck

SYNONYM *Dentinum umbilicatum* (Peck) Pouzar

Small to medium with orange, sharply sunken center on the cap; teeth cream or pinkish-buff, not decurrent; stem colored as cap or paler, bruising reddish-brown; usually under conifers

CAP 1–4 cm wide, convex then plane and mostly with sharply sunken center, like a belly button, margin frequently lobed, orange, orange-brown or reddish-brown, often subtly concentrically zoned, surface finely felted or spongy, especially over the center, opaque. **FLESH** white or color of cap surface, turning reddish-brown when exposed. **TEETH** cream or pinkish-buff, not decurrent, somewhat close, slender awl-shaped, up to 7 mm long. **STEM** 2–5.5 cm long, 4.5–7 mm broad, off-center (eccentric) or central, mostly equal, similar color to cap or paler and off-white, turning pale reddish-brown when bruised, with a white felted coating, then glabrous. **SPORE PRINT** white. **ODOR** not distinctive. **TASTE** mild.

Habit and habitat single, scattered, or clustered on the ground in wet areas, often boggy areas, under conifers like spruce, balsam fir and arborvitae, but also found under mixed hardwoods. July to September. **RANGE** widespread.

Spores ovoid, smooth, colorless in KOH, not amyloid, 7.5–9 × 7–8.5 µm.

Comments This species is most often confused with *Hydnum rufescens*, which is larger in size and lacks the umbilicate pileus.

Hericium americanum Ginns

BEAR'S HEAD TOOTH

Large white clusters of tight branches with long hanging teeth at the tips of each branch; growing on decaying hardwood logs and stumps or on standing and living trees

Fruit body 15–30 cm wide, tightly branched structure arising from a common base, with heads producing hanging teeth, white, but with age yellowish or brownish. **FLESH** white. **TEETH** white, close or crowded, stout, round, pointing downward, 5–20(–40) mm long. **STEM** absent or much reduced. **SPORE PRINT** white. **ODOR** not distinctive. **TASTE** not distinctive, or sweet.

Habit and habitat single or sometimes clustered on hardwood logs, stumps or standing and living trees. August to November. **RANGE** widespread.

Spores subglobose, very finely warty, colorless in KOH, amyloid, 5–7 × 4.5–6 μm.

Comments This species is commonly collected and considered a choice edible if carefully cleaned of debris and bugs. Submersing in salted water will help encourage the insects to leave. It freezes well for later use if first blanched or, better yet, lightly sautéed. Compare with *Hericium erinaceus*.

Hericium coralloides (Scopoli) Persoon, comb tooth, occurs in large white clusters with loose branching and short hanging teeth along the undersides of each branch. It too is considered a good edible once cleaned thoroughly. Occasionally a pink-colored form is collected. It may be a variant of this species or something not yet recognized as distinct. The branching on the pink form looks to be more compact and therefore different. If you find one, send me an email message and a dried sample to study.

Hericium coralloides

Hericium erinaceus (Bulliard) Persoon

BEARDED TOOTH

SYNONYMS *Hydnum erinaceus* Bulliard, *Hydnum caput-medusae* Bulliard, *Hericium caput-medusae* (Bulliard) Persoon

Large, white, unbranched ball with long hanging teeth covering the entire surface; rooted in wounds on living hardwood trees

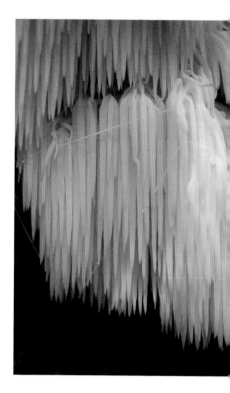

Fruit body 10–25(–40) cm wide, a ball-shaped mass covered with long fleshy teeth, arising from a single, tough, unbranched base. **FLESH** white, tough. **TEETH** white or creamy-white, with age turning yellow or brownish, especially over the tips, tightly packed, long and tapered downward, 10–40 mm long. **STEM** plug or stublike, tough, growing out of wounds on living trees. **SPORE PRINT** white. **ODOR** and **TASTE** not distinctive.

Habit and habitat single or sometimes in twos or threes on living hardwood trees such as beech, maples and oaks. August to November. **RANGE** widespread.

Spores subglobose or broadly ellipsoid, very finely warty or almost smooth, colorless in KOH, amyloid, 3–5 × 3–4 µm.

Comments It is easy to identify and considered a choice edible when young and fresh, but becomes sour tasting with age. The long hanging teeth also have prompted the use of the common name hedgehog mushroom, since it resembles the shaggy coat of that animal. See comments under *Hericium americanum* and *H. coralloides*, that differ by their branched fruit bodies with shorter teeth.

Sarcodontia setosa (Persoon) Donk

SYNONYMS *Hydnum setosum* Persoon, *Mycoacia setosa* (Persoon) Donk, *Hydnum earleanum* Sumstine, *Hydnum luteocarneum* Secretan

Crustlike on the undersides of apple tree branches, producing pale yellow or bright yellow, long, tapered, tightly packed teeth that turn wine-reddish from bruising or with age, texture tough, waxlike, odor fruity but unpleasant

Fruit body 3–10 cm wide, 5–20 cm long or more, completely attached to the undersides of living but weakened or decaying branches on living apple trees, irregular in outline, no upper surface visible (resupinate). **FLESH** yellow, tough, waxlike. **TEETH** pale or becoming bright yellow, discoloring wine red from bruising or with aging, tightly packed, tough-waxy, long with tapered tips, 5–12 mm long. **STEM** absent. **SPORE PRINT** white. **ODOR** fruity but unpleasant. **TASTE** mild.

Habit and habitat elongated patches on the undersides of damaged branches on living apple trees. July to October. **RANGE** widespread, but rarely collected.

Spores subglobose or slightly teardrop-shaped, smooth, colorless in KOH, not amyloid, 5–6 × 3.5–4 µm; with thick-walled convoluted cells (sclerocysts) in the flesh under the spines.

Comments This interesting, rarely collected species is restricted to apple trees in Europe, and perhaps also here in North America. Some accounts in northeastern North America report it from logs and standing wood of fruit trees, however the type of fruit tree was not indicated. (How about pear, plum, or peach?) Currently it is only definitely associated with old apple trees in unkempt, neglected orchards.

Climacodon septentrionalis
(Fries) P. Karsten

NORTHERN TOOTH

SYNONYMS *Hydnum septentrionale* (Fries) Banker, *Steccherinum septentrionale* (Fries) Banker

Large mass of thick, fan-shaped, overlapping, white, hairy caps; teeth long, pliant, crowded; on living maples or beech most often

CAP 10–30 cm wide, convex becoming plane or often undulating, fan-shaped and densely overlapping, several caps deep, off-white or buff, becoming cream color with age and eventually brown, surface hairy-shaggy and roughened. **FLESH** white, thick (2–4 cm), tough. **TEETH** white becoming cream, eventually brown, densely packed, soft-pliant, up to 10–20 mm long.

STEM absent, caps fused to a thick flattened mass. **SPORE PRINT** white. **ODOR** not distinctive when young, then unpleasant. **TASTE** mild when young, then bitter.

Habit and habitat single large overlapping clusters of many caps, on trunks of standing hardwoods, most often sugar maple but also beech, parasitizing the tree. Other hardwoods can be attacked as well. July to November. **RANGE** widespread.

Spores elliptical, smooth, colorless in KOH, not amyloid, 4.5–5 × 2–3 µm; cystidia fusoid, thick-walled and crystal-encrusted.

Comments This tree parasite can become quite massive and is often found growing well above ground level, so it can be challenging to examine closely. It looks like a polypore, but the teeth indicate otherwise.

Steccherinum ochraceum
(Persoon) Gray

SYNONYMS *Hydnum ochraceum* Persoon, *Climacodon ochraceus* (Persoon) P. Karsten, *Hydnum decurrens* Berkeley & M. A. Curtis, *Hydnum rhois* Schweinitz, *Steccherinum rhois* (Schweinitz) Banker

Crustlike peachy-orange patches on the undersides of downed decaying hardwood stems with small, short, crowded spines; thin, dry flesh

Fruit body 2–5 cm wide, 5–7 cm long, attached to the undersides of decaying branches, usually on the ground, lobed around the margin, upper surface when visible light gray-brown or dull white, zoned by color and velvety-hairy grooves and ridges. **FLESH** white, tough, dry. **TEETH** peach-orange or pale orange or tawny, fading to yellowish-salmon or brownish, crowded, short, blunt tipped, 1–2 mm long, often several fusing to near the tip, tips finely hairy. **STEM** usually absent. **SPORE PRINT** white. **ODOR** and **TASTE** not distinctive.

Habit and habitat as discrete circular or larger irregular patches on downed decaying hardwood branches or on stumps. August to November. **RANGE** widespread.

Spores ellipsoid or oblong-ellipsoid, smooth, colorless in KOH, not amyloid, 3–5 × 2–2.5 μm; cystidia on the spines, colorless, thick-walled, projecting beyond the basidia and usually with colorless encrusting material.

Comments Turning over moderately large decaying hardwood branches will reveal this short-toothed, colorful, resupinate species, growing flat against the undersides of the branches. It is not as common as the pure white *Irpex lacteus*, a polypore that looks like a tooth fungus and also tends to colonize smaller hardwood branches.

Club, Coral, and Fan Fungi

These fungi are grouped together because of their upright stature, like the corals of a marine reef. They can be unbranched or branched, and sometimes the branches are flattened and leaflike. As a group, they are not closely related to each other. Compare with the chanterelles.

Unbranched

Clavariadelphus Medium or large, fleshy-solid, club-shaped, mostly with brown colors; spores white or ochre

Clavulinopsis and **Clavaria** Unbranched mainly, but densely clustered or single, if bright yellow see *Clavulinopsis*, if white see *Clavaria*

Typhula (Macrotyphula) Tall, hollow, narrowly club-shaped, brown

Multiclavula Very small, white, mainly unbranched, on green-algae-covered wood or bare soil

Physalacria Small, white, cap inflated-hollow and bladderlike, densely clustered on decaying wood

Branched

Clavulina Small to medium, densely branched with upper branches flattened and cock's comblike with conical terminal teeth, white or grayish

Clavaria Small to medium, moderately branched, amethyst or purple, spores white

Artomyces (Clavicorona) Medium to large, on decaying wood, tips shaped like crowns, spores white

Ramaria Medium to large, profusely branched, usually colored (not white), spores golden-yellow or ochre-brown, on wood or on the ground

Ramariopsis Medium, white sometimes turning pink, profusely branched, branches very fragile, spores white

Branches flattened, leaflike

Sparassis Large, like a head of lettuce or cauliflower with long flattened branches bunched together, tough, white or buff colored, spores white

Thelephora Medium to large if many fused together, thick-fleshed, tough, dark brown or gray-brown, with smooth or bumpy reproductive surface, spores purple-brown

Clavariadelphus americanus (Corner) Methven

SYNONYM *Clavariadelphus pistillaris* var. *americanus* Corner

Broadly club-shaped with a rounded apex, orange-buff, vertically wrinkled; on the ground under oak and pine

Fruit body 1–3 cm wide, 3–15 cm high with up to 4 cm inserted into the substrate, unbranched and narrowly or broadly club-shaped with rounded apex, smooth becoming vertically wrinkled and grooved, orange-buff becoming orange-brown or cinnamon-brown, turning darker brown where bruised. **FLESH** white, turning erratically brown when exposed, thick. **STEM** not easily distinguished from reproductive tissues, white at base, with radiating white cords. **SPORES** white. **ODOR** and **TASTE** not distinctive.

Habit and habitat mostly in small clusters or scattered on the forest floor under mixed oak and pine. July to September. **RANGE** widespread.

Spores ovate or almond-shaped, smooth, colorless, not amyloid, 8–12 × 4–6 µm.

Comments *Clavariadelphus pistillaris* (Linnaeus) Donk is indistinguishable from *C. americanus*, except it grows mycorrhizally with beech instead of oak and pines, is bitter tasting, and the spores are larger (10.5–14 × 6–7.5 µm).

Clavariadelphus cokeri V. L. Wells & Kempton differs by its pinkish-buff or rose-pink fruit bodies that grow in bouquetlike clusters under hemlock.

Clavariadelphus flavidus Methven might also be confused with *C. americanus* since it grows in mixed coniferous forests, but the fruit bodies are bright yellow or lemon-yellow and the spores are much smaller, 7–9 × 4–5.5 µm.

Clavariadelphus truncatus Donk, flat-topped coral, has a large yellow or orange-yellow club-shaped fruit body with flattened, but often wrinkled top, and a smooth or wrinkled stem and reproductive surface turning cinnamon-brown or darker, with white base. It grows under conifers, looks somewhat like a chanterelle, and is edible.

Clavariadelphus truncatus

Typhula fistulosa (Holmskjold) Olariaga

SYNONYMS *Clavaria fistulosa* Holmskjold, *Clavariadelphus fistulosus* (Holmskjold) Corner, *Macrotyphula fistulosa* (Holmskjold) R. H. Petersen

Tall, stiffly erect, orange-brown or dark cinnamon-brown, narrowly club-shaped, hollow; on decaying hardwood branches on the forest floor

Fruit body 0.5–1 cm wide, 3–20 cm high, unbranched, narrowly club-shaped with rounded or somewhat tapered apex, stiffly erect but easily bent, smooth, ochre-yellow becoming reddish-orange-brown or dark cinnamon-brown, with white base. **FLESH** pale yellow next to outer rind, but mostly hollow. **STEM** similarly colored but clearly marked and distinct from the reproductive surface, glabrous, white at base. **SPORES** white. **ODOR** and **TASTE** not distinctive.

Habit and habitat scattered or in small or large clusters on decaying stems of mostly hardwoods, yellow birch and alder often. September to October, often after early snow fall. **RANGE** widespread but not common.

Spores teardrop-shaped or elliptical with tapered ends, smooth, colorless, not amyloid, 10–16 × 5.5–8.5 µm.

Comments One might confuse *Clavariadelphus ligula* (Schaeffer) Donk with *Typhula fistulosa* since they are both slender and club shaped, but *C. ligula* has white solid flesh, is much shorter, and grows under conifers, binding the needle duff with copious mycelial mats. *Clavariadelphus sachalinensis* (S. Imai) Corner, also found under conifers, can only be separated reliably from *C. ligula* by its larger spores (18–24 µm long vs. 12–16.5 µm for *C. ligula*).

Multiclavula mucida
(Persoon) R. H. Petersen

WHITE GREEN-ALGAE CORAL
SYNONYMS *Clavaria mucida* Persoon, *Lentaria mucida* (Persoon) Corner

Tiny, white, cylindrical, coral-like, single fruit bodies, often with tips turning brown; clustered on green-algae-covered wet wood or bare soil

Fruit body 1–1.5 mm wide, 5–15 mm high, erect, single, cylindrical or slightly inflated, but tapered to the tip, occasionally forked into one or more tapering branches in the upper half, white or pale cream or buff when fresh, the very tip becoming brown, or entirely reddish-brown with age. **FLESH** white, waxy-pliant, tough. **STALK** present but not obvious. **SPORES** white. **ODOR** and **TASTE** not distinctive.

Habit and habitat clustered on wet, green-algae-covered barkless wood or on green-algae-covered bare soil on banks. August to September. **RANGE** widespread.

Spores narrowly elliptical or cylindrical, colorless, not amyloid, 5.5–7.5 × 2–3 µm.

Comments *Multiclavula vernalis* (Schweinitz) R. H. Petersen is the other species found in the Northeast. It is frequently an alpine species, differing by its larger size, with individuals 10–30 mm high, and its orange inflated-club-shaped form with contrasting white stem. It is typically found on green-algae-covered soil.

Clavulinopsis fusiformis
(Sowerby) Corner

SPINDLE-SHAPED YELLOW CORAL
SYNONYMS *Clavaria fusiformis* Sowerby, *Ramariopsis fusiformis* (Sowerby) R. H. Petersen, *Clavaria compressa* Schweinitz, *Clavaria platyclada* Peck
Bright yellow or orange-yellow, tightly clustered and fused at the base, unbranched or very occasionally so; spores white

Fruit body 0.5–1.5 cm wide, 5–15 cm high, unbranched or sparingly so at the base or even near the apex, cylindrical or becoming flattened, often with a groove or channel, mostly pointed on the apex, bright yellow or orange-yellow, tips may become brownish with age, branches fragile and easily breaking. **FLESH** pale yellow, thin, becoming hollow. **STEM** not easily recognized, similar in color to the fertile portion or white at the very base. **SPORES** white. **ODOR** not distinctive. **TASTE** bitter.

Habit and habitat tightly clustered and fused at the bases, on humus under conifers or hardwoods. July to October. **RANGE** widespread.

Spores subglobose, smooth, colorless, not amyloid, 5.5–7(–9) × 4.5–7(–9) µm; basidia with 4 sterigmata, clamp connections present.

Comments This bright yellow clustered coral is very common and difficult to miss in the woods.

The easily overlooked *Clavulinopsis laeticolor* (Berkeley & M. A. Curtis) R. H. Petersen is similar in color but very small with a distinct yellow stem, and the fruit bodies occur as individuals scattered on humus.

Clavulinopsis gracillima (Peck) R. H. Petersen is similar but with pale apricot, pinkish-orange, or pale yellowish-buff fruit bodies.

Clavulinopsis laeticolor

Clavaria fragilis Holmskjold

WHITE WORM CORAL

SYNONYMS *Clavaria cylindrica* Bulliard, *Clavaria vermicularis* Swartz, *Clavaria eburnea* var. *fragilis* (Holmskjold) Persoon

White, clustered, erect, and coral-like, unbranched fruit bodies that are frequently curved; very fragile; on humus or soil

Fruit body 0.1–0.5 cm wide, 3–12 cm high, unbranched or sparingly so at the base or even near the apex, cylindrical or narrowly club-shaped, standing upright or undulating, round or becoming flattened, smooth, apex rounded or pointed, white or translucent, branches fragile and easily breaking, tips turning yellowish-brown. **FLESH** translucent-white, thin. **STEM** indistinct. **SPORES** white. **ODOR** and **TASTE** not distinctive.

Habit and habitat loosely or tightly clustered, on humus or soil under conifers or hardwoods. July to October. **RANGE** widespread.

Spores ellipsoid, smooth, colorless, not amyloid, 4.5–7 × 2.5–4 µm; basidia with 4 sterigmata, clamp connections absent.

Comments A very commonly found coral, this species is the white counterpart to the also commonly found, bright yellow *Clavulinopsis fusiformis*. When one tries to make a collection, the individual fruit bodies tend to break apart readily. To accurately tell the difference between the genera *Clavaria* and *Clavulinopsis*, you need a microscope to see whether clamp connections are present or absent.

Clavaria zollingeri Léveillé

SYNONYM *Clavaria lavandula* Peck
Vibrant purple or violet, small coral-like fruit bodies with rounded, antlerlike upper branches; on mossy beds under hardwoods

Fruit body 2–4(–7) cm wide, 3–8(–10) cm high, sparingly branched in the upper portions, often forking and antlerlike, branches 1–3(–6) mm wide, round, smooth, deep purple or violet or amethyst overall, fading with age, apex rounded or pointed, branches fragile and easily breaking. **FLESH** purple, thin. **STEM** not easily distinguished from branches. **SPORES** white. **ODOR** not distinctive. **TASTE** not distinctive, or of radish or cucumber.

Habit and habitat single or in small numbers on mosses and humus under hardwoods. July to September. **RANGE** widespread.

Spores ellipsoid or subglobose, smooth, colorless, not amyloid, 4–7 × 3–5 μm;
basidia with 4 sterigmata, clamp connections absent.

Comments *Clavulina amethystina* (Bulliard) Donk, also listed in some North American guides as *Clavaria amethystina* Bulliard, is similar in color and form to *Clavaria zollingeri* and can really only be separated by confirming the basidia bear only 2 sterigmata and the hyphae are endowed with clamp connections as is typical for *Clavulina* species. If the colors are tan, muted-lilac, or muted-purple and the branching is sparse with thick bumpy surfaces, the fungus is *Clavulina amethystinoides* (Peck) Corner. *Alloclavaria purpurea* (Fries) Dentinger & D. J. McLaughlin is a dull purplish or purplish-brown, unbranched, clustered coral found in coniferous forests, that produces long cylindrical sterile cells, or cystidia, projecting out from the basidia. It also has long, elliptical spores (8.5–12 × 4–4.5 μm).

Artomyces pyxidatus
(Persoon) Jülich

CROWN-TIPPED CORAL
SYNONYMS *Clavaria pyxidata* Persoon,
Clavicorona pyxidata (Persoon) Doty,
Merisma pyxidatum (Persoon) Sprengle,
Clavaria coronata Schweinitz

Large white, cream, or pale tan coral with
crownlike branch tips; on decaying logs,
common in hardwood forests

Fruit body 2–6(–10) cm wide, 6–10(–15) cm
high, erect, repeatedly branched, branches
1–5 mm wide, cylindrical or slightly
inflated, with tips forming cup-shaped
crowns bearing 4–8 conical points around
each rim, white or pale cream or buff when
fresh, becoming brown or with flesh-pink
hues, glabrous. **FLESH** white, pliant, tough.

STALK 1–3 cm long, 5–10 mm wide, colored
as branches or darker, covered with fine
hairs. **SPORES** white. **ODOR** not distinctive,
or faintly of raw potato. **TASTE** not distinctive, or slowly peppery.

Habit and habitat single, scattered, or
clustered on decaying logs or other woody
debris of hardwoods, especially willow,
poplar, aspen, maple, tulip trees. June to
September. **RANGE** widespread.

Spores elliptical, smooth, colorless, amyloid, 3.5–5 × 2–3 µm; shiny, refractive
hyphae (gloeopleurous) present in Melzer's reagent or phloxes-stained tissues.

Comments It is listed as edible and
can appear in large numbers on hardwood logs. Older field guides assign it
to *Clavicorona*.

Clavulina coralloides
(Linnaeus) J. Schröter

CRESTED CORAL

SYNONYMS *Clavaria coralloides* Linnaeus, *Clavulina cristata* (Holmskjold) J. Schröter, *Clavaria cristata* (Holmskjold) Persoon, *Ramaria cristata* Holmskjold

Small white, pale yellow or pale pinkish-tan branched coral with clusters of white, toothlike conical tips; branches fragile, easily breaking; on the ground in mixed woods with conifers; spores white

Fruit body 2–5(–10) cm wide, 2–8(–10) cm high, branched sparingly, with branch apex jagged from clusters of small tooth-like conical tips, arranged somewhat like a cock's comb and flattened, white or pale cream or pale pinkish-tan on the main branches, but the tips and the stem base white, branches fragile and easily breaking. **FLESH** white, fragile. **STEM** 0.5–3 cm long, 3–5 mm wide, white, glabrous. **SPORES** white. **ODOR** and **TASTE** not distinctive.

Habit and habitat single or scattered or clustered on humus under conifers or mixed conifers and hardwoods. June to October. **RANGE** widespread.

Spores subglobose, smooth, colorless, not amyloid, 7–11 × 5.5–7(–10) μm; basidia with only 2 incurved sterigmata, clamp connections present.

Comments Previously known as *Clavulina cristata*. *Clavulina cinerea* (Bulliard) J. Schröter is dark gray overall. *Clavulina rugosa* (Bulliard) J. Schröter has white bumpy-undulate branches and lacks the fringed apices. All three species can be infected in the stem base by a black ascomycete parasite, producing tiny black bumps, or perithecia, in the stem, turning the coral stem and lower branches progressively black.

Ramaria stricta (Persoon)
Quélet

STRAIGHT-BRANCHED CORAL
SYNONYMS *Clavaria stricta* Persoon, *Clavaria stricta* f. *fumida* (Peck) R. H. Petersen, *Lachnocladium odoratum* G. F. Atkinson

Tall, straight, multiple buffy-brown-colored branches with yellow tips; on downed logs; anise odor and bitter taste

Fruit body 4–10 cm wide, 4–14 cm high, abundantly and repeatedly branched, branches long, straight, parallel, compact, the tips yellow when young and the remainder mostly pale yellowish-buff, becoming orange-buff or brownish with spore maturation, bruising purplish-brown. **FLESH** white, turning brown when exposed, tough. **STEM** 0.2–2 cm long, 3–15 mm wide, sometimes not obvious, colored as branches, fuzzy over the base, also with white cords radiating into the substrate. **SPORES** dark golden-yellow. **ODOR** like anise or fragrant or not distinctive. **TASTE** bitter or astringent-bitter.

Habit and habitat single or scattered or clustered on well-decayed downed logs of hardwoods or conifers. June to October. **RANGE** widespread.

Spores elliptical, warted-bumpy, pale yellow-tan, 7–10 × 3.5–5.5 μm.

Comments A common wood-inhabiting coral fungus in the Northeast that appears earlier than other corals found on wood. *Ramaria concolor* (Corner) R. H. Petersen is similar, except the branch tips are the same color as the lower branches. *Ramaria rubella* (Schaeffer) R. H. Petersen [syn. *R. acris* (Peck) Corner] differs by its darker vinaceous-cinnamon-colored branches overall, its strongly acrid taste, and the white cords at the base turning bright pink in 10% KOH. If the branch tips, axils, or stem base have green colors, the taste is slightly acrid, the basal white cords do not turn pink in KOH, then it is *R. apiculata* (Fries) Donk. If the tips are shaped like little crowns and the branches are more yellow than brownish, see *Artomyces pyxidatus*.

Ramariopsis kunzei (Fries)
Corner

WHITE CORAL
SYNONYMS *Clavaria kunzei* Fries,
Ramaria kunzei (Fries) Quélet, *Clavulina
kunzei* (Fries) J. Schröder, *Clavulinopsis
kunzei* (Fries) Jülich, *Clavaria asperula*
G. F. Atkinson, *Clavaria velutina* Ellis &
Everhart

White, abundantly branched coral that
develops a pink flush, with branches fragile, easily breaking; on the ground in mixed
woods; white spores

Fruit body 2–12 cm wide, 2–10 cm high,
abundantly and repeatedly branched,
white but often with pink tints, especially
in age, branches fragile and easily breaking. **FLESH** white, fragile. **STEM** 0.5–3 cm
long, 3–10 mm wide, white or darker
pink than upper branches, fuzzy over
the base. **SPORES** white. **ODOR** and **TASTE**
not distinctive.

Habit and habitat single or scattered or
clustered on humus under hardwoods
or mixed with conifers. July to October.
RANGE widespread.

Spores broadly elliptical or subglobose,
minutely spiny, colorless, 3–5 × 3–4.5 μm.

Comments This can be a very common
species in summer and fall. It looks like it
could be a species of *Ramaria* since they
are usually highly branched, but *Ramaria*
species produce golden-yellow or ochre-
brown spores and the branches usually
have at least some color. See also *Clavulina coralloides*.

Sparassis spathulata
(Schweinitz) Fries

EASTERN CAULIFLOWER MUSHROOM

SYNONYMS *Merisma spathulatum* Schweinitz, *Sparassis simplex* D. A. Reid, *Sparassis herbstii* Peck

Large bouquetlike fruit body of white or pale yellow, densely clustered, vertical, flattened but wavy branches with colored zones on the tips; fleshy but tough; on the ground under oaks

Fruit body 10–45 cm wide, 10–40 cm high, a dense cluster of vertical, flat, undulating long branches, emerging from a buried base, white or pale yellow or yellowish-buff and each flattened branch with distinct colored zones on the tips. **FLESH** white, somewhat tough. **SPORES** white. **ODOR** and **TASTE** mild.

Habit and habitat single on the ground at the base of hardwood trees, especially oak, which it is parasitizing. June to October. **RANGE** widespread.

Spores elliptical, smooth, colorless, not amyloid, 6–8 × 5–6 µm; clamp connections absent.

Comments *Sparassis spathulata* is considered edible, but use the tender tips of the branches. *Sparassis americana* R. H. Petersen, formerly called *S. crispa* (Wulfen) Fries or the American cauliflower mushroom in our field guides in North America, differs by the thinner, crispy, nonzonate branch tips. DNA evidence has revealed that *S. cripsa* does not occur in North America, so *S. americana* has been described for our eastern North American taxon that is similar to the European *S. crispa*.

Thelephora terrestris Ehrhart

COMMON FIBER VASE

SYNONYM *Thelephora crustosa* Lloyd
Dark brown, fibrous-scaly cap; grayish-brown, bumpy undersurface; stem not obvious; tough-pliant; under conifers

Fruit body composed of one or more laterally fused often overlapping caps 2–5 cm wide and high, these also in rings 12 cm wide or more, individual caps flat but sunken centrally, radially wrinkled or fibrous-scaly, or both, margin white or cream when fresh, otherwise brown or dark brown and darker with age. **FLESH** brown, tough. **UNDERSURFACE** grayish-brown, radially wrinkled or folded, covered with small bumps or smooth. **STALK** when present short and tapered to base, covered with reproductive bumpy surface. **SPORES** purple-brown. **ODOR** moldy. **TASTE** mild.

Habit and habitat single or scattered or often clustered on the ground, in thick moss, in needle duff, under conifers with which it is mycorrhizal. June to December. **RANGE** widespread.

Spores angular-elliptical-nodulose and with short projecting spines (0.5 μm), pale brown in 3% KOH, not amyloid, 7–10 × 5–6 μm; clamp connections present.

Comments Of the several species of *Thelephora* in our region, the one most similar to *T. terrestris* is *T. americana* Lloyd. It is mycorrhizal with hardwoods and has smaller spores (6.5–8 × 4.5–6.5 μm) with longer spines (to 1 μm). *Thelephora anthocephala* (Bulliard) Fries is tough-pliant, coral-like, and grayish-brown with white tips. *Thelephora vialis* Schweinitz looks like a pagoda with many overlapping branches in a rosette fashion. *Thelephora palmata* (Scopoli) Fries is another coral-like form with fetid odor, that grows under conifers. These are the more commonly encountered ones.

Physalacria inflata (Schweinitz) Peck

BLADDER STALKS

SYNONYMS *Leotia inflata* Schweinitz, *Mitrula inflata* (Schweinitz) Fries, *Spathularia inflata* (Schweinitz) Cooke

Densely clustered, small white collapsing bladders on long white fuzzy stems; on decaying wood

CAP 1–2(–4) cm high, 0.1–1 cm wide, soft, rounded bladderlike, irregularly collapsed and folded or flattened, white or pale cream, smooth. **FLESH** thin, membranous, hollow inside. **STEM** 0.3–2 cm long, 0.2–0.5 mm wide, equal and round, minutely fuzzy, white, firm. **SPORE PRINT** white. **ODOR** and **TASTE** not distinctive.

Habit and habitat typically densely clustered on bark less wood of hardwoods. July to September. **RANGE** widespread.

Spores ellipsoid, smooth, colorless, 4–6 × 2–3.5 µm.

Comments *Physalacria* is phylogenetically related to the gilled mushrooms, but since it looks somewhat coral-like, I have placed it here among similar-looking fungi. Compare with the dry grayish-colored jelly fungus *Phleogena faginea* that can also be found on standing dead beech trees. Although *Phleogena* may resemble a large slime mold sporangium, it lacks the distinctive fibrous mesh (capillitium) of a slime mold.

Puffballs, Earthstars, Stinkhorns, Bird's Nest Fungi, and Allies

This group is also known as the gasteromycetes, or stomach fungi, since they produce and then release their spores inside a saclike structure within the fruit body, the "stomach," before the spores are exposed to air, water, or insects for dispersal. Most all are saprobes, using decaying organic materials like wood chips and humus to obtain their food. However, a few are mycorrhizal (*Calostoma*, *Pisolithus*, and *Scleroderma*) and are found growing with certain types of trees. The agroforest industry uses *Pisolithus* to help develop seedlings and young trees of conifers for reforestation plantings.

Some true puffballs have good flavor and are worth learning if you enjoy foraging for wild "mushroom" foods, for example, some species of *Lycoperdon* and *Calvatia*. Some Asian cultures use stinkhorn eggs and other parts of the fruit bodies as food, but the odor may give you pause.

In this group of fungi, insects often play a role in spore dispersal. The odor of stinkhorns attracts carrion flies that come to consume the sticky, stinky, mucous-like gel in which the spores mix, and unknowingly they fly away carrying the spores stuck to their bodies. These hitchhiking spores then spread to other woody substrates when the flies land, rubbing off the spores. In the bird's nest fungi, raindrops initially eject the spore packets, called peridioles, from the splash cup, but insects then aid in releasing the spores from the peridioles by chewing on these lentil-sized structures, exposing the spores to wind. The rest need some type of mechanical pressure on the spore sac—such as raindrops, falling twigs, or the feet of mice or larger animals—to puff the spores up into the air where they can be carried away on air currents.

Puffballs, earthstars, earthballs, and stalked puffballs

SPORE SAC OPENING BY AN IRREGULAR SPLITTING OR FISSURE ON THE TOP

Calvatia Medium to very large (basketball or larger), spore sac thin, constantly soft, fleshy, spores olive-brown

Bovista Medium (baseball-sized), inner spore sac thin, papery, dark colored, spores chocolate

Pisolithus Medium to large, on a thick stem, sac thin, lentil-sized peridioles producing cinnamon-brown spores

Scleroderma Medium or smaller, skin thick, white, spores black

SPORE SAC OPENING BY A PORE

Geastrum and *Astraeus* Stemless, but outer skin splitting into starlike rays, exposing sac with pore

Calostoma Sac with red puckered "mouth," sitting on a buried fibrous stalk, all wrapped in gel at first

Lycoperdon With or without a squat, fleshy stem, outer skin of sac granular or with soft spines

Bird's nest fungi

Crucibulum Peridioles white in a firm cup

Cyathus Peridioles gray in a firm cup

Nidularia Peridioles dark brown in a woolly sac

Stinkhorns

Lysurus Claws short, pink, free at tips, slime on the insides, at top of long, white spongy stem

Pseudocolus Claws long, red, fused at the tips, slime on the insides, stem short

Mutinus Stem tapered to tip, cap when visible, completely fused with stem, covered with smelly olive-brown slime

Phallus Cap margin free from stem, covered with slime, stem long, thick, spongy

Bovista pila Berkeley & M. A. Curtis

TUMBLING PUFFBALL

Grassy areas, round fruit bodies with a shiny bronze-brown or grayish-brown papery-brittle covering that splits open irregularly; extremely light and easily blown about

Fruit body 3–6(–9) cm, spherical or ellipsoid, the skin two layered, outer layer white, turning flesh-pink when handled, thin and flaking off revealing an inner brown or grayish-brown, glossy, thin, paperlike covering splitting open irregularly at the top, attached at the base by an easily broken cord, very light when mature and easily moved by air currents. **FLESH** white, then chocolate-brown and filled with dense fibrous mesh (capillitium). **STEM** absent. **SPORE COLOR** dark chocolate-brown. **ODOR** and **TASTE** not distinctive.

Habit and habitat single or scattered on the ground in grassy areas, pastures, around horse stables and cattle barns, in open woodland areas. July to October. **RANGE** widespread, persistent, year-round.

Spores globose, smooth, brown in KOH, not amyloid, 3.5–4.5 µm in diameter, often with short, colorless, cylindrical projection (pedicel); fibrous mesh (capillitium) dark brown, thick-walled, highly branched with sharply tapered tips.

Comments *Bovista plumbea* Persoon is a smaller (to 3 cm broad), less commonly collected species, characterized by the bluish-gray to purplish fruit body with numerous fibers anchoring it to the soil. *Bovista pila* can also be grayish, causing confusion. The spores for *B. plumbea* are larger and ovoid (5–7 × 4.5–6 µm) with long (to 14 µm) cylindrical colorless projections. These latter features are the best for separating these two species.

Calvatia gigantea (Batsch) Lloyd

GIANT PUFFBALL
SYNONYMS *Lycoperdon giganteum* Batsch, *Langermannia gigantea* (Batsch) Rostkovius, *Bovista gigantea* (Batsch) Gray
In grassy or sparsely wooded areas, very large, white, rounded fruit bodies with soft tanned leather feel when fresh; inside white at first, becoming yellowish then olive-brown with developing powdery spores

Fruit body 20–60 cm or more broad and high, rounded, soccer or basketball like, sometimes broader than tall, white, smooth like soft tanned leather, eventually cracking irregularly, attached by a rootlike cord. **FLESH** white, eventually yellowish then olive-green or olive-brown from spores maturing. **STEM** absent. **SPORE COLOR** olive-brown. **ODOR** and **TASTE** mild.

Habit and habitat single or scattered or in arcs or large circles on the ground in grassy areas, pastures, edges of meadows, under brushy areas near pastures, in open wooded areas. June to October. **RANGE** widespread.

Spores globose, minutely warted, pale brown in KOH, not amyloid, 3–5.5 μm in diameter; fibrous mesh (capillitium) thick-walled with pinhole pits in the walls.

Comments Most mycophagists consider this to be a fine edible fungus. The flesh must be completely white or the flavor will be off, and the firmer the flesh, the better. In some previous field guides this species is listed as *Langermannia gigantea*. See *Calvatia candida* for more information on other *Calvatia* species that can be found in the Northeast.

Calvatia candida (Rostkovius)
Hollós

ORANGE STAINING PUFFBALL
SYNONYMS *Langermannia candida* Rostkovius, *Lycoperdon candidum* (Rostkovius) Bonorden, *Calvatia rubroflava* (Cragin) Lloyd, *Lycoperdon rubroflavum* Cragin

In grassy areas, rounded fruit bodies with thick, tapered base, white but quickly staining bright yellow, eventually turning orange to reddish-brown; spore mass olive; odor disagreeable

Fruit body 2–10 cm broad, 2–5 cm high, rounded above, but base tapered and furrowed or pleated or pockmarked, becoming flattened, white with flesh-pink hues mixed in, turning yellow or reddish-orange when handled, eventually darkening with age, outer skin thin, a single layer, eventually falling off. **FLESH** white, turning yellowish-orange when exposed, eventually olive-green. **STEM** thick, tapered, sterile base of cottony tissue. **SPORE COLOR** olive. **ODOR** disagreeable with age. **TASTE** not distinctive.

Habit and habitat single or scattered or in circles on the ground in grassy areas, pastures. July to November. **RANGE** widespread in the southern areas of the Northeast.

Spores globose, weakly ornamented, pale brown in KOH, not amyloid, 3–5 µm in diameter, with short, colorless, cylindrical projection (pedicel); fibrous mesh (capillitium) with walls pitted.

Comments Earlier field guides identify this attractive but unpleasantly smelly puffball as *Calvatia rubroflava*. *Calvatia cyathiformis* (Bosc) Morgan, a choice edible when the flesh is pure white, produces purple-brown spores with age, is larger and tan-colored at first, does not stain yellow, and lacks a distinctly tapered stem base. *Calvatia craniformis* (Schweinitz) Fries is also a choice edible that is similar to *C. cyathiformis* but differs by its yellowish-green or yellowish-brown spores and the flattened patches covering the violet-brown skin that does not stain yellow.

Lycoperdon perlatum Persoon

GEM-STUDDED PUFFBALL

SYNONYMS *Lycoperdon gemmatum* Batsch, *Lycoperdon lacunosum* Bulliard

White, shaped like an upside-down pear, with detachable conical-granular spines when young, becoming bronze-brown with numerous circular scars from detached spines; the spore sac with olive-brown powdery spores and a fine mesh with age

Fruit body 2–6(–9) cm broad, 3–7(–9) cm high, pear-shaped, with stem bearing a rounded or abrupt inflated turban-shaped head, white becoming buff or pale brown, eventually bronze-brown, at first covered with brown tipped, white, granular, conical spines or warts that detach leaving round scars, spore sac opening by apical pore. **FLESH** white, then yellowish, or olive-green and eventually olive-brown from spores and fine fibrous mesh (the capillitium). **STEM** stout, equal or tapered to base, covered with smaller granular-conical spines, fleshy-chambered inside and spongy. **SPORE COLOR** olive-brown. **ODOR** and **TASTE** mild.

Habit and habitat single, scattered, or clustered on the ground in conifer or hardwood forests, also may be found in urban parks. July to October. **RANGE** widespread.

Spores globose, minutely spiny, pale brown in KOH, not amyloid, 3.5–4.5 µm in diameter, lacking a short hyaline peg; fibrous mesh (capillitium) thick-walled and with small circular pits in the walls.

Comments This is a choice edible when pure white throughout; if discoloring has begun, it is not recommended for the table. *Lycoperdon pyriforme*, another very common puffball in the Northeast, is brown and grows on wood. *Lycoperdon molle* Persoon has violet or purplish-gray-tinted spores, a well-developed stem, and finely granular spines on the surface of the spore sac.

Lycoperdon pyriforme
Schaeffer

PEAR-SHAPED PUFFBALL
SYNONYM *Morganella pyriformis* (Schaeffer) Kreisel & D. Krüger
Brown smooth puffball with olive-brown spores; attached to decaying wood by white cords

Fruit body 2–4(–5) cm broad, 2–3(–5) cm high, pear-shaped or subglobose, pale brown or almost white at first, soon darker brown, smooth at first or with tiny pale brown or nearly white granular spines or warts that disappear with maturity, opening by apical pore. **FLESH** white, becoming yellowish, then olive-green and eventually olive-brown from spores and fine fibrous mesh (the capillitium). **STEM** short, tapered or absent, when present, white and chambered inside, attached to substrate by white cords. **SPORE COLOR** olive-brown. **ODOR** mild. **TASTE** mild.

Habit and habitat single, scattered, or clustered on decaying wood or woody debris in conifer or hardwood forests. July to November. **RANGE** widespread.

Spores globose, smooth, pale brown in KOH, not amyloid, 3–4.5 μm in diameter, lacking a short hyaline peg; fibrous mesh (capillitium) slightly thick-walled and sparingly branched.

Comments This is another choice edible puffball when pure white throughout. It was moved to *Morganella* in 2003 based on molecular studies, then in 2008 moved back to *Lycoperdon* after expanded molecular studies. Compare with *Lycoperdon subincarnatum*. Like all members of *Lycoperdon*, it develops an apical pore in the spore sac for spore dispersal, and the spore sac is filled with a fine fibrous mesh to aid in mechanical ejection of the spores when raindrops, twigs, or other objects like mice feet smash into the spore sac, puffing the spores into the air.

Lycoperdon subincarnatum
Peck

SYNONYM *Morganella subincarnata* (Peck) Kreisel & Dring

Flesh-brown or pinkish-brown with conical pyramidal spines at first; spines falling off, leaving pockmarked surface; connected by white, thick cords; on moss-covered hardwood logs

Fruit body 1–3 cm broad, round, sometimes pear-shaped, pale flesh-brown or pinkish-brown at first, with cinnamon-brown or reddish-brown spines that converge into pyramidal warts, spines fall off leaving a pitted surface, spore sac opening by apical irregularly shaped pore. **FLESH** white, becoming dark olive-green, eventually developing purple or pinkish-brown colors, lacking a well-developed fibrous mesh when mature. **STEM** short, tapered, or absent, when present, white inside, attached to substrate by conspicuous, thick, white cords. **SPORE COLOR** gray with pink tint or pale purplish-brown. **ODOR** and **TASTE** mild.

Habit and habitat single, scattered, or clustered on decaying moss-covered wood or woody debris of hardwoods. July to November. **RANGE** widespread.

Spores globose, distinctly warty-spiny, pale brown in KOH, not amyloid, 3.5–4 μm in diameter, lacking a short hyaline peg; fibrous mesh (capillitium) lacking, but colorless, thin-walled, septate filaments (pseudocapillitium) present.

Comments This puffball is common, found on moss-covered hardwood logs in the Northeast. The spores are deep olive-green before developing the characteristic grayish-pink color, a feature not noted in other field guides but mentioned in Peck's original description. The most frequently encountered wood-inhabiting puffball is *Lycoperdon pyriforme*, which is typically larger, with a distinct stem. It is crust-brown-colored with a smooth surface or with tiny brown granular warts and a yellow-green fibrous mesh in the spore sac when mature.

Lycoperdon marginatum
Vittadini

White round puffball with long slender fused pyramidal spines; outer layer sloughing off in patches or sheets, leaving a brown smooth spore sac with an apical pore

Fruit body 1–5 cm broad, 2–4 cm high, round or flattened, white occasionally brown, spines long white converging into pyramidal warts of 2–4 spines fused at the tips, warts falling off in patches or sheets, exposing an olive-brown, smooth or granular and sometimes pitted, papery spore sac with an apical pore. **FLESH** white, becoming brown with a well-developed fibrous mesh (capillitium). **STEM** absent or short and rooting, with white sterile, chambered tissue supporting the spore producing tissue (the gleba), attached to substrate at a single point. **SPORE COLOR** olive-brown. **ODOR** and **TASTE** mild.

Habit and habitat single, scattered, or clustered on hard packed sandy soil, footpaths, trails, roadsides, may be found on the ground under hardwoods and conifers as well. July to November. **RANGE** widespread.

Spores globose, smooth or minutely punctate from tiny channels in the spore walls, pale brown in KOH, not amyloid, 3.5–4.2 μm in diameter, often with a short hyaline peg; fibrous mesh (capillitium) present, slender, thick-walled.

Comments *Lycoperdon curtisii* Berkeley looks similar in the early stages of development, but the spines fall off individually and not in patches or sheets. Another similar species, *L. echinatum* Persoon has long white spines that turn brown, then fall off individually, leaving pockmarks on the spore sac. Its spores are purple-brown.

Pisolithus arhizus (Scopoli) Rauschert

DYE-MAKER'S FALSE PUFFBALL
SYNONYMS *Lycoperdon arrizon* Scopoli, *Pisocarpium arhizum* (Scopoli) Link, *Pisolithus arenarius* Albertini & Schweinitz, *Scleroderma tinctorium* Persoon, *Pisolithus tinctorius* (Persoon) Coker & Couch

Large, dusty reddish-brown cankerous growth, thrusting out of sandy or hard-packed soil; in early stages with white, yellow or brownish lentil-sized pellets developing in the spore sac

Fruit body 4–10(–20) cm broad, 5–18(–30) cm high, pear-shaped or top-shaped with thick tapering base in the substrate, occasionally globose and lacking a rootlike base, surface olive-yellow or ochraceous-brown or very dark brown to black, skin thin, shiny, easily breaking apart exposing white or yellow to brownish lentil-shaped units (peridioles) that are 1–2 mm thick and 4 mm long, embedded in a black jelly, these units eventually release powdery spores, maturing in waves from the top to the bottom. **FLESH** purplish-brown or black, with paler colored peridioles packed tightly in, eventually entirely powdery and reddish-brown. **STEM** (pseudostem) equal or tapering somewhat downward, colored as the spore sac or darker, with yellow cords often attached to the base. **SPORE COLOR** reddish-cinnamon. **ODOR** and **TASTE** mild.

Habit and habitat single, scattered, or clustered in sandy soil, pastures, lawns, hard packed soil of roadsides, mostly around pine trees. July to October. **RANGE** widespread.

Spores globose, spiny, brown in KOH, not amyloid, 7–12 μm in diameter including the length of the spines that can be up to 2 μm long.

Comments Not the prettiest fungus, but it is known for aggressively partnering with forest trees, especially pines. The agroforestry industry uses it to rapidly develop seedlings for outplanting and forest reclamation. The mycorrhizal partnership provides significant advantage to the trees, especially in poor soils or other poor growing conditions. The fruit bodies can be used to produce colorful dyes as well, hence the common name.

Scleroderma citrinum Persoon

PIGSKIN POISON PUFFBALL

SYNONYMS *Scleroderma aurantium* var. *macrorhizum* (Fries) Šebek, *Scleroderma vulgare* Hornemann, *Scleroderma lycoperdoides* var. *reticulatum* Coker & Couch

Medium to large yellow-brown, scaly, cracked earthball, with a thick white "skin" and a dark purplish-black spore mass inside; opening by irregular splitting of the sac

Fruit body 2–10 cm broad, 2–4 cm high, round but soon flattened, firm or hard, yellow between the brown scales or yellowish-brown with darker brown scales, like a cracked mudflat, with smaller brown warts on each scale, wall of the spore sac thick, white, turning pink when fresh, eventually opening by irregular splitting. **FLESH** white, but soon dark purple-black in the spore sac from spores maturing, lacking a fibrous mesh (capillitium). **STEM** absent, attached to substrate from nub producing white cords. **SPORE COLOR** black or purplish-black. **ODOR** and **TASTE** mild.

Habit and habitat single, scattered, or clustered on the ground with mosses and often on well-decayed wood under hardwoods and conifers. July to November. **RANGE** widespread.

Spores globose, reticulate, dark brown in KOH, not amyloid, 8–13 μm in diameter; fibrous mesh (capillitium) absent, but colorless, thin-walled filaments present (pseudocapillitium).

Comments This species is the most commonly found earthball. It forms a mycorrhizal symbiosis with trees and is a bolete relative. One can find a small bolete, *Pseudoboletus parasiticus*, occasionally parasitizing and growing directly on the fruit bodies of this earthball. If the "skin" is flexible, not hard, and the scales are very small like dots, then you have found *Scleroderma areolatum* Ehrenberg that has spiny, not reticulate spores.

Scleroderma septentrionale
Jeppson

Sandy areas, especially dunes, a globose warty spore sac attached to a long, knobby, rooting, false stem; spore sac splitting open irregularly releasing dark grayish-black powdery spores

Fruit body 2–6 cm broad, round spore sac above, deeply rooting below, spore sac firm, pale yellow or with ochraceous-brown or brown colors mixed in, turning slowly red when injured, the surface brown scaly at first, becoming smooth with maturity, the wall of the sac 1–3 mm thick, white, opening by irregular splitting, sometimes with star-shaped fringe. **FLESH** white, soon dark gray or purplish-brown from spores maturing, lacking a fibrous mesh (capillitium). **STEM** (pseudostem) 3–9 cm long, 2–4 cm thick, equal or enlarged downward, cracked and lumpy, covered with sand grains, colored as the spore sac where visible, yellow or brownish in the cracked areas. **SPORE**

COLOR grayish-black or purplish-brown. **ODOR** and **TASTE** mild.

Habit and habitat single, scattered, or clustered in sandy soil under conifers, such as pitch pine and white pine, in areas with sand dunes. July to November. **RANGE** Limited to either oceanic or isolated inland areas with sand esker-conifer ecosystems.

Spores globose, reticulate-spiny, brown in KOH, not amyloid, 10–16 μm in diameter without spines, spines 2–4 μm long.

Comments *Scleroderma meridionale* Demoulin & Malençon, described from the Mediterranean region in Europe, is similar and found in the southern areas of North America. It differs by the smooth or finely felted spore sac, the intense sulfur-yellow color of the pseudostem, and the shorter spines, 1–2 μm, of the spores. When the fruit bodies of *S. septentrionale* are young and not yet producing spores, the warted, cracked spore sac reminds one of *S. citrinum*.

Calostoma cinnabarinum
Desvaux

STALKED PUFFBALL-IN-ASPIC
SYNONYM *Mitremyces cinnabarinus* (Desvaux) De Toni

Bright red, gel-covered, deeply rooting, stalked puffball, with raised scarlet-red puckered "mouth" on top of the red spore sac; mountainous areas with oaks

Fruit body entirely gelatinous covered at first, with a 1–2 cm broad round spore sac on deeply rooting stem, gelatinous covering translucent but with embedded reddish-orange seedlike pieces, the gel and "seeds" falling away from a bright red spore sac revealing a raised and puckered, deep scarlet-red "mouth" on the top of the spore sac that will eventually rupture for spore dispersal, spore sac slowly losing the red powdery covering and turning yellow with age. **FLESH** white then cream color. **STEM** (pseudostem) 2–4 cm long, 1–2 cm thick, equal, shaggy or fluted or lumpy, covered with translucent gel, colored as the spore sac. **SPORE COLOR** cream. **ODOR** and **TASTE** not distinctive.

Habit and habitat single, scattered, or clustered on the ground under hardwoods with oaks present, often in moss beds. April to October. **RANGE** widespread, in mountainous areas.

Spores oblong or elliptical, pitted-reticulate, colorless in KOH, not amyloid, 14–20 × 6.5–8.5 μm.

Comments *Calostoma lutescens* (Schweinitz) Burnap is similarly gelatinous-covered, but it produces a yellow spore sac with a bright red, puckered "mouth" and has a distinct collar at the base of the spore sac after the gel has dropped off. *Calostoma ravenelii* (Berkeley) Massee also has a yellow spore sac with a bright red "mouth" but lacks the gelatinous outer covering.

Geastrum saccatum Fries

ROUNDED EARTHSTAR

SYNONYM *Geastrum lloydianum* Rick

Round spore sac with circular ring around the pore, sac nestled down in thick, white, recurved rays; fruit body attached at one point

Fruit body 1–3 cm broad unopened, rounded or egg-shaped with pointed beak 1–2 mm high, buff or pale cinnamon-brown, surface felted, rubbery outer skin splitting into 4–9 white triangular rays curling back and under, revealing a rounded, pale gray spore sac, 0.5–2 cm broad, with a central apical pore surrounded by a paler colored, circular depression or ridge. **FLESH** white, solid, then powdery and brown at maturity. **STEM** lacking, attached at a point. **SPORE COLOR** brown. **ODOR** and **TASTE** not distinctive.

Habit and habitat clustered on rich humus and often around decaying woody debris in hardwood or coniferous forests. May to November. **RANGE** widespread.

Spores globose, with truncate warts, brown in KOH, not amyloid, 3.5–4.5 µm in diameter; fibrous mesh (capillitium) with thick, yellow-brown walls encrusted with debris.

Comments *Geastrum fimbriatum* Fries is similar but lacks the ring around the pore on the spore sac, and the spores are smaller (3–3.5 µm).

Geastrum fornicatum (Hudson) Hooker has its spore sac on a short pedestal and the rays point straight down, balancing all on the tips of the rays.

Geastrum pectinatum Persoon differs by the long slender stalk at the base of the spore sac and the beaked and lined pore at the top of the spore sac.

Astraeus hygrometricus (Persoon) Morgan, barometer earthstar, opens in wet weather and closes in dry conditions. The rays are dark brown and cracked.

Geastrum triplex Junghuhn, collared earthstar, has an egg-shaped fruit body with a well-developed conical beak and opens to reveal a spore sac sitting on a fleshy saucerlike collar surrounded by fleshy rays. No other species produces a fleshy collar.

Geastrum triplex

Crucibulum laeve (Hudson)
Kambly

WHITE-EGG BIRD'S NEST
SYNONYMS *Peziza laevis* Hudson, *Crucibulum vulgare* Tulasne & C. Tulasne, *Cyathus crucibulum* Persoon, *Nidularia crucibulum* Fries, *Nidularia laevis* (Hudson) Hudson

Very small yellowish-brown cups with tiny white eggs; usually clustered on woody debris

Fruit body resembling minute bird's nests, 5–12 mm wide, 5–10 mm high, squat vase-shaped cup with flared margin, externally yellow or tan or cinnamon-brown, at first covered by a woolly yellow or yellowish-brown lid (epiphragm), but this soon disintegrating, exposing the smooth inner shiny, white or pale ashy-gray surface, filled with smooth, tan or buff, tough, lentil-shaped eggs, 1–2 mm broad, eggs attached to the cup wall by a thin cord. **FLESH** of cup single layered. **SPORE COLOR** white or pale cream. **ODOR** not distinctive. **TASTE** not known.

Habit and habitat usually clustered on woody debris like sticks, leaves, nut shells, bark, also on wood chips frequently. July to November. **RANGE** widespread.

Spores elliptical, smooth, colorless, 7–10 × 3–6 µm.

Comments The eggs, actually peridioles, containing the spores are ejected from their nests, or splash cups, by raindrops. The cords attached to the eggs have a sticky pad at the opposite end, that can attach to woody substrates, causing the eggs to smash into the woody substrate, breaking the package open and releasing the spores to the wind. If the eggs are dark gray or black, you have a species of *Cyathus. Cyathus stercoreus* (Schweinitz) De Toni has golden to rusty-brown hairs outside the smooth inner cup, while *C. striatus* (Hudson) Willdenow is fuzzy gray to brown outside, and distinctly grooved or lined inside the cup. These two are the most common *Cyathus* species in the Northeast.

Nidularia pulvinata
(Schweinitz) Fries

PEA-SHAPED NIDULARIA
SYNONYM *Cyathus pulvinatus* Fries
Very small, soft, woolly brown balls, easily torn open, revealing dark gray-brown eggs packed into a sticky gel

Fruit body 1.5–10 mm wide, rounded, ball- or cushion-shaped, smooth at first, but somewhat lumpy with age from eggs pressing outward on the soft walls, brown or grayish-brown, fading to nearly white, shaggy-woolly or velvety, tearing open irregularly, revealing dark gray-brown, smooth, tough, lentil-shaped eggs, about 1 mm broad, embedded in a sticky gel. **FLESH** of cup insubstantial. **SPORE COLOR** pale, white or cream. **ODOR** not distinctive. **TASTE** not usually tasted.

Habit and habitat scattered or clustered on the upper sides of decaying barkless logs and other woody substrates. August to October. **RANGE** widespread.

Spores elliptical, smooth, colorless, 6–10 × 4–7 µm.

Comments Of the four genera of bird's nest fungi, two have a cord attached to the egg that aids in spore dispersal, *Cyathus* and *Crucibulum*, while the other two, *Nidularia* and *Nidula*, have eggs that lack a cord but are covered with sticky gel. Like *Cyathus* and *Crucibulum*, *Nidula* uses a splash cup mechanism (see *Crucibulum laeve* for a description of this process), but *Nidularia* lacks this standard cup. *Nidula* seems to be rarely collected in the Northeast but common in the Northwest.

Lysurus cruciatus (Leprieur & Montagne) Hennings

LIZARD'S CLAW

SYNONYMS *Aserophallus cruciatus* Leprieur & Montagne, *Anthurus cruciatus* (Leprieur & Montagne) E. Fischer, *Anthurus borealis* Burt, *Lysurus borealis* (Burt) Hennings

White, spongy, round stem bearing 5–7 short, erect, white or pinkish claws that are coated on the inside with a smelly olive-brown slime; coming out of white, saclike cups buried in the substrate

Fruit body starting as white 5-cm-long eggs, these producing long spongy stems with short erect triangular claws at the apex. **CAP** composed of 5–7 hollow, triangular, erect, free, unfused "claws," 1–2 cm wide, 2–5 cm high, claws white or pink colored, flattened or grooved externally, but on the inside they are rounded and covered with olive-brown slime. **STEM** 10–15 cm long, 1–1.5 cm broad, round, white, sponge-like and perforated, internally hollow, with a white saclike cup at the base, with radiating white cords. **FLESH** white or pale yellow, fragile, chambered. **SPORE COLOR** olive-brown. **ODOR** unpleasant. **TASTE** unknown.

Habit and habitat may be single but usually clustered on woody debris, straw, humus, on lawns, in gardens, under various trees. August to September. **RANGE** widespread.

Spores subcylindrical, smooth, olive-greenish, 3–4 × 1–2 µm; 5–8 spores per basidium.

Comments The short claws usually remain erect, only rarely opening outward like petals on a flower. The color of the claws is variable from off-white to pale cream or pinkish. *Lysurus mokusin* (Linnaeus) Fries is somewhat similar but smaller, with claw tips fused and a pink, angular stem. *Pseudocolus fusiformis* (E. Fischer) Lloyd has a short stem and long claws fused at the tips.

Mutinus ravenelii (Berkeley & M. A. Curtis) E. Fischer

SYNONYMS *Corynites ravenelii* Berkeley & M. A. Curtis, *Dictyophora ravenelii* (Berkeley & M. A. Curtis) Burt, *Ithyphallus ravenelii* (Berkeley & M. A. Curtis) E. Fischer

Dark pink or rosy-red, spongy, cylindrical stems bearing abruptly marked-off darker reddish, tapered heads, covered with green smelly slime; on wood chips and woody debris

Fruit body starting as white 2–4-cm-long ovate eggs, producing spongy stems bearing smelly slime over the upper tapered tip. **CAP** not obvious, but upper 3–5 cm tapered and abruptly marked-off by darker red color and texture, covered with olive-brown smelly slime, except for very apex with small pore, dark pink or rosy-red under the slime and spongelike with minute holes. **STEM** 4–8(–15) cm long, 0.5–2.5 cm broad, cylindrical, pink, with larger holes, hollow, surrounded by a white saclike cup. **FLESH** colored as surface, thin. **SPORE COLOR** olive-brown. **ODOR** strongly unpleasant, stinky. **TASTE** unknown.

Habit and habitat usually in clusters in wood-chip beds, sawdust piles, gardens, cultivated areas, pastures, lawns, parks. July to September. **RANGE** widespread.

Spores elliptical, smooth, 3.5–5 × 1.5–2.5 µm.

Comments *Mutinus elegans* (Montagne) E. Fischer, the elegant stinkhorn, can be similar in color or even with orange hues, but it differs by lacking a clearly marked "head," with the slime running further down the curved and more gradually tapered apex. *Mutinus caninus* (Hudson) Fries, the dog stinkhorn, has the abruptly marked-off head covered in spore slime, but the colors of the head and stem are various shades of orange, some quite pale.

Phallus rubicundus (Bosc) Fries

DEVIL'S STINKHORN

SYNONYMS *Satyrus rubicundus* Bosc, *Ithyphallus rubicundus* (Bosc) E. Fischer, *Phallus gracilis* (E. Fischer) Lloyd, *Ithyphallus rubicundus* var. *gracilis* E. Fischer

Tall, scarlet or pinkish-orange, spongy stalks, bearing sharply conical caps with olive-brown smelly slime; often on wood chips in hot, humid weather

Fruit body white 1–3 cm thick eggs, developing into long tapered cylinders with thimblelike caps. **CAP** 3–4 cm high, sharply conical or narrowly bell-shaped with a swollen perforate ring at the top, the base flaring out from the stem, covered with olive-brown, smelly slime until flies have removed this layer, then scarlet or orange or pinkish and smooth or wrinkled. **FLESH** colored as surface, thin. **STEM** 9–15(–20) cm long, 1–2.5 cm broad, cylindrical but often tapering at the apex, colored as cap, paler toward the base, coarsely spongelike with elongate holes, hollow, arising from a white to brownish saclike cup. **SPORE COLOR** olive-brown. **ODOR** quite unpleasant, like a poorly maintained outhouse. **TASTE** unknown.

Habit and habitat clustered in wood-chip beds, sawdust piles, gardens, cultivated areas, pastures, lawns, parks, and so on. July to September. **RANGE** limited to southern area in the Northeast.

Spores oblong, smooth, 3.5–5 × 1.5–2.5 µm.

Comments Resembling *Mutinus elegans* somewhat, *Phallus rubicundus* clearly differs by the free conical cap. Previously known from the southern states but now more frequently found in the Northeast, from Illinois to upstate New York, on wood-chip mulch used to dress up spaces around trees, walkways, and flower gardens. This is a subtropical fungus that is slowly invading northward.

Phallus ravenelii Berkeley & M. A. Curtis, Ravenel's stinkhorn, is a large, erect, white, spongy stem emerging from a pink saclike cup, tipped with a conical, smelly, olive-brown, slime-covered cap that flares out from the stem.

Phallus impudicus Linnaeus is the same size as *P. ravenelii* with similar features, except the cap is reticulate-ridged. See also *P. duplicatus*.

Phallus ravenelii

Phallus duplicatus Bosc

NETTED STINKHORN

SYNONYM *Dictyophora duplicata* (Bosc) E. Fischer

Large, erect, white, spongy stem, tipped with a conical, smelly, olive-brown, slime covered, reticulate-ridged cap that flares out from the stem; white lacey skirt hanging down between cap and stem

Fruit body as pink or off-white, ovate eggs, producing long spongy stems bearing smelly slime-covered caps. **CAP** 3–4 cm wide, 3–5 cm high, conical, attached to a white doughnut-shaped, perforate ring, hanging down, flaring out and free on the lower edge, covered with olive-brown smelly slime, ridged-reticulate underneath. **SKIRT** (indusium) white, lacey or fishnetlike, hanging down between the cap and stem for 3–6 cm, often flaring out. **STEM** 5–17 cm long, 3–4.5 cm broad, cylindrical, white, spongy with large holes, hollow, surrounded by a saclike cup, with white cords radiating from the cup. **SPORE COLOR** olive-brown. **ODOR** unpleasant, stinky. **TASTE** not usually tasted.

Habit and habitat single but usually in clusters in wood-chip beds, sawdust piles, gardens, cultivated areas, pastures, lawns, in parks, on stumps in wooded areas. July to October. **RANGE** widespread.

Spores elliptical, smooth, colorless, 3.5–4.5 × 1–2 μm.

Comments The lacey hanging veil, the indusium, can be quite attractive when it flares out like a flowing skirt, but the odor may keep you from handling such a beauty. If you can find the eggs—they do not smell bad—take some back and plant them in a flowerpot with loose soil. Watch for the emergence over the next day or so, and be vigilant. The fruiting process only takes a day once it starts.

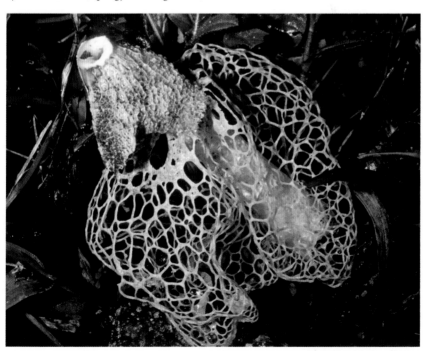

Jelly Fungi

The jelly fungi are distinguished from all other basidiomycetes by their septate, or partitioned, basidia, either horizontally septate, as in *Auricularia* and *Phleogena*, or vertically septate, as in *Exidia* and *Tremella*. Thus, for genera with vertically septate basidia, the basidia appear cruciate, or partitioned by a cross, when looking down on them from the top. This group also mostly produces gelatinous, jellylike fruit bodies, but some are tough and dry, as in *Phleogena* and *Tremellodendron*. *Tremella* species parasitize other fungi, as do *Syzygospora* species, while *Tremellodendron* and *Sebacina* form symbiotic mycorrhizae with plants, and the rest seem to be saprobic wood recyclers.

Dacrymyces, a saprobe on wood, is jellylike, but the basidia are not septate. *Dacrymyces* species produce a single-celled basidium shaped like a tuning fork, very different. A yellow species of *Tremella* will look quite similar to *Dacrymyces*, so look for the fungus being parasitized next to the *Tremella* to help make an identification in the field. If the decaying wood is bare, *Dacrymyces* is a reasonable guess.

Gelatinous, jellylike, yellow or orange-yellow

Dacrymyces Growing in bare wood with a rooting base
Tremella Growing on or near other fungi on wood, like polypores or *Stereum* species

Gelatinous, white or cream

Syzygospora Growing on mushrooms
Tremella Growing on other fungi on wood, like polypores or *Stereum*
Sebacina Growing on the ground and coral-like, or growing over the base of plant stems, usually green and herbaceous plants

Gelatinous, black or dark reddish-brown

Exidia Brainlike, disclike with ridges,
Tremella Leaflike, in bouquetlike clusters

Dry or flexibly waxy, coral-like, white, yellow, or gray

Calocera Small, yellow, sparingly branched
Tremellodendron Medium, white or cream, densely branched, tough
Phleogena Tiny, single unbranched, round cap, dry, tough

Gelatinous, jellylike, with cap and teeth

Pseudohydnum Colorless or pale brown on cap

Dacrymyces chrysospermus
Berkeley & M. A. Curtis

SYNONYMS *Dacrymyces palmatus* Bresadola, *Dacrymyces palmatus* Burt, *Tremella palmata* Schweinitz

Lobed or wrinkled, yellow or orange-yellow, slippery jelly, with white tough, rooting base; growing on decaying conifer logs

Fruit body 2–6 cm wide, 1–2.5 cm high, lobed or deeply wrinkled or brainlike, attached to the substrate at one point, orange-yellow or bright yellow, smooth, glabrous, gelatinous and slippery on the surface, with a white tough, rooting base, drying dark reddish and becoming a shiny-horny, brittle membrane. **FLESH** yellow, translucent, gelatinous. **SPORES** yellow. **ODOR** and **TASTE** not distinctive.

Habit and habitat single, scattered, or clustered on decaying conifer logs. May to November. **RANGE** widespread.

Spores broadly cylindrical and slightly curved, like fat sausages, smooth, colorless, having up to 7 cross walls, 17–25 × 6–8 µm; basidia cylindrical and Y-shaped or tuning-fork-shaped.

Comments *Tremella mesenterica* Retzius and *T. aurantia* Schweinitz look somewhat like *Dacrymyces chrysospermus* by the yellow, gelatinous, lobed fruit bodies, except they lack the white rooting base and both produce inflated, cruciate basidia. *Tremella mesenterica* parasitizes the wood-decay fungus *Peniophora* on hardwood branches, while *T. aurantia* parasitizes the bracket fungus *Stereum hirsutum*, also on hardwoods. *Tremella mesenterica* has thinner, more membranous folds than *T. aurantia* as well. *Tremella encephala* produces small pinkish brainlike balls with a tough white core while attacking *Stereum sanguinolentum* on decaying conifers.

Exidia glandulosa (Bulliard)
Fries

BLACK JELLY ROLL

SYNONYMS *Tremella glandulosa* Bulliard, *Exidia arborea* (Hudson) Saccardo

Small or medium-sized, black, brain-shaped, gelatinous masses that have bumps or are smooth; growing on dead hardwood branches

Fruit body 1–2 cm wide, fusing into larger masses, reaching 20–50 cm across, attached directly to the substrate, translucent colorless and lumpy at first, soon reddish-brown or blackish-brown and irregularly lobed, brainlike, smooth or warty overall, becoming flattened and crustlike in dry weather, but inflating again with moisture. **FLESH** thin, gelatinous, colored as surface. **SPORE COLOR** white. **ODOR** and **TASTE** not distinctive.

Habit and habitat small or large clusters on hardwood branches and sticks, especially oak, often on the ground. May to November. **RANGE** widespread.

Spores sausage-shaped, smooth, colorless, 10–16 × 4–5 µm; basidia inflated with internal longitudinal walls making 4 compartments, forming a cross from the top looking down.

Comments This black jelly fungus has been described many times under different names. Several other species of *Exidia* are in our area. *Exidia alba* (Lloyd) Burt, now classified as *Gloeotromera alba* (Lloyd) Ervin, remains pigmentless, pinkish, or even pale orange, and the fruit bodies are waxy gelatinous feeling and have gloeocystidia (cells containing oil droplets) in the tissues. *Exidia nucleata* (Schweinitz) Burt is vinaceous with small white, hard, granular spheres embedded in the gelatinous tissues. See further discussions under *Exidia recisa*.

Exidia recisa (Ditmar) Fries
AMBER JELLY ROLL
SYNONYM *Tremella recisa* Ditmar

Clusters of reddish or purplish-brown, jellylike cups or flattened lobes with ridges separating smooth concave surfaces; attached at a single point; on dead hardwood branches

Fruit body 1–4 cm wide, fusing into larger masses, attached at a single point to the woody substrate, often cuplike with raised edges and the concave depressions separated by ridges or the surface wrinkled, yellowish-brown or reddish-brown or purplish-brown, smooth on one side, with minute scales on the other, becoming flattened and crustlike in dry weather, but inflating again slowly with moisture. **FLESH** thin, firm-gelatinous, colored as surface. **SPORE COLOR** white. **ODOR** and **TASTE** not distinctive.

Habit and habitat small or medium-sized clusters, growing on hardwood branches and sticks, especially oak. May to November. **RANGE** widespread.

Spores sausage-shaped, smooth, colorless, 10–14 × 3–5 μm; basidia inflated with internal longitudinal walls making four compartments, forming a cross from the top looking down.

Comments This species and the black jelly fungus, *Exidia glandulosa*, are the two most common jellies found on downed decaying hardwood branches. If you find a brownish, tough-gelatinous, brainlike jelly on conifers, you most likely have *E. saccharina* Fries. See further comments under *E. glandulosa*.

Tremella foliacea Persoon

JELLY LEAF

SYNONYMS *Exidia foliacea* (Persoon)
P. Karsten, *Tremella ferruginea* Smith,
Tremella fimbriata Persoon, *Tremella
succinea* Persoon

Large cluster of tightly packed, cinnamon
or reddish-brown leaflike folds attached at
a single point; on decaying logs or various
types of woody debris

Fruit body 5–12(–20) cm wide, 5–10 cm
high, a tight cluster of gelatinous-rubbery,
leaflike folds, attached at a single point to
the woody substrate but lacking a distinct
stem, cinnamon-brown or reddish-brown
or purplish-brown, drying darker and
almost black, smooth on both surfaces.
FLESH thin, firm-gelatinous, colored as
surface. **SPORE COLOR** pale yellow. **ODOR**
and **TASTE** not distinctive.

Habit and habitat medium to large clusters
on decaying wood of hardwoods and conifers. July to November. **RANGE** widespread.

Spores subglobose or broadly elliptical,
smooth, colorless, 8–12 × 7–9 µm; basidia
inflated with internal longitudinal walls
making four compartments, and forming
a cross from the top looking down.

Comments Compare with *Exidia recisa*
that has very small lobes that are more
disclike than leaflike, but the colors can be
similar. Many species of *Tremella* can be
found in our area, but this one is very distinctive in form and color.

Sebacina sparassoidea
(Lloyd) P. Roberts

SYNONYMS *Tremella sparassoidea* Lloyd, *Corticium tremellinum* var. *reticulatum* Berkeley, *Tremella reticulata* (Berkeley) Farlow, *Tremella clavarioides* Lloyd

Coral-like, white, stubby, rubbery, hollow branches; on the ground in mixed moist woods

Fruit body 5–15 cm wide, 3–8 cm high or more, branching, coral-like, the branches thick, blunt, rubbery or firm-gelatinous and becoming hollow, white but becoming grayish-cream or ashy white or pale tan with age, smooth, often the branches with perforations or holes with age. **FLESH** thin, rubbery-elastic but fragile. **SPORE COLOR** white. **ODOR** and **TASTE** not distinctive.

Habit and habitat usually single on the duff in moist areas in conifer and mixed hardwood forests. July to October. **RANGE** widespread.

Spores elliptical, smooth, colorless, 9–11 × 5–7 μm; basidia inflated with internal longitudinal walls making four compartments, and forming a cross from the top looking down.

Comments In previous field guides as *Tremella reticulata*. However, since *Tremella* species are parasites on other fungi, and *Sebacina sparassoidea* is mycorrhizal, they really cannot be considered closely related. The species name *reticulata* was occupied in *Sebacina*, which dictates using the next available, oldest, name of a synonym, thus *sparassoidea*. Compare with *Tremellodendron schweinitzii*.

Tremellodendron schweinitzii (Peck) G. F. Atkinson

JELLIED FALSE CORAL

SYNONYMS *Thelephora schweinitzii* Peck, *Thelephora pallida* Schweinitz, *Sebacina pallida* Oberwindler, Garnica & K. Riess, *Tremellodendron pallidum* Burt

White, tough, coral-like fruit bodies, somewhat rubbery; on the ground under hardwoods

Fruit body 5–15 cm wide, 3–10 cm high, branching, coral-like, branches round at first, becoming flattened and tough-rubbery, fused laterally at the bases, white or buff, becoming yellowish or pale tan with age, smooth, often the branch tips becoming frayed with age. **FLESH** colored as surface, tough-pliant, solid. **STEM** composed of multiple fused branches of the same color, arising from matted white mycelium fusing leaves and debris in the substrate. **SPORE COLOR** white. **ODOR** not distinctive. **TASTE** not distinctive, or slightly bitter.

Habit and habitat single, scattered, or clustered on the duff, bare soil, mosses under hardwood forests. July to October. **RANGE** widespread.

Spores sausage-shaped, smooth, colorless, 7–11 × 4–6 µm; basidia inflated with internal longitudinal walls making 4 compartments, and forming a cross from the top looking down (tremelloid basidium).

Comments Though listed in previous field guides as *Tremellodendron pallidum*, the correct name is *T. schweinitzii*. A recent molecular study indicates more than one species may be displaying this morphotype, that is, there may be cryptic species that look alike but are genetically different. Detailed studies of populations will be necessary to sort out how many different species may be involved. In any case, they appear to be mycorrhizal, probably with oaks and perhaps other hardwoods.

Calocera cornea (Batsch) Fries

SYNONYMS *Clavaria cornea* Batsch, *Calocera palmata* (Schumacher) Fries, *Tremella palmata* Schumacher

Small, gelatinous, yellow or orange-yellow, unbranched or sparingly branched, coral-like fruit bodies with sharply tapered tips; on barkless decaying logs

Fruit body 2–3 mm wide, 1–2 cm high, coral-like with cylindrical or more often sharply tapered tips, single or sparingly branched, sometimes forked, orange-yellow or bright yellow, smooth, glabrous, gelatinous or somewhat waxy and flexible, typically with a white fuzzy base, or naked and inserted into the wood, bases often fused. **SPORES** yellow. **ODOR** not distinctive. **TASTE** not usually tasted.

Habit and habitat clustered and often in rows on barkless decaying logs, mostly hardwoods but occasionally conifers as well. July to September. **RANGE** widespread.

Spores cylindrical and slightly curved, smooth, colorless, often with 1 cross wall, 7–10 × 3–4 µm; basidia cylindrical and Y-shaped or tuning-fork-shaped.

Comments *Calocera viscosa* (Persoon) Fries is larger, up to 10 cm high, with repeatedly, evenly forked, round-tipped, slippery branches, occurring on conifer wood.

Syzygospora mycetophila
(Peck) Ginns

SYNONYMS *Tremella mycetophila* Peck, *Exobasidium mycetophilum* (Peck) Burt, *Christiansenia mycetophila* (Peck) Ginns & Sunhede, *Carcinomyces mycetophilus* (Peck) Oberwinkler & Bandoni

White or cream-yellow, gelatinous-waxy, saucer-shaped, or convoluted tumorous growths on the fruit bodies of *Gymnopus dryophilus*, the oak-loving collybia

Fruit body a tiny or massive tumorlike outgrowth on the mushroom *Gymnopus dryophilus*; tumors rounded, saucer-shaped, brainlike, growing single or clustered on the mushroom cap, gills, or stem; white or cream-yellow, gelatinous or somewhat waxy and flexible. **SPORES** white. **ODOR** not distinctive. **TASTE** not usually tasted.

Habit and habitat parasitic and occurring only on *Gymnopus dryophilus*. July to September. **RANGE** widespread.

Spores ellipsoid or cylindrical, smooth, colorless, 6–9 × 1.5–2.3 μm; basidia single-celled with 4 spores; clamp connections present.

Comments If such tumors are found on *Rhodocollybia butyracea*, then the mycoparasite is most likely *Syzygospora tumefaciens* (Ginns & Sunhede) Ginns, or if just on the lamellae of *R. butyracea*, then it is *S. norvegica* Ginns. *Syzygospora marasmoidea* Ginns is found completely deforming the pileus of *Marasmius pallidocephalus* Gilliam. Each of these species of *Syzygospora* differs by its microscopic features as well as its particular microhabitat. Five other species of *Syzygospora* are known, all are mycoparasites of *Gymnopus*, *Rhodocollybia*, or crust fungi on decaying wood.

Pseudohydnum gelatinosum (Scopoli) P. Karsten

JELLY TOOTH, JELLY FALSE TOOTH

SYNONYMS *Hydnum gelatinosum* Scopoli, *Steccherinum gelatinosum* (Scopoli) Gray, *Tremellodon gelatinosus* (Scopoli) Fries

White or gray or sometimes brownish, completely gelatinous fruit bodies; laterally stalked, with tongue-shaped cap and decurrent, conical, white teeth; on decaying conifer wood

CAP 1–7 cm wide, convex, becoming plane, tongue or kidney-shaped from the top, variable in color, translucent off-white or pale gray or brown, smooth or somewhat fuzzy, gelatinous but dry to touch. **FLESH** translucent, thick, gelatinous. **TEETH** white or grayish or with fain blue tint, decurrent, short, conical, 2–4 mm. **STEM** when present, 3–6 cm long, 5–20 mm broad, lateral, equal or tapered to the base, colored as cap or paler, gelatinous. **SPORE PRINT** white. **ODOR** and **TASTE** not distinctive.

Habit and habitat single, scattered, or in small clusters, sometimes overlapping, on decaying conifer wood. August to November. **RANGE** widespread.

Spores globose, smooth, colorless, 5–7 µm; basidia inflated with internal longitudinal walls making 4 compartments and forming a cross from the top looking down (tremelloid).

Comments True tooth fungi are fleshy-brittle and not gelatinous. They also have 1-celled basidia. The jelly tooth or jelly false tooth is listed as edible but really has no flavor of its own.

ASCOMYCETES

The ascomycetes produce their spores in microscopic cylindrical tubes or sacs, the asci, that act as water cannons, shooting the spores up into the air to be carried away and dispersed by air currents. The asci are arranged in fruit bodies of three main types: the apothecium, that can be shaped like a sponge, brain, tongue, or cup; the perithecium, that is a minute flask with a porelike opening (usually many are embedded in a more solid support tissue called the stroma); and the cleistothecium, also minute but a completely enclosed ball, typical of the powdery mildews on plants in your gardens and hedgerows in the fall. Morels, false morels, cup fungi, and earth tongues are examples of apotheciate ascomycetes, with their reproductive surface exposed or displayed openly for spore dispersal. The earth tongue allies, because they are erect like the earth tongues, for example, *Cordyceps* and *Podostroma*, are really more closely related to

other perithecium-producing ascomycetes, such as *Daldinia*, *Hypomyces*, and *Xylaria*. For convenience of identification, I have placed them together based on fruit body form, not because they are closely related.

Of the more than 100,000 species of fungi described so far, 64,000 are ascomycetes. They produce moderately large fruit bodies, though not as large as some of the basidiomycetes, and they produce some of the smallest fruit bodies, about the size of a pinhead. In this group we find the most expensive food on the planet, the truffles of Italy and France, and some fine gourmet morels from our local woods, but the group mainly includes lots of wood-decaying fungi, plant parasites, and even animal parasites (ringworm). Many also form symbioses with plants (mycorrhizae) and algae (the lichens). First I present the morels and false morels, followed by the cup fungi, earth tongues, and their allies.

Morels and False Morels

Morels and false morels are ascomycetes that produce an exposed apothecium on a stem. The shape of the reproductive apothecium, and in some cases the color, along with the types of spores, determine the genus. *Morchella* species are choice edible fungi, whereas most *Gyromitra* and *Helvella* species produce toxins and should not be consumed.

Cap poroid or spongelike

Morchella Apothecium poroid, spongelike with pits and ridges

Cap brainlike or saddle-shaped, brown or white

Gyromitra Apothecium either folded and brainlike or saddle-shaped, brown- or white-colored.

Cap saddle-shaped or cup-shaped, gray or black

Helvella Apothecium saddle-shaped, but gray or black or cup-shaped

Morchella americana Clowez & Matherly

YELLOW MOREL

SYNONYMS *Morchella californica* Clowez & D. Viess, *Morchella claviformis* Clowez, *Morchella populina* Clowez & Lebeuf

Only in the early spring, on the ground under hardwoods, with oval, spongelike, gray becoming yellowish caps; stem whitish with a granular surface, hollow, fragile

CAP 7–11 cm high, 4–5 cm wide, oval or broadly conical or irregularly cylindrical, attached to the stem, spongelike with pits and ridges, pits rounded or elongated vertically or laterally, generally pale grayish or pale brown when young, becoming cream-yellow or yellowish-tan with age, hollow, inside colored as surface, granular inside. **FLESH** firm but fragile-brittle. **STEM** 2–12 cm long, 20–90 mm wide, equal or enlarged downward, off-white or buff or sordid cream-buff, densely granular overall, longitudinally ribbed, hollow and like the cap inside. **ODOR** and **TASTE** not distinctive.

Habit and habitat single, scattered, or clustered on the ground in soil, under white ash, in apple orchards, standing dead elms, mixed hardwoods, old abandoned railroad beds, steep banks, burned areas and along stream sides. April to May. **RANGE** widespread.

Spores ellipsoid, smooth, colorless, with spherical droplets on the outer ends, 17–24 × 11–15 µm; asci with thin even walls, and apex with flattened or collapsing "lid" (operculate), not turning blue amyloid.

Comments After the black morels have fruited, the yellow morels appear, usually within just a few weeks the succession begins. Former names for this species are *Morchella esculenta* and, more recently, *M. esculentoides*. It is a highly prized edible when prepared properly and is a lot of fun to hunt for, once the snow has departed. Many new species of morels have been described in North America based on DNA evidence, only some of the common species will be considered here.

Morchella angusticeps Peck

BLACK MOREL

Only in the early spring, on the ground under conifers or hardwoods, with conical, spongelike, black ridged cap, on whitish stem with a granular surface, hollow, fragile

CAP 2–9 cm high, 2–6 cm wide, conical, sometimes oval-shaped, attached to the stem with a flaring rim, spongelike with pits and well-defined ridges, ridges black or dark brown, pits yellow-brown or grayish-brown and elongated longitudinally, hollow, buff colored and granular inside. **FLESH** firm but fragile-brittle. **STEM** 5–10 cm long, 20–40 mm wide, equal or enlarged downward, off-white or buff or sordid cream-buff, densely granular overall, longitudinally ribbed, hollow, like the cap inside. **ODOR** and **TASTE** not distinctive.

Habit and habitat single, scattered, or clustered on the ground in sandy or mineral soil, under conifers or hardwoods or in mixed forests, often associated with dead standing elms. April to May. **RANGE** widespread.

Spores ellipsoid, smooth, colorless, with spherical droplets on the outer ends, 22–28 × 11–15 μm; asci with thin even walls, and apex with flattened or collapsing "lid" (operculate), not turning blue amyloid.

Comments When the lilacs begin to break bud or before, look for this very early fruiting species. It is edible and choice with thorough cooking. Dried and used later to make sauces, it provides excellent flavor. Stomach upset has been reported for some, especially if not cooked thoroughly or if consumed with alcohol. Most individuals do not have these problems, but when first trying any new wild "mushroom," eat small amounts to test your digestive system.

Morchella punctipes Peck

HALF-FREE MOREL

Only in the early spring, on the ground under hardwoods, with short, conical, half-free, yellowish-brown spongelike cap, with ridges becoming darker; stem whitish with a densely granular surface, hollow, fragile

CAP 1–4 cm high, 1–4 cm wide at the base, short conical and mostly free from the stem, attached about the midpoint, spongelike with vertically elongated pits and ridges, generally yellowish-brown, ridges may become darker brown to black, hollow. **FLESH** firm but fragile-brittle. **STEM** 2–15 cm long, 10–20 mm wide or up to 40 mm in base, equal or enlarged downward, off-white or buff or sordid cream-buff, densely granular overall, longitudinally ribbed, hollow, like the cap inside. **ODOR** and **TASTE** not distinctive.

Habit and habitat single, scattered, or clustered on the ground in soil, under hardwoods, wherever other species of morels are found. April to May. **RANGE** widespread.

Spores ellipsoid, smooth, colorless, with spherical droplets on the outer ends, 24–30 × 12–15 µm; asci with thin even walls, and apex with flattened or collapsing "lid" (operculate), not turning blue amyloid.

Comments This species formerly has been labelled *Morchella semilibra* in older field guides, but DNA evidence indicates that species is only found in Europe. Our eastern half-free morel is fairly common in the early spring. Listed as good, not choice, the flesh is not as substantial as that of the black and yellow morels.

Gyromitra esculenta
(Persoon) Fries

CONIFER FALSE MOREL
SYNONYM *Helvella esculenta* Persoon

Cap brown, wrinkled brainlike, chambered on inside; stem white, hollow but often chambered on inside; under conifers

CAP 4–10 cm high, 3–12 wide, convoluted-folded, brainlike or occasionally saddle-shaped, also wrinkled over surface, margin inrolled, underside not visible, reddish-brown or with some yellow mixed in, dark bay-brown or even nearly black with age, smooth, dry. **FLESH** pale tan, brittle, chambered. **STEM** 3–9 cm high, 10–35 mm wide, white or pale cream-yellow, round or longitudinally compressed, or fluted and occasionally bumpy, with small branlike dots or glabrous, hollow with 1–2 chambers. **ODOR** and **TASTE** not distinctive.

Habit and habitat scattered or clustered on the ground in needle litter under conifers or poplar or aspen. April to June. **RANGE** widespread.

Spores long ellipsoid or somewhat fusiform, smooth, colorless but often with two large shiny oil droplets inside, 17–22 × 7–9 µm; asci with thin even walls, and apex with flattened or collapsing "lid" (operculate), not turning blue amyloid.

Comments Poisonous. It produces a toxin that can cause death. As the name *esculenta* implies, it can only be eaten after treating the fruit bodies to remove the toxin, a compound similar to one used to manufacture rocket fuel. The treatment involves drying then rehydration, or several rounds of boiling and discarding the cooking water each time. This may or may not remove all the volatile toxin. Compare with species of *Morchella* that have a spongelike cap and are excellent edibles.

Gyromitra infula (Schaeffer) Quélet

SADDLE-SHAPED FALSE MOREL
SYNONYM *Helvella infula* Schaeffer

Brown saddle-shaped cap with smooth or slightly wrinkled surface; stem paler than cap, smooth or irregular, but not ribbed; often on decaying conifer wood

CAP 2–13 cm high, 2–10 wide, saddle-shaped with two or occasionally three upraised lobes, smooth or wrinkled, not brainlike, margin where may be connected to stem at one point, underside when visible white or brown, variable in color, tan or more often brown or reddish-brown, dry. **FLESH** white or brown, brittle. **STEM** 3–8 cm long, 20–30 mm wide, paler than cap, often white or pale buff, equal or enlarged downward, smooth or irregular but not ribbed or folded. **ODOR** and **TASTE** not distinctive.

Habit and habitat single or scattered on rotting wood or on the ground in humus under conifers and or mixed forests. July to October. **RANGE** widespread.

Spores ellipsoid, smooth, lacking bumps at either end, colorless, often with two shiny oil droplets inside, 17–24 × 7–12 µm; with thin even walls, and apex with flattened or collapsing "lid" (operculate), not turning blue amyloid.

Comments Like *Gyromitra esculenta*, this species contains toxins that can make you very ill. *Gyromitra ambigua* has red and purple hues in the cap, longer spores, 22–33 × 8–12 µm, and distinctive external bumps at either end of the spores. It is found less frequently in the northernmost areas and is also toxic.

Gyromitra korfii (Raitviir) Harmaja is a large often square, yellowish or chestnut-brown cap with longitudinal folds and multichambered inside; stem white with longitudinal folds, stocky. It is our most

Gyromitra korfii

common false morel found under hardwoods in the early spring. It was named for Richard "Dick" Korf, an ascomycete taxonomist at Cornell University.

Helvella lacunosa Afzelius

FLUTED BLACK HELVELLA

SYNONYMS *Helvella mitra* Schaeffer, *Helvella costata* Berkeley, *Helvella sulcata* Afzelius

Dark gray, brown or black irregularly lobed or saddle-shaped cap, underside of cap gray, smooth, ribbed; stem white then gray, paler than cap, deeply ribbed and fluted

CAP 1–5 cm high, 1–5(–10) wide, variously shaped, irregularly lobed or saddle-shaped or like a miter cap, dark brown or dark gray or black, smooth but often wrinkled, the margin attached to the stem in places, the underside gray, smooth, often ribbed. **FLESH** thin, brittle. **STEM** 2–15 cm long, 10–30 mm wide, equal, deeply ribbed and fluted, may be white when young, then gray, surface glabrous, hollow and chambered. **ODOR** and **TASTE** not distinctive.

Habit and habitat single, scattered, or clustered on the ground or among mosses on wet ground under conifers and in boggy areas. June to November. **RANGE** widespread.

Spores broadly ellipsoid, smooth, colorless with one oil droplet inside, 14–22 × 10–14 µm; asci thin walled, operculate, not turning blue amyloid.

Comments The colors and the fluted stem are distinctive. However, the shape of the cap and the color combinations of cap and stem vary a great deal.

There may be many cryptic species involved here. For instance, the former *Helvella lacunosa* on the West Coast is now divided into *H. vespertina* N. H. Nguyen & Vellinga and *H. dryophila* Vellinga & N. H. Nguyen, the former associated with conifers, the latter with oak trees.

Helvella crispa (Scopoli) Fries is similar to *H. lacunosa* with the fluted stem, but the cap is white.

Helvella elastica Bulliard, smooth-stalked helvella, is grayish-brown, shallowly saddle-shaped cap with inrolled margin, underside of cap pale tan or white and glabrous; stem paler than cap, mostly cream-buff.

Helvella stevensii Peck is similar to *H. elastica* in color and shape of cap and stem, but the underside of the

Helvella elastica

cap is hairy and the margin of the cap rolls up and back over the surface.

If the cap and stem are dark grayish or grayish-black and the undersurface of the cap is fuzzy, you may have *Helvella pezizoides* Afzelius.

Helvella macropus (Persoon)
P. Karsten

LONG-STALKED GRAY CUP
SYNONYMS *Peziza macropus* Persoon, *Cyathipodia macropus* (Persoon) Dennis, *Macroscyphus macropus* (Persoon) Gray, *Macropodia macropus* (Persoon) Fuckel

Grayish-brown hairy cups on long, slender, hairy, grayish-brown stems

CUP 1–4 cm wide, bowl-shaped or flat and platter-shaped, glabrous above except for hairy, fringed margin, surface dull grayish-tan or grayish-brown, underside grayish-brown, dotted with small mounds of grayish hairs. **FLESH** thin, brittle. **STEM** 1–7 cm long, 3–5 mm wide, equal, grayish-brown, coated with tufts of grayish hairs. **ODOR** and **TASTE** not distinctive.

Habit and habitat single or scattered on the ground in humus under conifers or hardwoods or in mixed forests, also from rotting wood. June to November. **RANGE** widespread.

Spores fusiform or elongate ellipsoid, finely bumpy, colorless with a central shiny oil droplet inside, 18–25 × 10.5–12.5 µm; asci operculate, not turning blue amyloid.

Comments *Helvella corium* O. Weberbauer has a fruit body that is entirely black and lacks the hairy covering on the outside of the cup and stem. This species is usually found in a boreal forest ecosystem. *Helvella queletii* Bresadola has a white or cream-colored, ribbed stem that does not extend up onto the underside of the grayish-brown cup. *Helvella acetabulum* (Linnaeus) Quélet is bulky with the ribs on the stem extending onto the bottom of the cup, nearly to the cup margin.

Cup Fungi, Earth Tongues, and Pyrenomycete Allies

The cup fungi, or discomycetes, are placed in two different orders, depending upon the type of ascus that is produced—either operculate (or suboperculate) or inoperculate. You need a microscope, however, to make the precise determination. Fungi in the order Pezizales produce either operculate or suboperculate asci, while the Helotiales produce inoperculate asci. Generally, the large-cupped species produce operculate asci, while the smaller-cupped ones produce inoperculate asci. However, there are exceptions. *Scutellinia* is a genus with small cups, but it produces operculate asci. The earth tongues and their pyrenomycete allies all produce inoperculate asci, the earth tongues on soft-fleshed club-shaped or rounded caplike apothecia on stems, the pyrenomycetes in minute, flask-shaped perithecia embedded in a firm supporting tissue referred to as the stroma. The stroma can be flat and growing over something, like a mushroom or bolete that is being parasitized, or the stroma can become more complex, consist of one or more layers, and produce tongue- or clublike structures bearing the perithecia, as in species of *Cordyceps* and *Xylaria*, or dead man's fingers.

The following guide will help you more rapidly select the proper genus to consider, or it will at least give you a starting point. Be sure to read the comments sections carefully for each species entry, as they will give you clues to many additional species that cannot be imaged here.

Cup fungi

CUPS SCARLET, RED, OR DARK ORANGE
Sarcoscypha Scarlet red cups, white fuzzy outside or hairless
Microstoma Scarlet red cups, with long hairy margins on outside of cup
Scutellinia Scarlet reddish-orange cups, with brown eyelash hairs around margin, small, on wood
Aleuria Orange overall, inside cups and externally, large, brittle, lacking fuzz or hairs, on mineral soil or sand

CUPS WHITE, BROWN, OR BLACK
Peziza, *Pachyella* Brown, medium to large cups or saucers
Urnula, *Bulgaria* Black, gelatinous cups
Wynnea Large, rabbit-ear-shaped, dark brown or black outside, orange-brown inside
Humaria White deep cups, with brown eyelash hairs covering the outside

CUPS PINK, PURPLE, GREEN, BLUISH-GREEN, YELLOW, OR TAN
Ascocoryne, *Ascotremella* Purple, rubbery, or jellylike
Neobulgaria Translucent, pale pink, rubbery, or jellylike
Chlorociboria Deep bluish-green overall, small, staining wood blue-green
Bisporella Bright yellow, very small, with short stem
Galiella Orange-tan cups, dark brown or black outside, gelatinous
Chlorencoelia Olive-green or olive-brown, flat, fragile cups, outside fuzzy, on wood (inoperculate asci)

Plicariella Dark olive-green cups with brown hairy-pyramidal external warts, on rich humus (operculate asci)

Earth tongues

Cudonia Cap mushroomlike, cream-yellow, stem often ribbed, on leaves
Leotia Cap rounded yellow or dark green cap and jellylike, stem yellow
Microglossum Cap club-shaped, yellow, with scaly stem
Spathulariopsis Cap paddle-shaped, with brown fuzzy stem
Trichoglossum Cap tongue or lance-shaped, black, stem and cap bristly
Mitrula Cap inflated, round or elongate yellow or pinkish, in fens or bogs
Vibrissea Cap rounded, convex, yellow or orange, stem black-dotted, on sticks in water

Pyrenomycete allies

GROWING ON WOOD, SOMETIMES BURIED AND NOT OBVIOUS

Podostroma Large club-shaped, ochre-yellow or brownish, bumpy, stem short buff
Daldinia Grape- or golf-ball-sized, hard, reddish-black, zoned inside, carbonlike
Hypoxylon Pea- or marble-sized, hard, bumpy, carbonlike
Xylaria Club or finger or coral-like, hard, often bumpy, carbonlike crust around white, tough, supporting center

GROWING ON BURIED "FALSE" TRUFFLES OR INSECT LARVAE

Tolypocladium Growing on underground trufflelike ascomycetes, such as *Elaphomyces*
Cordyceps and *Ophiocordyceps* Growing on insect larvae, usually buried in decaying wood

PARASITIZING MUSHROOMS OR BOLETES

Hypomyces Producing colored crustlike stroma with embedded or superficial perithecia

Sarcoscypha dudleyi (Peck)
Baral

SCARLET CUP
SYNONYM *Peziza dudleyi* Peck

Large scarlet-red cups with white outside, on fallen decaying branches in hardwood forests in the spring

Fruit body a deep, then shallow cup 2–6.5 cm wide, dark scarlet-red inside, smooth, margin incurved, splitting with age, outside white or off-white or pale pinkish-red showing through from the inside pigments, minutely hairy. **FLESH** thin, tough. **STEM** when present, short, 1–3 cm, tapered, white hairy. **ODOR** and **TASTE** not distinctive.

Habit and habitat single, scattered, or clustered on branches of hardwoods lying on the forest floor. March to May, sometimes later. **RANGE** widespread.

Spores ellipsoid, smooth, colorless, with two large oil droplets inside, 26–40 × 10–12 μm; asci cylindrical, with walls even, with a flattened or collapsing "lid" just below the apex (suboperculate), not turning blue amyloid.

Comments Found mostly in the spring just before the morels start to appear or at the same time. The colorful cups are hard to miss, unless they are covered by fallen leaves.

Sarcoscypha austriaca (Beck ex Saccardo) Boudier is similar, mainly differing by spores with truncate ends and filled with numerous small oil droplets.

Sarcoscypha occidentalis (Schweinitz) Saccardo, stalked scarlet cap, has small bright scarlet-red cups, smooth and hairless on the outside, and white, hairless stems. It grows on hardwood sticks and branches.

Sarcoscypha occidentalis

Microstoma floccosum
(Schweinitz) Raitviir

SHAGGY SCARLET CUP
SYNONYMS *Peziza floccosa* Schweinitz, *Sarcoscypha floccosa* (Schweinitz) Saccardo, *Plectania floccosa* (Schweinitz) Seaver, *Anthopeziza floccosa* (Schweinitz) Kanouse

Small bright scarlet-red cups, with white silky tufted hairs on the outside; perched on white hairy stems and growing from hardwood sticks and branches

Fruit body a cup on a stem, cup 0.5–1 cm wide, goblet-shaped with a tiny opening at first, becoming cup-shaped, bright scarlet-red inside and outside, smooth inside, densely white silky hairy outside, hairs sticking together in tufts especially around the margin of the cup. **FLESH** red, thin. **STEM** 3–5 cm long, 1–5 mm wide, equal, cylindrical, white or pale ashy translucent, densely white hairy, fleshy but tough. **ODOR** and **TASTE** not distinctive.

Habit and habitat scattered or clustered on decaying hardwood branches and woody debris, often the wood is buried. June to August. **RANGE** widespread.

Spores narrowly ellipsoid, smooth, colorless, 20–35 × 15–17 µm; asci cylindrical, with walls even, and with flattened or collapsing "lid" or operculum just below the apex (suboperculate), not turning blue amyloid.

Comments Compare with *Sarcoscypha occidentalis* that is clearly stalked and *Scutellinia scutellata* that is not stalked. The latter is a very common cup fungus on all types of decaying hardwood.

Urnula craterium (Schweinitz) Fries

DEVIL'S URN

SYNONYMS *Peziza craterium* Schweinitz, *Sarcoscypha craterium* (Schweinitz) Bánhegyi

Black goblet-shaped, rubbery, with grayish-brown scales on the outer surface; on decaying branches, early spring

Fruit body 3–8 cm wide, 4–6 cm high, goblet-shaped or urn-shaped, closed at first, opening to reveal deep cup with triangular flaps on the margin, inside dark brown or black, smooth, outside dark brown or black, densely covered with reddish-brown or grayish-brown hairy scales. **FLESH** rubbery or gelatinous when fresh, becoming leathery and tough. **STEM** 3–4 cm long, equal or tapered downward, black. **ODOR** and **TASTE** not distinctive.

Habit and habitat usually clusters on decayed woody branches and logs of hardwoods, often the cups sticking out from decaying leaf litter over the branches. March to May. **RANGE** widespread.

Spores elongate ellipsoid or spindle-shaped, smooth, colorless, 25–35 × 12–14 µm; asci cylindrical, with walls even, with a flattened or collapsing "lid" just below the apex (suboperculate), not turning blue amyloid.

Comments Only found in the early spring during morel season.

Wynnea americana Thaxter
MOOSE ANTLERS

Large cups, rabbit ear–like with inner surface orange or pinkish-red, outer surface dark brown or reddish-brown, clusters on a short stem arising from an underground ball of tough tissue

Fruit body 2–14 cm wide, 6–13 cm high, many erect, rabbit-ear-shaped fruit bodies with inrolled margins, tightly clustered together, arising from a common stalk, inner surface orange or pinkish-red or brownish-orange, smooth, outside dark brown, reddish-brown or almost black and roughened with small rounded bumps or wrinkled. **FLESH** brown, tough. **STEM** 2 cm long, brown, attached to an underground ball of tough tissue (a sclerotium). **ODOR** and **TASTE** not distinctive.

Habit and habitat clusters single or scattered on the ground under hardwoods.

July to October. **RANGE** widespread but not commonly collected.

Spores ellipsoid-fusoid, slightly curved, some with nipplelike points at either end, longitudinally lined, colorless, 32–40 × 15–16 µm; asci cylindrical, with walls even, with a flattened or collapsing "lid" just below the apex (suboperculate), not turning blue amyloid.

Comments *Wynnea sparassoides* Pfister, another infrequently seen ascomycete, has the truncated, earlike fruit bodies fused into a large bouquet of beige or yellowish-brown cups arising from a thick and long rooting stem. It looks a little like cauliflower, hence the common name stalked cauliflower fungus. *Sparassis spathulata*, the eastern cauliflower mushroom, is a basidiomycete that is more frequently encountered.

Disciotis venosa (Persoon) Arnould

VEINED CUP

SYNONYMS *Peziza venosa* Persoon, *Discina venosa* (Persoon) Fries

Yellow or reddish-brown, large cups on soil or humus, with wrinkled or veined inner surface, outside white with brown scales; early spring

Fruit body 5–20 cm wide, cup-shaped, inner surface pale yellow-brown or reddish-brown, smooth, becoming wrinkled or veined with age, outside surface dingy white or tan, often dotted with small brown dots or scales. **FLESH** thin, brittle, pale brown. **STEM** 0.5–1 cm long and broad, stout and buried in the ground, or absent. **ODOR** and **TASTE** not distinctive.

Habit and habitat scattered or clustered on soil or humus under hardwoods. March to May. **RANGE** widespread.

Spores broadly ellipsoid, smooth, pale yellowish, 19–25 × 12–15 μm; asci cylindrical, with thin even walls, and apex with flattened or collapsing "lid" (operculate), not turning blue amyloid.

Comments You need a microscope to reliably differentiate the several large brown cup fungi that may appear early in the spring. However, if your specimen has a wrinkled, obviously veined surface, it is most likely this species. Compare with *Peziza varia*, a common large brown cup fungus on wood chips, that has a smooth inner surface and asci that turn blue in Melzer's reagent.

Peziza varia (Hedwig) Albertini & Schweinitz

SYNONYMS *Octospora varia* Hedwig, *Aleuria varia* (Hedwig) Boudier, *Galactinia varia* (Hedwig) Le Gal, *Humaria varia* (Hedwig) Saccardo

Large yellowish-brown cups with white fuzzy coating on outer surface; on well-decayed wood of hardwoods

Fruit body 2–15 cm wide, cup-shaped or almost flat with age, margin incurved, flat or bent backward, often splitting, inner surface light brown or yellowish-brown or grayish-brown, mostly smooth but often puckered in the center, outside white from dense coating of fuzzy hairs, becoming similar color as the inside once the hairs collapse or are rubbed off. **FLESH** pale brown, thin, brittle. **STEM** absent. **ODOR** and **TASTE** not distinctive.

Habit and habitat single, scattered, or clustered on well-decayed wood of hardwoods, or wood chips or soil with collapsed woody debris. July to October. **RANGE** widespread.

Spores ellipsoid, smooth, colorless, 14–17.5 × 8–10.5 µm; asci with thin even walls, apex with flattened or collapsing "lid" (operculate), turning blue amyloid over most of the ascus.

Comments With recent DNA studies, *Peziza repanda* Wahlenberg, *P. cerea* Sowerby ex Fries, and *P. micropus* Persoon appear to be synonyms of *P. varia*, though these conclusions formally have not been made taxonomically. *Peziza arvernensis* Roze & Boudier, formerly known as *P. sylvestris*, is very similar but differs by its finely warted spores. There are several other medium- to large-sized brown cup fungi, and a microscope is necessary to confirm identifications for most. Compare with *Pachyella clypeata*, also found on decaying hardwood logs.

Peziza badia Persoon

SYNONYMS *Plicaria badia* (Persoon) Fuckel, *Scodellina badia* (Persoon) Gray

Large dark brown cups on sandy soil or needle duff under conifers; summer and fall

Fruit body 3–10 cm wide, deeply cup-shaped, inner surface dark reddish-brown or dark purple-brown or dark olive-brown, smooth, outside reddish-brown or purplish-brown, granular from minute fuzzy particles. **FLESH** brown, thin, brittle. **STEM** absent. **ODOR** and **TASTE** not distinctive.

Habit and habitat scattered, or more often clustered on sandy soil or needle duff or decaying woody litter under conifers. July to October. **RANGE** widespread, but rarely collected.

Spores ellipsoid, with fine irregular reticulum, colorless, 17–22 × 8–10 μm; asci with thin even walls, apex with flattened or collapsing "lid" (operculate), turning blue amyloid over most of the ascus.

Comments *Peziza phyllogena* Cooke (syn. *Peziza badioconfusa* Korf) is very similar but fruits in the early spring and has warted spores.

Aleuria aurantia (Persoon) Fuckel

ORANGE PEEL

SYNONYMS *Peziza aurantia* Persoon, *Otidea aurantia* (Persoon) Massee, *Peziza coccinea* Hudson

Bright orange, clustered cups, on bare soil

Fruit body cup-shaped, 2–10 cm wide, becoming flattened, sometimes deeply split down one side, clustered and overlapping, bright orange and smooth inside, fading to pale yellowish-orange, outside white fuzzy at first, but soon smooth and orange. **FLESH** orange, brittle. **STEM** absent. **ODOR** and **TASTE** not distinctive.

Habit and habitat on the ground in disturbed areas, roadsides, paths, gardens, with clay or sandy soil, usually in clusters of overlapping cups. May to October. **RANGE** widespread.

Spores ellipsoid, bumpy and reticulate, colorless, 18–22 × 9–10 μm; ascus tip broad, rounded, thin-walled (operculate), not amyloid.

Comments This species is edible and, with the bright colors, easy to identify. The problem is gathering enough to make it worth your time. It certainly is photogenic, because of the brilliant colors and the size. Compare with *Sarcoscypha dudleyi*, scarlet cup.

Pachyella clypeata (Schweinitz) Le Gal

SYNONYMS *Peziza clypeata* Schweinitz, *Discina clypeata* (Schweinitz) Saccardo

Medium-sized, dark brown, saucer-shaped, thick gelatinous fruit bodies, firmly attached to wet, well-decayed hardwood logs in seepage or swampy areas

Fruit body 2–4(–8) cm wide, round, saucer-shaped and completely attached to the woody substrate, surface brown or dark brown with red or purplish tints, mostly smooth but wrinkling with age, slippery or sticky when fresh and wet, outside surface not visible since it is attached to the substrate. **FLESH** rubbery or gelatinous when fresh. **STEM** absent. **ODOR** and **TASTE** not distinctive.

Habit and habitat single, scattered, or small clusters on wet well-decayed wood of hardwoods, most often lying in standing water or very mucky depressions. June to October. **RANGE** widespread.

Spores ellipsoid, smooth, colorless, 18–25 × 13–16 µm; asci with thin even walls, apex with flattened or collapsing "lid" (operculate), turning blue amyloid over most of the ascus.

Comments There certainly are other species of *Pachyella*, however this species seems to be the most commonly encountered, along with a much smaller species now placed in the genus *Adelphella* based on molecular and morphological features. *Adelphella babingtonii* (Saccardo) Pfister, Matočec & I. Kušan is smaller (to 2 cm at most), pale yellow-brown with translucent, watery or gelatinous flesh, and produces slightly smaller, compact, ellipsoid spores. *Pachyella punctispora* Pfister has finely warted spores.

Plicariella flavovirens (Fuckel) Van Vooren & Moyne

SYNONYMS *Plicaria flavovirens* Fuckel, *Peziza flavovirens* (Fuckel) Cooke, *Sphaerospora flavovirens* (Fuckel) Saccardo, *Scabropezia flavovirens* (Fuckel) Dissing & Pfister

Small cups, directly attached to wet, well-decayed hardwoods, inside dark olive-green, outside brown and covered with thick conical-hairy, pyramidal warts

Outer surface

Fruit body a cup 0.5–1.6 cm wide, dark olive-green or green inside, smooth, chestnut-brown or with yellowish hues on tips of hairs on the outside of the cup, composed of mostly thick, conical-hairy, pyramidal warts making up the surface. **FLESH** dark green, producing a wheylike, white juice when cut. **ODOR** and **TASTE** not distinctive.

Habit and habitat single or scattered on well-decayed wood of hardwoods, the one imaged here from a stump of American chestnut, or on bare soil with decaying wood. June to September. **RANGE** widespread.

Spores globose, uniformly warted, colorless, 14–16 μm; asci with thin even walls, apex with flattened or collapsing "lid" (operculate), turning blue amyloid over most of the ascus.

Inside surface

Comments These small dark cups can be hard to find because of the size and the colors. Examine well-decayed logs and stumps that are almost completely broken down, or bare ground around such objects.

Galiella rufa (Schweinitz) Nannfeldt & Korf

HAIRY RUBBER CUP

SYNONYMS *Bulgaria rufa* Schweinitz, *Gloeocalyx rufa* var. *magna* Peck

Pale orange or tan inner cup, dark reddish-brown to black, hairy outer cup; flesh thick and gelatinous or rubbery; on decaying hardwood debris

Fruit body 2–3 cm wide, goblet-shaped at first, becoming cup-shaped, inner surface pale orange or orange-tan or tan, smooth, outer surface black or very dark reddish-brown, matted hairy, margin of cup with toothlike triangular flaps, creamy-tan on the inside, black on the outside. FLESH gelatinous, rubbery-tough. STEM 1–2 cm long 3–5 mm wide, tapered to a basal point, attached by dense black hairs. ODOR and TASTE not distinctive.

Habit and habitat scattered or clustered on sticks, branches and logs of hardwoods. July to September. RANGE widespread.

Spores ellipsoid or subfusiform, thick-walled, with tiny bumps, colorless in 3–5% KOH, 17–21 × 8–10 µm; asci cylindrical, walls even, with a flattened or collapsing "lid" just below the apex (suboperculate), not turning blue amyloid.

Comments *Bulgaria inquinans* (Persoon) Fries is similar in size, rubbery texture, and occurrence on decaying hardwoods. However, it differs by the black inner surface of the cup, the scaly outer surface, its lack of toothlike triangular flaps on the cup margin, and by having inoperculate asci with four brown and four colorless spores.

Humaria hemisphaerica
(F. H. Wiggers) Fuckel

BROWN-HAIRED WHITE CUP
SYNONYMS *Peziza hemisphaerica* F. H.
Wiggers, *Lachnea hemisphaerica* (F. H.
Wiggers) Gillet, *Peziza hispida* Sowerby

Small cups on the ground, on mosses or
well-decayed wood with white or pale gray
inner surface, contrasting sharply with
the bristly brown hairs covering the outer
surface

Fruit body goblet-shaped then deeply
cup-shaped, 1–3 cm wide, 0.5–1 cm high,
white or pale grayish and smooth on the
inner surface, outside brown and covered
with bristly brown hairs making a fringe
around the cup rim. **FLESH** thin. **STEM**
absent. **ODOR** and **TASTE** not distinctive.

Habit and habitat scattered or clustered on
soil or mixed in the decaying leaves and
needles or among mosses or sometimes
on wet well-decayed wood. July to August.
RANGE widespread.

Spores ellipsoid, with flattened ends
with fine bumps in Melzer's reagent,
colorless, 22–27 × 10–15 µm; asci with
thin even walls, apex with flattened or
collapsing "lid" (operculate), not turning
blue amyloid.

Comments *Trichophaea hemisphaerioides*
(Mouton) Graddon looks very similar but
is smaller and flatter and occurs on newly
burned forest areas often accompanied by
a small rosette moss (*Funaria*). The ascospores of *T. hemisphaerioides* are smaller,
13–18 × 7–8 µm. *Jafnea semitosta* (Berkeley
& M. A. Curtis) Korf, is somewhat similar, but the fruit bodies are much larger
(1.5–7 cm wide) with a sparsely hairy
brownish outer surface and off-white inner
surface, but with a false stem projecting
into the humus from the base of the cup.

Scutellinia scutellata
(Linnaeus) Lambotte

EYELASH CUP

SYNONYMS *Peziza scutellata* Linnaeus, *Patella scutellata* (Linnaeus) Morgan, *Lachnea scutellata* (Linnaeus) Gillet

Bright red or orange-red, small discs with brown eyelash hairs around the margin, clustered together on decaying woody substrates

Fruit body shallowly cup-shaped, 0.5–2 cm wide, bright red or orange-red and smooth on the upper surface, outside covered with brown hairs that project beyond the cup rim, eyelashlike. **FLESH** thin. **STEM** absent. **ODOR** and **TASTE** not distinctive.

Habit and habitat clustered on wet wood, among mosses on wood, on well-decayed wood on the ground. June to November. **RANGE** widespread.

Spores ellipsoid, bumpy, colorless, 18–19 × 10–12 µm; asci with thin, even walls, apex with flattened or collapsing "lid" (operculate), not turning blue amyloid.

Comments You need a microscope to definitively identify the other species of *Scutellinia*. If you find a small eyelash cup, 1.5–5 mm wide, on wood with pale orange-yellow colors, you could have *S. erinaceus* (Schweinitz) Kuntze. If you think you have collected *S. scutellata* but your microscope shows that the spores are reticulate, check the underside of the cups. If they lack hairs, you most likely have *S. pennsylvanica* (Seaver) Denison.

Ascocoryne sarcoides (Jacquin) J. W. Groves & D. E. Wilson

PURPLE JELLY DROPS

SYNONYMS *Lichen sarcoides* Jacquin, *Bulgaria sarcoides* (Jacquin) Dickson, *Coryne sarcoides* (Jacquin) Tulasne & C. Tulasne, *Peziza sarcoides* (Persoon), *Tremella sarcoides* (Jacquin) Fries

Purple or reddish-purple, small, jellylike discs on decaying hardwood logs

Fruit body cup-shaped or flat disc or saucer-shaped, 0.5–1 cm wide, usually several clustered together, purple or reddish-purple and smooth on the upper surface, outside only slightly paler. **FLESH** purple, jellylike. **STEM** when present, very short, sometimes ribbed. **ODOR** and **TASTE** not distinctive.

Habit and habitat clustered on well-decayed or at least barkless hardwood logs and stumps, also known to grow on downed conifers. September to October. **RANGE** widespread.

Spores fusiform (canoe-shaped) and with 1 or occasionally 2 cross walls, colorless, 10–21 × 3–5 µm; asci cylindrical, with thickened walls at apex with a pore (inoperculate), turning blue amyloid.

Comments *Ascocoryne cylichnium* (Tulasne) Korf looks very similar and is found in similar habitats, but it is distinguished by the multiple cross walls in the spores that are also larger, 18–27 × 4–6 µm.

Chlorociboria aeruginascens
(Nylander) Kanouse ex C. S. Ramamurthi, Korf & L. R. Batra

GREEN STAIN
SYNONYMS *Peziza aeruginascens* Nylander, *Chlorosplenium aeruginascens* (Nylander) P. Karsten

Blue-green, small cups on stems, with associated green-stained wood; on downed logs and woody debris of hardwoods

Fruit body cup-shaped or becoming plane, 0.2–0.5 cm wide, blue-green on the smooth upper surface, similar color or paler on the outside. **FLESH** blue-green, thin. **STEM** same color as cup, 1–2 mm long, central of off-center. **ODOR** and **TASTE** not distinctive.

Habit and habitat single, scattered, or clustered on downed, decaying hardwood logs, especially oak but also beech, staining the wood green. June to November. **RANGE** widespread.

Spores subfusiform or cylindrical, small oil droplet at each end, colorless, 6–10 × 1–2 μm; asci cylindrical, with thickened walls at apex with a pore (inoperculate), turning blue amyloid, with smooth cylindrical spiraling cells on outside skin of the cup.

Comments You will most likely see the green-stained wood before you actually see or find the fruit bodies. *Chlorociboria aeruginosa* (Oeder) Seaver ex C. S. Ramamurthi, Korf & L. R. Batra is very similar and can only be separated with assurance microscopically since the spores are 9–15 × 1.5–2.5 μm and the cylindrical spiraling cells on the ectal excipulum are roughened.

Bisporella citrina (Batsch) Korf & S. E. Carpenter

YELLOW FAIRY CUPS

SYNONYMS *Peziza citrina* Batsch, *Helotium citrinum* (Hedwig) Fries, *Calycella citrina* (Hedwig) Boudier

Bright yellow or lemon-yellow, small discs clustered together on decaying woody substrates like logs, stems, stumps, cut decaying lumber, particle board; very common

Fruit body cup-shaped or flat disc or saucer-like, 0.1–0.3 cm wide, usually a large number clustered together or in a well-defined patch, bright lemon-yellow or yellow and smooth on the upper surface, outside similar. **FLESH** pale yellow. **STEM** when present, same color as cup, very short, tapered. **ODOR** and **TASTE** not distinctive.

Habit and habitat clustered on typically barkless logs and stumps, on particle board or cut pieces of lumber that are decaying. July to November. **RANGE** widespread.

Spores ellipsoid and with 1 cross wall, making a 2-celled spore (hence the genus name), colorless, 9–14 × 3–5 μm; ascus tip tapered, thick-walled with a pore (inoperculate) faint blue amyloid at tip.

Comments A similar, but less often encountered species is *Bisporella sulfurina* (Quélet) S. E. Carpenter. It is bright sulfur-yellow, with smaller somewhat translucent cups only reaching 1.5 mm in diameter. These tiny bright yellow disc fungi are found growing on wood or on the old black crust of the fruit bodies of carbon balls and other carbonaceous pyrenomycetes. Besides the size, color, and habitat, they also differ by narrower spores, reaching only 2 μm wide. If a well-formed stipe is present, you could have a species of *Hymenoscyphus*. There are a number of quite small-stalked discomycetes.

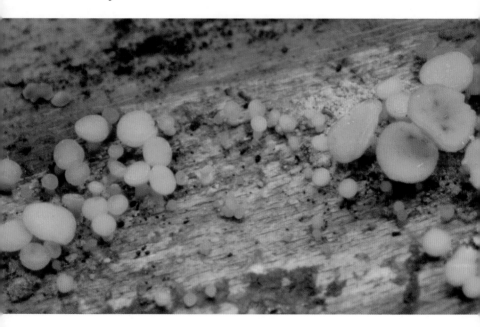

Cudonia lutea (Peck) Saccardo

YELLOW CUDONIA

SYNONYMS *Vibrissea lutea* Peck

Small yellow smooth cap and yellow smooth or wrinkled stem; cap margin inrolled or incurved, smooth on the underside, without gills

Fruit body mushroomlike with a cap and stalk. **CAP** 1–2 cm wide, convex but often convoluted, margin strongly inrolled, dull cream-yellow or ochre-buff, smooth, dry, smooth on the underside. **FLESH** firm or somewhat leathery. **STEM** 2–6 cm high, 2–5 mm wide, yellow or cream-yellow, smooth or longitudinally wrinkled or lined. **ODOR** and **TASTE** not distinctive.

Habit and habitat scattered or clustered on decaying leaves under beech and other hardwoods. July to September. **RANGE** widespread.

Spores needle-shaped with many cross walls, smooth, colorless, 40–70 × 1.5–2 μm; ascus tip tapered, thick-walled with a pore (inoperculate), not amyloid at tip.

Comments This small ascomycete can look like a mushroom, but the underside of the cap is smooth. Compare with the gelatinous *Leotia lubrica* that can be somewhat similar in color. *Cudonia circinans* (Persoon) Fries grows under conifers and has brownish hues in cap and stem.

Leotia lubrica (Scopoli) Persoon

OCHRE JELLY CLUB

SYNONYMS Helvella lubrica Scopoli, Leotia viscosa Fries

Gelatinous, yellowish-ochre, bumpy or wrinkled slippery cap; stem similar in color and texture, also minutely scaly

CAP 1–4 cm wide, variable in shape but mostly convex and bumpy or wrinkled, margin rolled down and inward, glabrous, slippery, gelatinous texture, buff or ochre or yellow or dull olive, some dark green with age, underside smooth. **FLESH** gelatinous. **STEM** 2–8 cm long, 5–10 mm wide, equal or enlarged downward, similar color as cap or paler, minutely scaly, slippery when fresh, gelatinous texture. **ODOR** and **TASTE** not distinctive.

Habit and habitat scattered or clustered on the ground in humus, in mosses, on rotting wood, under conifers or hardwoods or in mixed forests. July to October. **RANGE** widespread.

Spores long narrow ellipsoid or canoe-shaped, smooth, colorless, with 3–8 cross walls, 16–25 × 4–6 µm; ascus tip tapered, thick-walled with a pore (inoperculate), not amyloid at tip.

Comments Leotia viscosa Fries has a dark green cap with yellow stem, but some consider this a synonym of L. lubrica. A possible third species, L. atrovirens with an all-green cap and stem, is now thought to be L. lubrica that has been infected or parasitized by a mold, causing the flesh to turn pea-green or bluish-green. Also known as jelly babies, they are fairly common in the fall months.

Microglossum rufum
(Schweinitz) Underwood

ORANGE EARTH TONGUE

SYNONYM *Geoglossum rufum* Schweinitz

Bright yellow or orange-yellow, club-shaped with head often compressed; stem similar in color and scaly-granular

Fruit body 2–7 cm high. **CAP** 0.3–1.5 cm wide, variable in shape but mostly tongue-shaped or club-shaped, yellow or orange-yellow, flattened and then grooved in the middle, glabrous, soft in texture, fragile. **FLESH** firm. **STEM** 1–4 cm long, 1.5–4 mm wide, equal, cylindrical, yellow or orange-yellow like head, densely scaly or granular from pale yellow scales, fleshy. **ODOR** and **TASTE** not distinctive.

Habit and habitat scattered or more often clustered on decaying leaves and needle beds, well-rotted wood, and in *Sphagnum* and other mosses, under hardwoods and conifers. July to September. **RANGE** widespread.

Spores cylindrical and curved (sausage-shaped), smooth, colorless, with 7–15 cross walls, 18–40 × 4–6 μm; asci elongate clavate, with walls thickened at apex and with a pore (inoperculate), turning blue amyloid.

Comments *Neolecta irregularis* (Peck) Korf & J. K. Rogers is only vaguely similar because of the yellow colors. The fruit bodies are irregularly contorted and short-branched over the upper two-thirds, with a stem finely powdered or with no recognizable stem at all, and it occurs under conifers.

Spathulariopsis velutipes
(Cooke & Farlow) Maas Geesteranus

VELVETY FAIRY FAN

SYNONYM *Spathularia velutipes* Cooke & Farlow

Fan-shaped, yellowish, flattened cap on a brown velvety stem; bright orange mycelium at base

Fruit body fan or paddle-shaped, 2–6 cm high, 1–3 cm wide. **CAP** flattened, smooth, with rounded edges, attached and running down stem on opposite sides, yellow or cream color or with brown tints. **FLESH** thin, brittle. **STEM** 2–6 cm long, 10–15 mm wide, equal and round or somewhat flattened, orange-brown or brown, velvety or fuzzy overall, with bright orange mycelium radiating into substrate from the base. **ODOR** and **TASTE** not distinctive.

Habit and habitat scattered or clustered on the ground or on well-decayed wood of hardwoods or conifers in mixed forests. July to September. **RANGE** widespread.

Spores long needle-shaped, with many cross walls, smooth, colorless, 33–43 × 1.5–3 µm; asci cylindrical, with thickened walls at apex with a pore (inoperculate), not turning blue amyloid.

Comments *Spathularia flavida* Persoon has a paler buff-colored flattened cap, lighter colored nonvelvety stem, and white mycelium at the base of the stem. Also compare with *Microglossum rufum*.

Podostroma alutaceum
(Persoon) G. F. Atkinson

SYNONYMS *Sphaeria alutacea* Persoon, *Hypocrea alutacea* (Persoon) Cesati & De Notaris, *Xylaria alutacea* (Persoon) Gray, *Cordyceps alutacea* (Persoon) Quélet

Large club-shaped with not well-differentiated cap and stem, cap ochre-yellow, bumpy with minute holes; stem short buff-white; on often buried, decaying wood

Fruit body a column or club (stroma) with an obscure cap and a short stem. **CAP** 1–3 cm high, 6–10 mm wide, just the upper portion or even two-thirds of the club darker yellow or ochre-yellow, becoming brownish with age, covered with tiny darker ochre-yellow bumps with minute holes (perithecia), surface smooth or undulate, slightly broader than the base or stem of the club. **FLESH** white, tough. **STEM** 1–2 cm long, slightly less wide than the head, buff or off-white, equal, glabrous. **ODOR** and **TASTE** not distinctive.

Habit and habitat single, scattered, or clustered on buried wood most often in hardwood forests. June to September. **RANGE** widespread but rarely encountered.

Spores long ellipsoid, with cross walls, breaking up into smaller part spores that are globose or subglobose, finely bumpy, colorless, 4–4.5 × 3–4 μm; asci narrowly cylindrical, with walls at apex thickened into a cap with a channel or pore (inoperculate), not turning blue amyloid.

Comments Compare with *Clavariadelphus americanus*, since this basidiomycete has somewhat similar colors and shape. However, *C. americanus* is easily distinguished by its soft, more fragile tissues and by the lack of darker colored perithecia embedded in its flesh.

Trichoglossum farlowii
(Cooke) E. J. Durand

SYNONYM *Geoglossum farlowii* Cooke

Dark grayish-black tongue- or lance-shaped head on a dark black velvety or bristly stem; on the ground in leaf litter or mosses, under hardwoods

Fruit body 3–8 cm tall. **CAP** 2–5 mm wide, mostly lance-shaped with a rounded tip or narrow and tonguelike, dark grayish-black and mostly decurrent on the rounded stem edges. **FLESH** black, soft-brittle. **STEM** round or flattened compressed, especially near the head, equal, dark black, minutely bristly or velvety. **ODOR** and **TASTE** not distinctive.

Habit and habitat scattered or most often clustered on the ground in leaf litter, well-decayed wood, or mosses, under hardwoods. July to September. **RANGE** widespread.

Spores very long cylindrical, with 0–5 cross walls, brown, 45–90 × 6–7 µm; asci cylindrical, with thickened walls at apex with a pore (inoperculate), blue amyloid at tip.

Comments True earth tongues are fleshy and mostly dark grayish-black or black. *Trichoglossum* species are all blackish with dark brown or black sharp-pointed setae in the tissues of the stem and head (apothecium), which usually can be seen with a hand lens as minute bristles. Species of *Geoglossum* lack such setae. There are many species in each genus here in the Northeast, and microscopic features are used to identify them. *Thuemenidium atropurpureum* (Batsch ex Fries) Kuntz [syn. *Geoglossum atropurpureum* (Batsch) Persoon], also frequently encountered, is dark purplish-brown and lacks setae. Compare also with *Tolypocladium ophioglossoides* that is tough and attached to underground false truffles by bright yellow cords.

Mitrula elegans Berkeley

SWAMP BEACON

In wet mucky areas on decaying leaves, caps small, bright yellow, waxy or jellylike, fragile; stem translucent or dingy white, often translucent

CAP 0.2–1.2 cm wide, 0.6–2 cm high, variable in shape but mostly ellipsoid or club-shaped or pear-shaped, sometimes irregularly wrinkled or somewhat folded longitudinally, lacking a free margin, clear butter-yellow or orange-yellow, glabrous, soft jellylike in texture, fragile. **FLESH** gelatinous. **STEM** 2–4 cm long, 1–3 mm wide, equal, colorless translucent or off-white, glabrous, slightly slippery, gelatinous texture. **ODOR** and **TASTE** not distinctive.

Habit and habitat scattered or clustered on decaying leaves, needles, twigs in immersed or mucky wet areas, especially boggy areas or fens. May to September. **RANGE** widespread.

Spores long narrow ellipsoid or fusiform, smooth, colorless, with 3–8 cross walls, 10–18 × 2.5–5 μm; asci narrowly club-shaped, with walls thickened at apex and with a pore (inoperculate), turning blue amyloid.

Mitrula lunnulatospora

Comments *Mitrula lunnulatospora* Redhead shares the same habitat as *M. elegans*, but has a pinkish or yellowish pink-colored cap. The crescent-shaped spores of *M. lunnulatospora* present the best way to confirm identification.

Vibrissea truncorum (Albertini & Schweinitz) Fries

WATER CLUB

SYNONYM *Leotia truncorum* Albertini & Schweinitz

Small orange-yellow heads on white or blackish stems, attached to immersed decaying branches in cold water streams or boggy areas

Fruit body 1–2 cm high. **CAP** 1.5–5 mm wide, convex with inrolled margin, glabrous, soft in texture, yellow or orange-yellow or reddish-orange. **FLESH** firm. **STEM** 0.6–1.5 cm long, 1–1.5 mm wide, equal, cylindrical, white or pale ashy translucent, becoming blackish-dotted from base upward, densely but minutely hairy, fleshy. **ODOR** and **TASTE** not distinctive.

Habit and habitat scattered or clustered on immersed, decaying, small branches in cold water streams and boggy areas, usually at higher elevations. July to September. **RANGE** widespread.

Spores needlelike, smooth, colorless, with multiple cross walls, 120–250 × 1–1.5 µm; asci narrowly club-shaped, with walls thickened at apex and with a pore (inoperculate), not turning blue amyloid.

Comments It is odd to see a fungus growing immersed in water, but this is a favored habitat for this small ascomycete. Look around carefully and you may also find a *Mitrula* where it is a little drier.

Tolypocladium ophioglossoides (J. F. Gmelin)
Quandt, Kepler & Spatafora

GOLDENTHREAD CORDYCEPS

SYNONYMS *Sphaeria ophioglossoides*
J. F. Gemlin, *Cordyceps ophioglossoides*
(J. F. Gmelin) Fries, *Elaphocordyceps*
ophioglossoides (J. F. Gemlin) G. H. Sung,
J. M. Sung & Spatafora

Dark brown or black, tongue-shaped and
bumpy head; stem black above ground,
yellow below ground with yellow cords
attached to a false truffle

Fruit body tonguelike cap on a stem (the
stroma). **CAP** 1–1.5 cm wide, 2–2.5 cm
high, dark reddish-brown, eventually
black, smooth at first, then finely roughened with beaks of tiny flask-shaped spore
producing vessels (perithecia) that eject
white powdery spores, pores evident under
a lens. **FLESH** white, tough. **STEM** 2–8 cm in
length, 3–10 mm wide, equal, bluish-black
or black above ground, bright golden-
yellow below ground, with thick yellow
cords running from the base, attached to
one or more false truffles. **ODOR** and **TASTE**
not distinctive.

Habit and habitat single, scattered, or
clustered on the ground and deeply rooting down to a false truffle, *Elaphomyces*.
August to October. **RANGE** widespread.

Spores threadlike, breaking into smaller
segments (part spores), part spores ellipsoid, smooth, colorless, 2.5–5 × 1.5–2 μm;
asci with walls thickened at apex and with
a pore (inoperculate), not amyloid at tip.

Comments A commonly encountered
cordyceps, this is the one that most closely
resembles the earth tongues, *Geoglossum*
and *Trichoglossum*. The cap of *Trichoglos-*
sum is bristly, but not roughened with
perithecial beaks like cordyceps. Rub your
fingers over the cap. If it feels rough, dig

down carefully to find the false truffle. Just
follow the yellow cords. Molecular phylogenetic research has significantly changed
our understanding of the evolution of these
parasites, thus all the new names.

Tolypocladium longisegmentum (Ginns) Quandt, Kepler & Spatafora

SYNONYMS *Cordyceps longisegmentis* Ginns, *Elaphocordyceps longisegmentis* (Ginns) G. H. Sung, J. M. Sung & Spatafora Small brown or black, very solid, rounded and roughened head; on a firm, yellow becoming grayish or sooty-black stalk attached directly to an underground false truffle

Fruit body firm round cap on a long stem (together the stroma). **CAP** 1–2 cm wide, olive-brown or dark brown or black, finely roughened with emerging beaks of tiny flask-shaped spore producing vessels (perithecia), pores evident under a lens. **FLESH** very firm. **STEM** to 13 cm in length, 13–14 mm wide, equal, bright golden-yellow at first, soon with grayish to sooty-black colors, but always yellow over the base, attached directly to the false truffle. **ODOR** and **TASTE** not distinctive.

Habit and habitat single, scattered, or clustered on the ground and deeply rooting down to and attached to a tuber-shaped false truffle, *Elaphomyces*. July to November. **RANGE** widespread.

Spores very long, threadlike, breaking into smaller segments (part spores), part spores rodlike, 40–65 × 4–5 μm; asci long cylindrical with walls thickened at apex and with a pore (inoperculate), not amyloid at tip.

Comments *Tolypocladium capitatum* (Holmskjold) Quandt, Kepler & Spatafora looks very similar and can only be reliably separated by measuring the part spores (15–20 μm long).

Cordyceps militaris (Linnaeus) Fries

SYNONYMS *Clavaria militaris* Linnaeus, *Torrubia militaris* (Linnaeus) Tulasne & C. Tulasne

Small orange clubs with darker orange bumps embedded in the cap; stem ochre-yellow arising from large, dark brown, segmented insect larvae buried in the substrate; on the ground or well-decayed wood

Fruit body club-shaped. **CAP** 1–2 cm high, 0.5 cm wide, cylindrical, orange-yellow, covered with darker orange dots of tiny flask-shaped spore producing structures (perithecia) embedded in the swollen tissue (stroma), individual perithecia and their pores evident under a lens. **FLESH** pale orange, tough. **STEM** round, 3–4 cm high, 3–5 mm wide, yellow or ochre-yellow, smooth, arising from a larva that is segmented and dark brown. **ODOR** and **TASTE** not distinctive.

Habit and habitat single or clustered on the ground or well-decayed wood, attached to moth or butterfly larva. July to October. **RANGE** widespread.

Spores thread like, breaking into smaller segments (part spores), part spores brick-like, smooth, colorless, 3.5–6 × 1–1.5 μm; asci long cylindrical with walls thickened at apex and with a pore (inoperculate), not amyloid at tip.

Comments The bright orange colors attract collectors, so this fungus is frequently found in our forests. If the cap is rough, dig down into the ground with care to find the larva. A relative of this species from Asia, *Ophiocordyceps sinensis*, has quite a history and is worth looking up online because of the interest in medicinal use of the fungus.

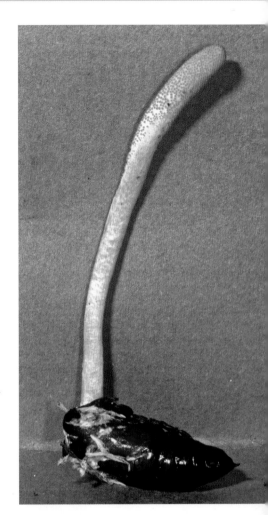

Ophiocordyceps variabilis
(Petch) G. H. Sung, J. M. Sung & Spatafora

SYNONYMS *Cordyceps variabilis* Petch, *Cordyceps viperina* Mains, *Cordyceps ithacensis* Balazy & Bujakiewicz
Yellow or ochre-yellow stems bearing clusters of flask-shaped spore producing structures behind the stem tip; stem attached to elongate, segmented fly larva in well-decayed wood

Fruit body firm, erect with only slightly swollen or often horizontally bent apex (snakelike head), on which one or more fertile cushions bear 10–20 tiny flask-shaped spore producing structures (perithecia) on the tip or more often on the stem below the tip, individual perithecia usually obvious with darker orange pores evident under a lens. FLESH tough. STEM round, 2–24 mm high, 0.2–3 mm wide, yellow or ochre-yellow, finely roughened with soft scales or patches, one or two arising from a wood eating fly larva, the host pupa cadavers, segmented and dark reddish-brown. ODOR and TASTE not distinctive.

Habit and habitat one or two on a dead larva of wood eating flies (*Xylophagus* sp.) in well-decayed wood. August to October. RANGE widespread.

Spores threadlike, breaking into smaller segments (part spores), part spores cylindrical, smooth, colorless, 5–10 × 1.5–3 µm; asci long cylindrical with walls thickened at apex and with a pore (inoperculate), not amyloid at tip.

Comments A very distinctive cordyceps that often resembles a snakelike pattern to the stromata, hence one of the synonyms, *Cordyceps viperina*.

Hypomyces lactifluorum
(Schweinitz) Tulasne & C. Tulasne

LOBSTER MUSHROOM
SYNONYM *Sphaeria lactifluorum*
Schweinitz

Orange or reddish-orange flat, hard, tissue with tiny bumps, covering the fruit bodies of mushrooms, appearing like a cooked lobster; on the ground in hardwoods

Fruit body a flat, firm skin covering the mushroom, orange or orange-red or deep purplish-red sheet of fungal tissue (stroma) with tiny bumps of reddish-orange or reddish-purple flask-shaped fruit bodies (perithecia) embedded in the stroma, perithecia typically darker in color, white spores eventually released, appearing as powdery white piles.

Habit and habitat parasite on fruit bodies of the mushrooms *Russula* and *Lactarius* species in hardwood forests of beech, birch, oak. July to October. **RANGE** widespread.

Spores canoe-shaped with a single cross wall in the middle, coarsely bumpy, colorless, 35–40 × 4.5–7 µm; asci cylindrical with walls thickened at apex and with a pore (inoperculate), not amyloid at tip.

Comments The lobster mushroom is prized as an edible of choice quality. However, most of the time it is almost impossible to know what host mushroom was attacked by this ascomycete parasite, because the host is covered so quickly and thoroughly. The lobster mushroom has been used as a delicacy in food preparation for hundreds of years without a documented problem. It seems to be a safe bet if you like gourmet dining.

Hypomyces lateritius (Fries) Tulasne & C. Tulasne is a parasite only on the gills of *Lactarius* species, where it shows up as

Hypomyces lateritius

tan or yellowish brown with darker brown bumps embedded in the sheetlike tissue covering the gills. In the photo, it is attacking *L. vietus*.

Hypomyces camphorati Peck

SYNONYM *Peckiella camphorati* (Peck) Seaver

Bright lemon-yellow tissue with dark olive-brown bumps, covering the gills of *Lactarius camphoratus*, but the cap looks normal

Fruit body a flat, firm skin covering only the gills, off-white at first but soon bright yellow or lemon-yellow, fading to beige, producing a firm sheet of fungal tissue (the stroma) with tiny bumps of dark olive-brown or brownish flask-shaped fruit bodies (perithecia) embedded in the stroma, perithecia typically darker in color than the stroma, white spores eventually released, appearing as powdery white piles.

Habit and habitat parasite on the gills of *Lactarius camphoratus* in hardwood forests of beech, birch, oak. July to October. **RANGE** widespread but rare.

Spores canoe-shaped with tapered projections at each end (apiculate), lacking a cross wall in the middle but inclusion making it appear there may be one (pseudoseptate), minutely bumpy, colorless, 15–20 × 3.5–5 µm; asci cylindrical with walls thickened at apex and with a pore (inoperculate), not amyloid at tip.

Comments The characteristic odor of *Lactarius camphoratus* was absent or much reduced with this ascomycete parasite covering the gills. Again, the mushroom has to be turned over to even see the lemon-yellow stroma on the gills.

Hypomyces luteovirens

Hypomyces luteovirens (Fries) Tulasne & C. Tulasne is found on the gills and upper stem of red- and purple-capped *Russula* species as a green coating with tiny dark olive-green fruit bodies embedded in the tissue covering the gills and upper stem of the host.

Hypomyces chrysospermus
Tulasne & C. Tulasne

GOLDEN HYPOMYCES
SYNONYM *Sepedonium chrysospermum*
(Bulliard) Fries
Growing on boletes, white cottony, then bright yellow powdery, eventually firm reddish-brown layer covering entire fruit body and dotted with tiny protruding beaks producing white ascospores

Fruit body a skin growing over the entire bolete, with three growth stages; soft and white at first producing colorless asexual 1-celled spores (10–30 × 5–12 µm) on verticillate whorls of spore producing cells, looking like spokes of a wagon wheel (this stage forms the mold *Verticillium*), the outline and structures of the bolete fruit body still obvious; soon becoming bright lemon-yellow, soft, powdery and mostly covering the entire bolete fruit body, obscuring structures and producing asexual, 1-celled, yellow, thick-walled warted spores, called chlamydospores (10–25 µm in diameter); the final phase producing a firm reddish-brown sheet of fungal tissue with tiny, flask-shaped fruit bodies (perithecia) embedded in the tissue (stroma), spores eventually released from pores in perithecia and obvious as a white powdery covering.

Habit and habitat growing on various species of boletes. June to September.
RANGE widespread.

Spores (sexual spores) canoe-shaped with a single cross wall in the middle, finely bumpy, colorless, 25–30 × 5–6 µm; asci cylindrical with walls thickened at apex and with a pore (inoperculate), not amyloid at tip.

Comments The image provided shows the last two stages of the ascomycete parasitism on the bolete as described above. Although a few of the common *Hypomyces* species are covered in this guide, many others attack different mushrooms. For example, a common white parasite on *Amanita* is *H. hyalinus* (Schweinitz) Tulasne & C. Tulasne.

Daldinia concentrica (Bolton) Cesati & De Notaris

CARBON BALLS

SYNONYMS *Sphaeria concentrica* Bolton, *Hypoxylon concentricum* (Bolton) Greville Hard carbonlike texture, pinkish-gray or reddish-brown, becoming black; zoned on inside; on downed logs, stumps

Fruit body 2–4 cm wide, nearly round or hemispherical, pale grayish, then pinkish-gray or vinaceous-gray or reddish-brown, eventually black from the spores, finely dotted with miniscule openings of the flask-shaped fruit bodies (perithecia), and black spores eventually released out of these holes. **FLESH** (stroma) hard, carbonlike, dark grayish-brown or black, zoned, with upper zone filled with perithecia, lower zones are sterile tissue. **STEM** absent. **ODOR** and **TASTE** not distinctive.

Habit and habitat usually clustered on downed hardwood logs or on stumps, especially ash and beech. June to September. **RANGE** widespread.

Spores ellipsoid but with one flat side, smooth, black, 12–17 × 6–9 µm; asci cylindrical with thickened, tapered apex with a pore (inoperculate), blue amyloid at the tip.

Comments Some of the common names for this fungus are quite funny, especially the stories that go with the common names. For example, one alternative is cramp balls, because it was believed they would cure attacks of cramping muscles if one carried them as a personal item, perhaps placing them under one's arm to do the job. Compare with *Hypoxylon* that lacks zones in the stroma.

Hypoxylon fragiforme
(Persoon) J. Kickx f.

RED CUSHION HYPOXYLON
SYNONYM *Sphaeria fragiformis* Persoon
Small, round, reddish-colored carbon balls
with bumpy surface; on downed beeches

Fruit body 2–13 mm wide, 1.5–6 mm high,
nearly round or hemispherical, bumpy,
rust or dark brick red, finely dotted with
miniscule openings of the flask-shaped
fruit bodies (perithecia), and black spores
eventually released out of these holes.
FLESH (stroma) hard, carbonlike, orange-
red granular layer overlying a black,
unzoned layer. **STEM** absent. **ODOR** and
TASTE not distinctive.

Habit and habitat usually densely clustered on the bark of downed beech
logs or stumps. July to November.
RANGE widespread.

Spores ellipsoid, smooth, dark brown,
with germ slit running the length, 10.5–15
× 5–7 µm; asci cylindrical with thickened,
tapered apex with a pore (inoperculate),
blue amyloid at the tip.

Comments A microscopic examination is
generally necessary to separate the many
species of *Hypoxylon*. The strawberry-
shaped, bumpy, hard carbonlike, reddish
fruit bodies that grow on downed beech
trees make this particular species easy
to identify.

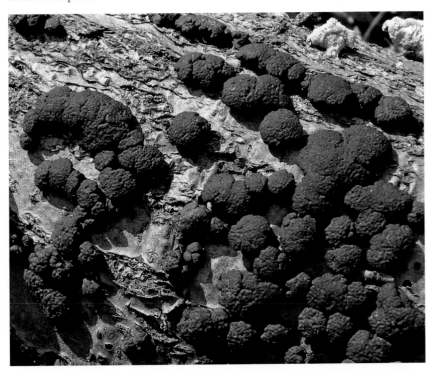

Xylaria longipes Nitschke

DEAD MAN'S FINGERS

SYNONYM *Xylosphaera longipes* (Nitschke) Dennis

Club-shaped, black, tough-pliant, white flesh with black round cavities lining the surface; on hardwood debris

Fruit body 2–11 cm high, 1–2 cm wide, club-shaped, ash-gray at first from asexual spores, soon black, surface cracked, finely dotted with miniscule openings of embedded fruit bodies (perithecia). **FLESH** (stroma) firm, white, covered with black perithecia embedded and appearing as minute round cavities. **STEM** black as upper portion, short or long. **ODOR** and **TASTE** not distinctive.

Habit and habitat single, scattered, or clustered on decaying hardwood logs, branches, various woody debris, especially of beech and sugar maple. June to November. **RANGE** widespread.

Spores ellipsoid or slight curved, smooth, dark brown or black, with germ slit spiraling around the entire length, 13–15 × 5–7 µm; asci cylindrical with thickened, tapered apex with a pore (inoperculate), blue amyloid cylinder at the tip.

Comments Three species in the Northeast look alike and have all been called dead man's fingers, or as I teach my students, cadaverous digits (it is more politically correct). *Xylaria longipes*, very frequently encountered, has a cracked surface and the short spores produce spiral germ slits that go completely around the spore from top to bottom. *Xylaria polymorpha* (Persoon) Greville and *X. schweinitzii* Berkeley and M. A. Curtis produce longer spores, 22–28 µm, with short germ slits. *Xylaria polymorpha* has straight germ slits, while *X. schweinitzii* produces short oblique or curving germ slits on most of its spores.

BASIC MICROSCOPY

To see small structures, good quality, high-powered optics are essential. A compound light microscope with a built-in light source, an adjustable substage condenser with an NA of 0.9 or 1.25, at least three achromatic lenses (10×, 40× NA 0.65 and 100× NA 1.25), and a pair of 10× eye pieces, one with an ocular micrometer, are the minimum components necessary to make accurate observations.

You will also need the following:

- pair of fine-tip jeweler's forceps
- pair of dissecting needles
- glass slides and cover glasses (18 or 22 mm square and 1 or 1.5 mm thick)
- several solutions in dropper bottles
  - distilled water
  - 80–95% ETOH
  - household ammonia
  - 3% KOH
  - saturated Congo red in ammonia
  - Melzer's reagent
  - cotton blue in lactic acid
- box of single-edged razor blades
- spot plate for preparing dried samples for sectioning
- pith sticks or foam packing peanuts for holding revived tissues for sectioning
- immersion oil for the 100× lens
- lens paper and cleaning fluid

Never leave oil on the lens. At a minimum, wipe off the oil with lens paper after each session. Clean the immersion lens with lens paper and cleaning fluid regularly. Do not use a paper towel or tissue paper, or you may scratch that expensive lens.

I strongly urge you to buy *How to Identify Mushrooms to Genus III: Microscopic Features* (Largent et al. 1977) and to contact your nearest mycological club or association, and see if they offer training in using the microscope for identifying fungi.

However, if you wish to go it alone, start your explorations using these simple techniques. Use the very edge of the gills, about 1 mm square, to look for cheilocystidia and pleurocystidia and determine spore morphology. If the fungus is fresh, make some sections that include the cap surface to determine orientation of the cap cells. If it is a white-spored mushroom, always examine in Melzer's reagent first to determine if the spores are amyloid, turning bluish-black, or not. Compare unstained tissue and spores in ammonia or KOH mounts to confirm amyloid reactions. A different mount in Congo red, destained with clear ammonia, will help with finding clamp connections and cystidia. You can purchase some reagents from a supermarket, such as distilled water and household ammonia (nonsudsing and unscented), the others are available from online sources, such as Fisher Scientific (fishersci.com).

For Melzer's reagent, try contacting your local college or university to see if someone in the science department can help make this reagent for you (for a donation to the institution, of course) as it contains one ingredient, chloral hydrate, that is now federally controlled. Another option is to use a compounding pharmacy as Lawrence Leonard suggests in a 2006 article published in *MacIlvainea* (vol. 16, no. 1, spring), which you can view online at namyco.org/docs/Melzer_Lugo.pdf. Using a compounder is a great suggestion, if you can get your medical physician to write a prescription for you. The formula for this reagent is published in *How to Identify Mushrooms to Genus III*.

GLOSSARY

acrid An intensely sharp or burning taste.

adnate (of gills) Attached to the stem over most of the gills' height, also bluntly attached.

adnexed (of gills) Gill edge curving upward on the stem leaving a small triangular space between the gill and stem, also in this guide as bluntly attached.

adnexed finely (of gills) Gill edge curving up and barely attached to stem, in this guide as *barely attached.*

amyloid (of spores and tissues) Staining blue or blue-black to blue-gray in Melzer's reagent.

annulus Ring of tissue left on the stem from the remains of a veil.

appendiculate (of cap margin) Hung with pieces of tissue of the partial veil.

appressed Flattened onto the surface of the cap or stem, as in appressed fibrils or scales.

apothecium (of fruit body in ascomycetes) The hymenium constantly exposed during development and spore release; can be as cups, spongelike, saddlelike, clublike, mushroomlike, paddlelike, and so on.

ascomycete One of a major group of fungi producing asci and ascospores; member of the class Ascomycetes.

ascospore The sexual spore produced from meiosis by ascomycetes; they are symmetrical and lack any attachment pegs; see *basidiospores.*

ascus (pl. **asci**) Saclike or cylindrical cell that produces and contains the ascospores of an ascomycete inside the sac as a product of meiosis.

attached (of gills) Reaching the stem and being attached to it.

basidiomycete One of a major group of fungi producing basidia and basidiospores; member of the class Basidiomycetes.

basidiospore The sexual spores produced from meiosis by basidiomycetes; they are generally asymmetrical and have a distinct attachment peg on the bottom, called the apiculus or hilar appendix.

basidium (pl. **basidia**) Club-shaped cell of basidiomycetes on which basidiospores are formed at the end of pegs or sterigmata, after meiosis.

bell-shaped (of cap) Having a raised rounded center with margin becoming straight or upturned slightly, resembling a church bell, campanulate.

biodiversity The variety of life; unfortunately, fungal biodiversity for most of North America, and elsewhere on the planet, is still underexplored and not well understood.

bluntly attached (of gills) Gills attached straight into the stem, adnate, or slightly curving up the stem, emarginate or even adnexed.

bolete A fleshy mushroomlike fruit body with a removable tube layer on the undersurface of the cap; taxonomically placed in their own order Boletales.

bouquetlike Fruit bodies of mushrooms and other fleshy fungi that cluster together, joined at their stem bases like a bouquet of flowers, caespitose.

broad (of gills) Describes the distance between the gills' attachment to the cap and the gills' lower edge when it is equal to or greater than the thickness of the flesh of the cap.

brown rot Wood rot in which the fungus degrades the cellulose but not the lignin, leaving a brown, often cubical, firm residue; not as common as white rot.

bruising Changing color when handled, rubbed, or otherwise injured.

buff Very pale yellow-brown.

button (mushroom) The young unopened fruit body, with the veil intact and covering the gills, and the cap not yet expanded.

calyptrate (of spores) Having a delicate skirtlike covering that separates from the spore wall proper.

canescent (of cap) White and glistening from a grayish-white pubescence.

cap The upside-down cuplike part of a fruit body whose undersurface bears the basidia in a hymenium on gills, teeth, tubes, veins, or smooth surfaces.

capillitium (in puffballs and slime molds) The netlike threads found inside many puffballs that aid in spore dispersal.

cellulose The principal cross-linked polysaccharide in plant cell walls.

chlamydospores Thick-walled asexual spores of the mold phase of some ascomycetes, especially those that parasitize other fungi.

circumsessile (in *Amanita*) A tight roll of volva tissue that goes around the base of the stem.

clavate (of stem or basidia) Club-shaped.

close (of gills) Spaced intermediately between subdistant and crowded.

conical (of cap) Cone-shaped, more or less.

conifer Cone-bearing tree with needles or scales, such as hemlock, spruce, pine, fir, or arborvitae.

conk The common name for a large, woody, hoof-shaped polypore growing on trees.

convex (of cap) Rounded, shaped like an inverted bowl.

cortina (type of partial veil) A hairy, silky mass of spiderweb-like filaments attached from the cap margin to the stem.

crowded (of gills) So close together that spaces between the gills are hard to see.

cruciate (of basidia or basidiospores) Having a cross shape; in basidia of certain jelly fungi the longitudinal wall formation inside the basidium after meiosis makes four compartments that appear cross-shaped from the top of the basid-ium; an important character of the order Tremellales; basidiospores can produce long wings at the corners of angled spores, making the spore look like a cross.

cyanophilous (of granules in basidia of Lyophyllaceae) Staining deep blue when pieces of air-dried lamellae are soaked in a drop of cotton blue dye in lactic acid on a glass slide, then gently heated over an alcohol lamp before removing the stained tissues and placing them in a fresh drop of clear lactic acid, squashing the mount after the cover glass is placed on to separate out individual basidia for observations; same as siderophilous reaction.

cystidium (pl. cystidia) A sterile hyphal end cell with a distinctive shape or wall thickness, found on the edges and faces of the gills, also may be found on the cap and stem surfaces.

decurrent (of gills) Attached and running down the stem.

deliquescent (of gills) Autodigesting or liquefying at maturity, turning into a black inky liquid; as in the genus *Coprinus*.

depressed (of cap) With the central portion lower than the margin.

dextrinoid (of spores and tissues) Staining reddish-brown in Melzer's reagent.

disc The central part of the cap surface of a mushroom.

discomycete An ascomycete exposing the reproductive surface, the hymenium, on fruit bodies shaped like cups, discs, tongues, saddles, paddles, sponges, or brains.

distant (of gills) Spaced widely apart.

dots (of stem) Small white or colored points over the apex, or sometimes covering the entire stem, pruina; if sticky then referred to as glandular dots, as in boletes and the genus *Suillus* in particular.

duplex (of tissues in polypores and tooth fungi) A soft layer above or below a hard or firm layer.

eccentric (of stem attachment) Stem attached to the cap off-center.

ecosystem A biological community of interacting organisms and their physical environment.

egg Immature stage of amanitas and stinkhorns; also a peridiole of bird's nest fungi.

elliptical or ellipsoid (of spores) Rounded on ends and with curved sides; having the outline of an ellipse.

emarginate (of gills) Similar to adnexed but not going very far up the stem, making a tiny empty triangular space.

equal (of stem) Having a constant diameter from top to base.

fairy ring Ring of mushrooms growing from the periphery of a radially spreading, underground mycelium.

family A taxonomic group of related genera, ranking above genus and below order; named with the suffix –aceae, as in Amanitaceae.

farinaceous An odor or taste of freshly ground flour meal; mealy.

fibril An aggregation of hyphae forming a threadlike filament.

fibrillose (of surface of cap or stem) Having visible fibrils, often differently colored than the surface below them.

fibrous (of flesh of cap or stem) Composed of stringlike, rather tough tissue.

flesh The inner tissue of the cap or stem when viewed with the naked eye.

fleshy (of cap or stem) Soft, decaying readily.

floccose (of cap or stem) Composed of a cottony surface, resembling flannel.

forked (of gills and veins) Branching, irregularly or equally and dichotomously.

free (of gills) Not attached to the stem.

friable (of universal veil) Breaking up readily, crumbling.

fruit body The organized reproductive structure of a fungus that produces spores, such as the mushroom, coral, polypore, cup, perithecium, and so on.

fungus (pl. fungi) A nonphotosynthesizing, spore-producing organism made up of hyphae with chitinous cell walls and that produce enzymes and absorb food from their environment.

gasteromycete One of a group of diverse basidiomycetes that develop spores inside spore cases and do not actively discharge their spores, for example, puffballs, stinkhorns, and bird's nest fungi.

genus A group of similar species; the taxonomic rank below family and above species.

germ pore (of spores) The differentiated tip on a spore where the spore wall is so thin it appears as if a hole or opening is present; the place through which the germ tube often extends upon spore germination.

gills Platelike structures arranged radially on the underside of the mushroom cap, on which the hymenium and spores are formed.

gleba The spore-producing tissue, or spore mass, within the peridium of a gasteromycete.

globose Spherical, or nearly so.

glutinous (of cap or stem) Having a thick layer of mucouslike, sticky material that pulls away from the surface in long strands; see *viscid*.

granulose Covered with granules, like fine grains of salt.

gregarious A pattern of fruiting in which many mushrooms grow close together but are not attached to each other.

hardwood In the broad sense, denotes a nonconiferous tree, such as maple, beech, birch, oak, aspen, cherry, basswood, cottonwood, willow, alder, and so on.

humus A mixture of decayed vegetation on the forest floor.

hyaline Transparent, colorless.

hygrophanous Appearing one color when wet and then changing to a different, faded color when moisture is lost.

hymenium The spore-bearing layer of a fruit body.

hypha (pl. hyphae) A microscopic tubular filament that is the basic structural unit of the body of the mycelium and therefore the fruit body of a fungus.

incurved (of cap margin) Curved or bent inward, with edge pointing at the stem.

indusium (of stinkhorns) A lacy veil that hangs down from the cap.

inoperculate (of asci) Having a pore, not a lid, at the apex of the ascus, through which the spores are shot.

inrolled (of cap margin) Curved in toward the gills and rolled up.

KOH Potassium hydroxide; usually a 2.5–3% aqueous solution is used for reviving tissues and examining spore color.

lamella (pl. lamellae) Gill that reaches the stem.

lamellula (pl. lamellulae) Shortened gill that reaches only partway to the stem; often arranged in distinct tiers with one lamella in the first tier, two lamellulae in the second, shorter tier, and three even shorter lamellulae in the third tier, forming a triangular shape between lamellae with the shortest lamellulae on the outsides of the triangle.

latex Juice- or milklike fluid exuding from a cut or injured portion of some mushrooms, especially species of *Lactarius, but also Mycena.*

LBM Little brown mushroom, a term denoting small, brownish, hard-to-differentiate mushrooms.

lichen A dual organism whose body is made of a fungus, usually an ascomycete, and a blue-green alga or cyanobacterium.

lined (of cap) Having visible lines, ridges, or grooves on the margin; striate; if translucent-lined, then the margin is smooth and the gills from below the cap surface show through.

maculate (of cap or stem) Spotted with different colors or stains.

margin (of gills or cap) The edge; in the case of the cap, the area away from the disc toward and including the edge.

marginate (of gills) Having a different color at the edge than on the face of the gills.

mealy (texture) Appearing as if covered with coarse meal; (taste) like freshly ground flour.

Melzer's reagent An iodine solution used to test spores and tissue for amyloid or dextrinoid reactions.

membranous Resembling a membrane or thin skin.

mitriform (of spores) Shaped like the mitered cap worn by bishops during Catholic religious ceremonies; that is, swollen in the middle, tapered at the top sharply, but only somewhat tapered below, and with flattened (truncate) end with a broad germ pore.

mushroom Strictly, as used in this book, the fruit body of a fleshy fungus with gills; also a much less precise general term for any fruit body of a fungus.

mycelioid (of stem base of mushrooms) Resembling fluffy mycelium, often differently colored than the stem.

mycelium A mass of hyphae or fungus filaments; the assimilative portion of a fungus that is obtaining nutrients.

mycoflora The fungi characteristic of an area.

mycologist A scientist who studies fungi.

mycology The science of studying fungal biology.

mycophagist One who explores and forages for fungi to be used as food.

mycorrhiza (pl. mycorrhizae) The symbiotic association of fungal mycelium and the root tips of trees or other plants.

narrow (of gills) Describes the distance between the gills' attachment to the cap and the gills' lower edge when it is less than the thickness of the flesh of the cap.

nonamyloid (of spores and tissues) Remaining colorless or merely yellowish in Melzer's reagent.

notched (of gills) Having a notch at the point of attachment to the stem; sinuate.

ochraceous Ochre-colored, deep orange-brown or yellow-brown.

oil droplets (of spores) Apparently oily, spherical material inside a cell, especially spores, when viewed under a microscope.

operculate (of asci) Having a lid or plate-like opening at the top of the ascus, through which the spores are shot.

ornamentation (of spore surfaces) Having warts, ridges, lines, or wrinkles, not smooth.

ovate Having an outline or shape like the long axis of a chicken egg.

ovoid A solid, shaped like a chicken egg.

pallid Very pale, an indefinite whitish color.

papilla or papillate (of cap) A small, nipple-shaped projection.

parasitic The relationship of one organism, the parasite, living in or on another living organism, the host, and obtaining nourishment from the association, usually to the detriment of the host.

partial veil A membranous, weblike, or glutinous covering that extends from the cap margin to the stem, protecting the young gills or tubes.

peridioles The small lentil-shaped spore capsules produced in the splash cups of bird's nest fungi, called eggs.

peridium The wall surrounding the spore case in gasteromycetes such as puffballs.

perithecia (ascomycetes) Tiny flask-shaped or round fruit bodies that have a pore at the top through which the ascospores are discharged.

phylogenetics The study of evolutionary relationships among groups of organisms.

pileus The cap of a mushroom, bolete, polypore, tooth fungus, chanterelle, and so on.

plage (of mushroom spores) A depression or flat unornamented area on an ornamental spore surface next to the apiculus on the inner face of the spores, defining the genus *Galerina*.

plane (of cap) Flat, or nearly so, not curved.

plano-convex (of cap) Convex with a flat disc, or central area.

pleurotoid Type of mushroom fruit body with cap and gills, but either lacking a stem or having a highly reduced and lateral stem; as in *Pleurotus*.

polypore Common name for members of the family Polyporaceae, with firmly attached, thin tube layers on leathery or woody fruiting bodies.

pores The mouths or openings of the tubes in boletes and polypores.

pruina Small white or colored dots, often on the apex of the stem but sometimes covering the stem.

pruinose (of stem) Appearing powdered by fine dots, as if sprinkled with flour or sugar.

pseudorhiza A long, false rootlike structure; defines the genera *Hymenopellis* and *Phaeocollybia*.

pubescent (of cap or stem) Covered with fuzz or hairlike material.

punctate (of stem or spores) Having tiny, barely visible dots.

Q (of spores) A measurement derived from dividing the length by the width of each spore.

radially arranged Radiating from a central point, like the spokes of a wheel, as in differently colored or shaded fibrils on the cap of a fungus.

recurved (of cap margin or scales) Having an edge curved up and back.

resupinate (of fruit body) Lying flat, crustlike on substrate with hymenium facing outward, lacking a stem or well-defined cap.

reticulate (of stem surface or spores) Marked with a raised fishnet, veined, or netlike pattern.

rhizoid A rootlike structure attached to the stem base, usually white or brightly colored, soft but tough.

rhizomorph A rootlike structure attached to the stem base, black and hard, carbonaceous, often penetrating the substrate; common in *Armillaria* and some species of *Marasmius*.

ring Tissue left on the stem from the remains of a veil; annulus.

saclike (of volva) Shaped like a cup or sac around the base of the stem, saccate.

saprobe An organism that lives on dead or decaying organisms or organic material.

scaber Rough knobs of tufted hyphal ends, projecting from surface of stem; characteristic of the bolete genus *Leccinum*.

scaly (of surface of cap or stem) Having small, often tapered or pointed pieces of tissue, lying flat or standing up on the surfaces.

sclerotium A fleshy mass of hyphae of definite structure serving as a resting stage for a fungus.

scurfy (of cap or stem) Minutely roughened with small scales.

separable (of gills or stems) Easily separated from the cap.

septate (of hyphae) Having cross walls.

sessile (of fruit body) Stemless, attached directly to the substrate.

setae Sterile, thick-walled, pointed, stiff cells on a fruit body, usually having colored walls; found in ascomycetes and basidiomycetes.

sexual reproduction The fusion of nuclei of different mating types, followed by meiosis to produce spores.

siderophilous (of granules in basidia and spore walls in Lyophyllaceae) Staining purplish-black when treated with aceto-carmine dye in the presence of iron oxide and heated; same as cyanophilous reaction.

sinuate (of gills) Notched gills in which the gill edge becomes abruptly concave as it meets the stem, attached but leaving an obvious channel around the stem.

slimy (of cap or stem) Having a thick layer of mucouslike, sticky material that pulls away from the surface in long strands, glutinous.

slippery (of cap or stem) Having a thin layer of mucouslike material making the surface wet and hard to hold, viscid.

species A taxonomic group representing a population of individuals that have numerous morphological characteristics in common and whose DNA is highly similar; usually considered capable of interbreeding.

spines The pendant, toothlike, spore-bearing structures characteristic of the tooth fungi.

spore The microscopic reproductive and dispersive unit of a fungus.

spore print The visible deposit of basidiospores made by placing the cap of a fungus, removed from its stem, on white paper and covering it for a few hours.

squamules Minute, tiny scales.

stem The structure supporting the cap or head of a fungus; also called a stipe or stalk.

sterigma (pl. sterigmata) Tapered extension of the basidium that produces the basidiospores; when broken, leaves a small portion on the basidiospore, the apiculus or hilar appendix.

sterile Without reproductive spores; the opposite of fertile.

sterile base The sterile, chambered base below the gleba in certain gasteromycetes.

stipe The stem or stalk of a fruit body.

striate Having a surface marked with roughly parallel lines, grooves, or ridges, lined.

stroma (of ascomycetes) A compact mass of tissue, often black or highly colored, that has fruit bodies, typically perithecia, embedded in or on it.

subdecurrent (of gills) Running a short distance down the stem.

subdistant (of gills) Spaced intermediately between close and distant.

subglobose (of spores) Nearly globose, having a Q value slightly more than 1.

substrate The material on which the fruit body is found and from which the fungus obtains its nourishment.

subzonate (of cap) Faintly zoned in concentric rings by color, texture, or ridging.

tawny Rich yellowish-brown; the color of a lion.

taxonomy The science of systematically classifying organisms by emphasizing their relationships at the morphological, anatomical, and molecular levels.

teeth The pendant, spinelike, spore-bearing structures characteristic of the tooth fungi.

terrestrial Growing on the ground.

toadstool A term used to denote an inedible or poisonous, stemmed mushroom, or an alternative to mushroom in general.

translucent-lined (of cap margin) Having thin, translucent, smooth flesh that allows the gills to show through as lines or striations.

tremelloid (type of fruit body or basidium) Relating to the genus *Tremella*; having a jellylike fruit body and basidia with longitudinal cruciate septa.

troops A pattern of fruiting in which mushrooms grow close together in large numbers, but not so close as to be considered a cluster.

truncate (of cap, gills, or spores) Chopped-off in appearance, having a flat surface, for example, the center of the cap, the edge of lamellulae, the apex of spores.

tubes Hollow, cylindrical structures lined with basidia and open at one end as a pore; characteristic of boletes and polypores.

tuberculate (of spores) Roughened with rounded bumps.

type specimen or type A collection of fruit bodies from which the original concept of a species or other taxonomic group is derived.

umbo Protrusion or knob on the disc or center of the cap.

undulate (of cap margin) Broadly wavy.

universal veil A layer of tissue completely surrounding the developing fruit body, pieces of it sometimes remaining on the cap as warts or patches when the fruit body expands, also surrounding the base of the stem as a volva, often on both the cap and stem as differently colored and differently textured tissues when compared to the skin of the cap and stem.

veil A structure covering and protecting the hymenium from damage by environmental stresses or insects; see also *partial veil* and *universal veil*.

velvety (of cap or stem) Soft, smooth, having very short hairs, like velvet.

ventricose (of cystidia or stem) Swollen in the middle and tapered at either end.

verticillate Forming whorls around a common point, as leaves on a stem in the bedstraw plant.

vinaceous Having the color of red wine.

viscid (of cap or stem) Slippery to the touch when wet from a thin mucous layer, becoming sticky when losing moisture; see also *glutinous* and *slimy*.

volva Remnants of the universal veil left in various forms on or at the base of the stem.

warts (surface of cap or stem base) Small patches of universal veil remnants resembling warts; (surface of spores) small, rounded projections that look like warts.

waxy (of gills) Looking as if coated with wax, like waxed paper.

white rot Wood rot produced by basidiomycetes, that degrades both the cellulose and the lignin, leaving a white, stringy, soft residue.

zonate Having zones of different textures or colors, usually on the cap surface.

FURTHER READING

General Information

Largent, David L., and Daniel E. Stuntz. 1973, 1977 revised. *How to Identify Mushrooms to Genus I: Macroscopic Features*. Eureka, California: Mad River Press.

Largent, David L., and Harry D. Thiers. 1973. *How to Identify Mushrooms to Genus II: Field Key to Genera*. Eureka, California: Mad River Press.

Largent, David L., David Johnson, and Roy Watling. 1977. *How to Identify Mushrooms to Genus III: Microscopic Features*. Eureka, California: Mad River Press.

Largent, David L., and Timothy J. Baroni. 1988. *How to Identify Mushrooms to Genus VI: The Modern Genera*. Eureka, California: Mad River Press.

Specialized Information

Bessette, Alan E., Arleen R. Bessette, William C. Roody, and Steven A. Trudell. 2013. *Tricholomas of North America: A Mushroom Field Guide*. Austin: University of Texas Press.

Bessette, Alan E., William C. Roody, and Arleen R. Bessette. 2000. *North American Boletes*. Syracuse, New York: Syracuse University Press.

Beug, Michael W., Alan E. Bessette, and Arleen R. Bessette. 2014. *Ascomycete Fungi of North America: A Mushroom Reference Guide*. Austin: University of Texas Press.

Breitenbach, J., and F. Kränzlin. 1984–1995. *Fungi of Switzerland*. 6 vols. Lucerne, Switzerland: Verlag Mykologia.

Gilbertson, Robert L., and Leif Ryvarden. 1986. *North American Polypores*. 2 vols. Oslo, Norway: Fungiflora A/S.

Hansen, Lise, and Henning Knudsen. 2000. *Nordic Macromycetes*. 3 vols. Copenhagen, Denmark: Nordsvamp.

Kendrick, Bryce. 1992. *The Fifth Kingdom*. 2nd ed. Newburyport, Massachusetts: Focus Information Group.

Kluting, Kerri L., Timothy J. Baroni, and Sarah E. Bergemann. 2014. Toward a stable classification of genera within the Entolomataceae: a phylogenetic re-evaluation of the *Rhodocybe–Clitopilus* clade. *Mycologia* 106:1127–1142.

Knudsen, Henning, and Jan Versterholt, eds. 2008. *Funga Nordica: Agaricoid, Boletoid and Cyphelloid Genera*. Copenhagen, Denmark: Nordsvamp.

Moser, Meinhard. 1978. *Agarics and Boleti*. 4th ed. Stuttgart, Germany: Roger Phillips.

Roberts, Peter, and Shelley Evans. 2011. *The Book of Fungi: A Life-size Guide to Six Hundred Species from Around the World*. Chicago, Illinois: University of Chicago Press.

Seaver, Fred J. 1928. *The North American Cup-fungi (Operculates)*. New York: Seaver.

Stamets, Paul. 2005. *Mycelium Running*. Berkley, California: Ten Speed Press.

Field Guides Useful for Northeastern Fungi

Barron, G. 1999. *Mushrooms of Northeast North America*. Vancouver, Canada: Lone Pine. 336 p.

Bessette, A. E., Arleen R. Bessette, and David W. Fischer. 1997. *Mushrooms of Northeastern North America*. Syracuse, New York: Syracuse University Press. 582 p.

Binion, D. E., H. H. Burdsall, Jr., S. L. Stephenson, O. K. Miller, Jr., W. C. Roody, and L. N. Vasilyeva. 2008. *Macrofungi Associated with Oaks of Eastern North America*. Morgantown: West Virginia University Press. 467 p.

Kibby, G., and R. Fatto. 1990. *Keys to the Species of Russula in Northeastern North America*. Somerville, New Jersey: Kibby-Fatto Enterprises. 70 p.

Kuo, M. 2005. *Morels*. Ann Arbor: University of Michigan Press. 205 p.

Kuo, M., and A. Methven. 2010. *100 Cool Mushrooms*. Ann Arbor: University of Michigan Press. 210 p.

Kuo, M., and A. S. Methven. 2014. *Mushrooms of the Midwest*. Urbana: University of Illinois Press. 427 p.

Lincoff, G. H. 1992. *The Audubon Society Field Guide to North American Mushrooms*. New York, New York: Knopf. 926 p.

McNeil, R. 2006. *Le Grand Livre des Champignons du Quebec et de l'Est du Canada*. Waterloo, Canada: Editions Michel Quintin. 575 p.

Miller, O. K., Jr., and H. H. Miller. 2006. *North American Mushrooms: A Field Guide to Edible and Inedible Fungi*. Guilford, Connecticut: FalconGuide. 584 p.

Phillips, R. 1991. *Mushrooms of North America*. Boston: Little, Brown and Company. 319 pp.

Roody, W. C. 2003. *Mushrooms of West Virginia and the Central Appalachians*. Lexington: University Press of Kentucky. 520 p.

MYCOLOGICAL RESOURCES

Useful Online Sites

blog.mycology.cornell.edu—Cornell Mushroom Blog

indexfungorum.org—Index Fungorum, the best source for looking up the correct spelling of scientific names or seeing more species names for a given genus

mushroomexpert.com—Michael Kuo

mycobank.org—Mycobank, maintained by the International Mycological Association, another source for looking up names of fungi

mycoportal.org—Mycology Collections Portal

mycoquebec.org—Les champignones du Québec

namyco.org—North American Mycological Association, an excellent source for finding a mushroom club near you

tomvolkfungi.net—Tom Volk's Fungi

Northeast Mycology Organizations

For contact information see the website of the North American Mycological Association.

United States

Berkshire Mycological Society
Boston Mycological Club
Cape Cod Mushroom Club
Central New York Mycological Society
Central Pennsylvania Wild Mushroom Club
Connecticut Valley Mycological Society
Connecticut-Westchester Mycological Association (COMA)
Eastern Penn Mushroomers
Long Island Mycological Club
Maine Mycological Society
Michigan Mushroom Hunters Club
Mid Hudson Mycological Association
Mid York Mycological Society
Minnesota Mycological Society
Monadnock Mushroomers Unlimited
Mycological Association of Greater Philadelphia
New Jersey Mycological Association
New York Mycological Society
Pioneer Valley Mycological Association
Rochester Area Mycological Association
Susquehanna Valley Mycological Society
Western Pennsylvania Mushroom Club
Wisconsin Mycological Society
WNY Mycology Club
Wyoming Valley Mushroom Club

Canada

Cercle des Mycologues de Montréal
Foray Newfoundland and Labrador
Mycological Society of Toronto

PHOTO AND ILLUSTRATION CREDITS

All drawings are by Marjorie Leggitt.

All photographs are by the author except as listed below.

Howard E. Bigelow, *Macrolepiota procera, Lepiota cristata, Russula silvicola, Lactarius piperatus, Lactarius subpurpureus, Tricholoma aurantium, Tricholoma virgatum, Tricholoma odorum, Leucopaxillus albissimus, Lyophyllum multiforme, Marasmius strictipes, Mycena atkinsoniana, Mycena rosella, Hypsizygus ulmarius, Pluteus aurantiorugosus, Inocybe albodisca, Cortinarius caperatus, Coprinus comatus, Leccinum aurantiacum, Albatrellus ovinus, Phaeolus schweinitzii, Pycnoporus cinnabarinus, Turbinellus floccosus, Geastrum triplex, Lysurus cruciatus, Phallus duplicatus, Gyromitra esculenta, Helvella elastica, Microstoma floccosum, Urnula craterium, Peziza varia, Mitrula elegans.*

Nina Burghardt, *Calliderma indigofera.*

Gary Emberger, *Lenzites betulina.*

Michael Goldman, *Tricholomopsis sulphureoides, Neoalbatrellus caeruleoporus, Trichaptum biforme, Daedaleopsis confragosa, Daedalea quercina, Phellinus igniarius.*

Roy Halling, *Chlorophyllum molybdites, Marasmius rotula, Boletus subvelutipes, Xanthoconium purpureum, Leccinellum griseum.*

Rick Kerrigan, *Agaricus placomyces.*

Lance Lacey, *Geastrum saccatum.*

Renée Lebeuf, *Amanita flavorubens, Amanita fulva, Amanita multisquamosa, Chlorophyllum rachodes, Leucoagaricus americanus, Leucoagaricus leucothites, Tricholoma grave, Tricholomopsis rutilans, Melanoleuca alboflavida, Rugosomyces carneus, Clitocybe sinopica, Clitocybe robusta, Gymnopus dryophilus, Baeospora myriadophylla, Crinipellis campanella, Marasmius bellipes, Marasmius pallidocephalus, Mycetinis scorodonius, Marasmius sullivantii, Mycena pura,*

Neolentinus lepideus, Ossicaulis lignatilis, Phyllotopsis nidulans, Cheimonophyllum candidissimum, Schizophyllum commune, Pluteus petasatus, Trichopilus jubatus, Clitocella mundula, Crepidotus nyssicola, Tapinella panuoides, Gymnopilus luteus, Pholiota multifolia, Pholiota squarrosoides, Pholiota lenta, Inocybe tahquamenonensis, Inocybe rimosa, Cortinarius distans, Cortinarius gentilis, Melanophyllum haematospermum, Coprinopsis variegate, Coprinellus micaceus, Stropharia rugosoannulata, Psathyrella piluliformis, Hypholoma capnoides, Austroboletus gracilis, Chalciporus piperatoides, Strobilomyces strobilaceus, Tylopilus plumbeoviolaceus, Tylopilus rubrobrunneus, Retiboletus ornatipes, Suillus grevillei, Boletopsis grisea, Grifola frondosa, Laetiporus sulphureus, Polyporus arcularius, Royoporus badius, Cantharellus cinnabarinus, Polyozellus multiplex, Craterellus tubaeformis, Sarcodon scabrosus, Hydnum repandum, Climacodon septentrionale, Physalacria inflata, Helvella lacunosa, Sarcoscypha dudleyi, Sarcoscypha occidentalis, Pachyella clypeata, Humaria hemisphaerica, Spathulariopsis velutipes, Vibrissea truncorum, Tolypocladium longisegmentis.

Amanda Neville, *Mycena galericulata, Hypoxylon fragiforme.*

Joseph Nuzzolese, *Hydnum umbilicatum.*

Clark Ovrebo, *Tricholoma caligatum, Tricholoma portentosum, Tricholoma subsejunctum.*

Jack Ruggirello, *Hericium coralloides, Sebacina sparassoidea.*

INDEX

Bold pages indicate detailed accounts with photographs.

ABOUT THE AUTHOR

Timothy J. Baroni, Distinguished Professor of Biology at the State University of New York, College at Cortland, works globally on biodiversity research of macrofungi. He is the author or co-author of two books: *A Revision of the Genus* Rhodocybe *Maire*, and *How to Identify Mushrooms to Genus VI*. He has published 70 articles on taxonomy of macrofungi, describing 9 new genera and 100 new species and varieties of macrofungi. He brings to this work a passion for helping others learn how to identify species of macrofungi and a desire to display the beauty of the fungal organisms. He has served as president of the Mycological Society of America, has received numerous awards from the State University of New York for teaching and research, and was selected (2009) for the Distinguished Mycologist Award given by the Mycological Society of America.

PHOTO IN THE AUTHOR'S COLLECTION.

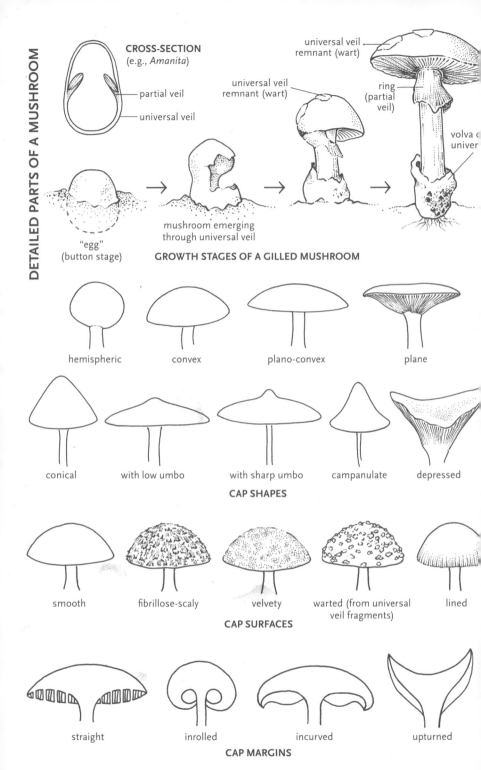

DETAILED PARTS OF A MUSHROOM

CROSS-SECTION
(e.g., *Amanita*)

partial veil

universal veil

universal veil remnant (wart)

universal veil remnant (wart)

ring (partial veil)

volva (universal...)

"egg" (button stage)

mushroom emerging through universal veil

GROWTH STAGES OF A GILLED MUSHROOM

hemispheric

convex

plano-convex

plane

conical

with low umbo

with sharp umbo

campanulate

depressed

CAP SHAPES

smooth

fibrillose-scaly

velvety

warted (from universal veil fragments)

lined

CAP SURFACES

straight

inrolled

incurved

upturned

CAP MARGINS